FRICTION STIR WELDING AND PROCESSING VIII

TMS2015

144th Annual Meeting & Exhibition

March 15-19, 2015 • Walt Disney World • Orlando, Florida, USA

FRICTION STIR WELDING AND PROCESSING VIII

Proceedings of a symposium sponsored by
the Shaping and Forming Committee of
the Materials Processing & Manufacturing Division of
TMS (The Minerals, Metals & Materials Society)
held during

TMS2015
144th Annual Meeting & Exhibition
March 15-19, 2015
Walt Disney World • Orlando, Florida, USA

Edited by:
Rajiv S. Mishra • Murray W. Mahoney
Yutaka Sato • Yuri Hovanski

Editors
Rajiv S. Mishra
Murray W. Mahoney
Yutaka Sato
Yuri Hovanski

ISBN 978-3-319-48604-8
ISBN 978-3-319-48173-9 (eBook)
DOI 10.1007/978-3-319-48173-9

Chemistry and Materials Science: Professional

Published by Springer International Publishers, Switzerland, 2016
Reprint of the original edition published by John Wiley & Sons, Inc., 2015, 978-1-119-08249-1

Printed on acid-free paper

This Springer imprint is published by Springer Nature
The registered company is Springer International Publishing AG
The registered company address is: Gewerbestrasse 11, 6330 Cham, Switzerland

TABLE OF CONTENTS
Friction Stir Welding and Processing VIII

High Temperature Materials I

High Temperature Materials II

Aluminum and Magnesium Alloys

Dissimilar Materials

Friction Stir Processing

Friction Stir Related Technologies

Simulations and Measurements

Poster Session

EDITORS

Rajiv S. Mishra is a Professor of Materials Science and Engineering in the Department of Materials Science and Engineering at the University of North Texas. He is also the UNT Site Director of the NSF I/UCRC for Friction Stir Processing and a Fellow of ASM International. His highest degree is Ph.D. in Metallurgy from the University of Sheffield, UK (1988). He has received a number of awards which include the Firth Pre-doctoral Fellowship from the University of Sheffield, the Brunton Medal for the best Ph.D. dissertation in the School of Materials from the University of Sheffield in 1988, the Young Metallurgist Award from the Indian Institute of Metals in 1993, Associate of the Indian Academy of Sciences in 1993, and the Faculty Excellence Awards from the University of Missouri-Rolla each year from 2001 through 2007. Dr. Mishra has authored or co-authored 266 papers in peer-reviewed journals and proceedings and is principal inventor of four U.S. patents. His current publication based h-index is 41 and his papers have been cited more than 7,000 times. He has co-edited a book on friction stir welding and processing, and edited or co-edited thirteen TMS conference proceedings. He is the chair of the TMS Structural Materials Division and serves on the TMS Board of Directors as the SMD Director. He serves on the editorial board of *Materials Science and Engineering A*, *Science and Technology of Welding and Joining*, and *Advances in Materials Science and Engineering*. He has recently published a book titled *Friction Stir Welding and Processing: Science and Engineering* with Springer and several short books with Elsevier.

Murray W. Mahoney: Prior, Manager/Senior Scientist, Structural Metals Department, Rockwell Scientific; B.S., physical metallurgy, University of California Berkeley; M.S., physical metallurgy, University of California Los Angeles. Mr. Mahoney has over 48 years of experience in physical metallurgy and related disciplines. Most recently, his work has centered on the development of innovative joining technologies for aerospace, transportation, and oil and gas applications and thermomechanical processes to control the microstructure in structural alloys to enhance specific properties. This work has led to the introduction of friction joining processes to join materials considered unweldable, improve superplasticity in structural alloys, facilitate room temperature forming of structural aluminum alloys, and enhance material properties such as strength, fatigue life, and corrosion resistance. These studies have resulted in a more complete metallurgical understanding of joining fundamentals, formability, and corrosion resistance. The primary research emphasis has been to improve manufacturing efficiency while reducing fabrication costs. Mr. Mahoney has authored or co-authored more than 120 published papers and has been awarded 21 United States patents, organized and hosted a number of Friction Stir Welding symposia, participated as co-editor/author on the first reference book on friction stir welding and processing, and been awarded the honor of Rockwell Scientific Technologist of the Year.

Yutaka Sato is currently an Associate Professor in the Department of Materials Processing at Tohoku University, Japan. He earned a Ph.D. in Materials Processing at Tohoku University (2001). His Ph.D. thesis was titled "Microstructural Study on Friction Stir Welds of Aluminum Alloys." He participated in friction stir research of steels at Brigham Young University for a year in 2003. He is a member of Sub-commission III-B WG-B4 at IIW, which is a working group to build international standardization of friction stir spot welding. His work has focused on metallurgical studies of friction stir welding and processing for more than a decade. He has obtained fundamental knowledge on development of grain structure, texture evolution, joining mechanism, behavior of oxide-layer on surface, properties-microstructure relationship, and so on. Recently, he has centered on developing friction stir welding of steels and titanium alloys, and new tool materials. He has received a number of awards including District Contribution of Welding Technology Award from Japan Welding Society in 2005, Kihara Award from Association for Weld Joining Technology Promotion in 2008, Prof. Koichi Masubuchi Award from AWS in 2009, Murakami Young Researcher Award from the Japan Institute of Metals in 2010, Aoba Foundation Award in 2010, and Honda Memorial Young Researcher Award in 2011. He has authored or co-authored more than 220 papers in peer-reviewed journals and proceedings.

Yuri Hovanski is a Senior Research Engineer at Pacific Northwest National Laboratory. He earned a B.S. Degree in Mechanical Engineering at Brigham Young University, and then completed his M.S. degree in Mechanical Engineering at Washington State University. He serves as the vice-chair of the TMS Shaping and Forming Committee, and is the Chair of the Industrial Advisory Board for the Center of Friction Stir Processing. He has participated in friction stir related research for more than 15 years investigating weld formability, abnormal grain growth, and the influence of post weld microstructure and texture on mechanical properties. More recently, he has focused on the development of low-cost solutions for friction stir welding, introducing cost efficient solutions for thermal telemetry, new tool materials and production techniques for friction stir spot welding tooling, and utilizing thermo-hydrogen processing to aid friction stir welding of titanium alloys. He continues this effort today enabling high speed friction stir welding to support high volume, cost-sensitive applications. He recently introduced and patented friction stir scribe technology that enables lap welding of highly dissimilar materials. He actively reviews friction stir related literature for several publications and has documented his work in more than 35 publications.

FRICTION STIR WELDING AND PROCESSING VIII

High Temperature Materials I

Friction Stir Welding and Processing VIII
Edited by: Rajiv S. Mishra, Murray W. Mahoney, Yutaka Sato, and Yuri Hovanski
TMS (The Minerals, Metals & Materials Society), 2015

A Study of Friction Stir Welding for Clad Pipelines

T. KATAYAMA[1], Y. KISAKA[1], F. KIMURA[1]
Y. S. SATO[2], H. KOKAWA[2]

[1] Nippon Steel & Sumikin Engineering CO., LTD
20-1, Shintomi, Futtsu-City, Chiba, 293-0011, Japan
[2] Department of Materials Processing,
Graduate School of Engineering, Tohoku University
6-6-02 Aramaki-aza-Aoba, Aoba-ku, Sendai 980-8579, Japan

Keywords: Clad pipe, Friction Stir Welding, Pipeline

Abstract

The demand on clad pipelines which are combined corrosion resistance alloy (CRA) with carbon steel has been increasing recently. Metal inert gas (MIG) welding with nickel based alloy wire is usually applied for the girth welding of stainless clad pipes. On the work-barge to lay offshore pipelines, however, it has some difficulties. To solve those assignments, FSW+ MAG welding process has been developed.
In this study, several FSW conditions were examined, and sound FSW joint without dilution between stainless steel and carbon steel were obtained. Then, corrosion resistance and mechanical properties of FSW+MAG welding joints were evaluated.

1. Introduction

Pipelines are typical ways to transport natural gas or oil economically and effectively in middle or long distance. In Southeast Asia, especially, the demand on offshore pipelines used clad pipes which are combined CRA with carbon steel has been increasing recently [1]. Constructions for offshore pipelines are performed on a special work-barge called pipe laying barge, "Kuroshio", which is owned by Nippon Steel & Sumikin Engineering Co., Ltd. The offshore construction with high productivity has been required because the operational cost for pipe laying barges is extremely expensive[2],[3],[4].
Figure 1 shows the longitudinal section of "Kuroshio". The barge consists of various stations, i.e., three welding ones, equipping automatic welding systems shown in Figure 2, two non-destructive testing ones, a repair one, and a coating one. The barge goes forward with a pipe length distance when the predetermined operations at all stations are finished. Pipelines are constructed by repeating those works, but if welding defects are detected, the barge cannot move until completing repair welding. Thus, to keep the high productivity of the offshore pipeline construction, not only high speed welding but also high quality welding is necessary.

In the case of construction for clad pipelines, MIG welding is usually employed for the girth welds as a welding method. Typically, when pipes which have API 5L X65 as base materials of carbon steel are welded, nickel based alloy wire is used as welding wire because of the following two reasons. Firstly, at this situation, carbon steel wire results in a metallurgical problem, i.e. brittle martensitic structure. Another one is that a weld metal should have higher strength than base materials.
On the other hand, MIG welding using nickel based alloy wire may cause some issues. For example, incomplete fusion due to cold lap is easy to occur because melting point of nickel

based alloy is lower than that of carbon steel. In other words, molten pool made of nickel based alloy wire is difficult to melt a bevel wall of carbon steel. In addition, the weld beads often have convex shape.
Therefore, such weld beads would reduce productivity because careful grinding which is additional work to prevent weld defects is needed after every welds. Thus, conventional MIG welding process has some problems in terms of high productivity as well as high quality. The main purpose of this study is to solve such assignments mentioned above.

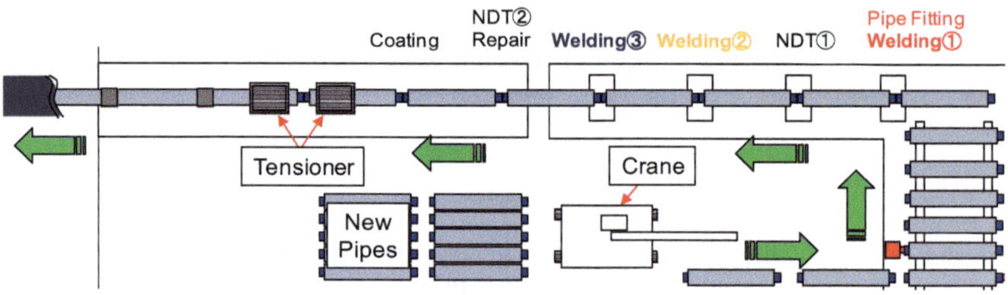

Fig.1 Longitudinal section of Kuroshio

Fig.2 Automatic welding system

2. FSW + MAG welding process

Figure 3 shows the comparison between conventional MIG welding process and developing process, FSW + dual torches MAG welding process. In the conventional process, two welding heads having single welding torch travel downward during welding along the guide rail which is set on the pipes. The sequence is repeated from root pass to cap pass. Nickel based alloy wire is used as welding wire in all passes as mentioned above, so some issues may appear.
In the developing process, on the other hand, FSW is conducted for root welding from inside of clad pipes, instead of MIG welding from outside of them. At the first step, the welding tool is inserted at bottom of pipes and then it is traveled along a groove. At this time, if welding without dilution between stainless steel (316L) and carbon steel (X65) has been achieved, it would not need to consider to the metallurgical problem which is caused by mixing of both the materials. Therefore, this suggested method might improve productivity of onsite welding for offshore pipeline constructions.

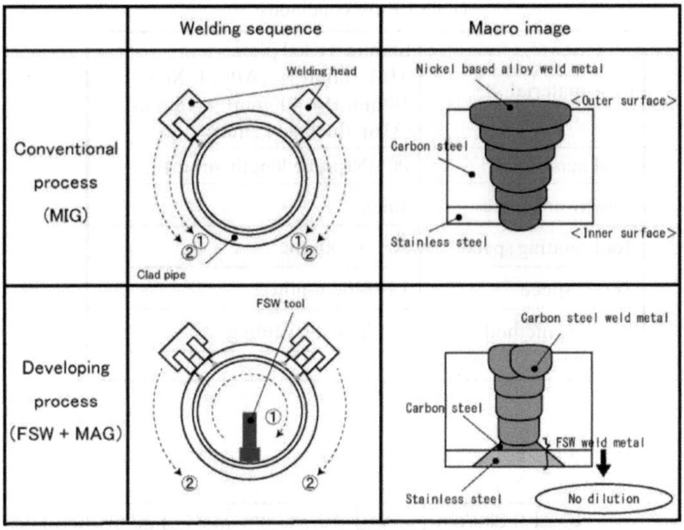

Fig.3 Schematic illustrations of the conventional and the developing methods

In this paper, several FSW conditions were preliminarily tried to obtain a good joint without dilution between stainless steel and carbon steel. Secondly, corrosion resistance on bead surface after FSW was investigated using pitting potential measurement test. And then, mechanical properties of "FSW+ MAG" joints were examined.

3. Experiment and discussion

3.1 Investigation of FSW condition without dilution between stainless steel and carbon steel

Several experiments had been performed to realize FSW + MAG welding process. Preliminarily, FSW trials were conducted using clad plates to certify whether joining can be done without dilution between stainless steel and carbon steel. In this experiment, FSW was performed under the conditions shown in Table 1, i.e. traveling speed and tool rotation speed were changed as the parameters.
Experimental results are shown in Figure 4. The sound FSW joints without dilution between stainless steel and carbon steel were obtained under some conditions. In order to confirm presence of peeling and mixing, microstructure observation by optical microscopy and EPMA analysis of major chemical compositions of stainless steel were conducted, as shown in Figure 5 and Figure 6. These results indicate that the sound joints have neither dilution nor separation between 316L and X65.

Table 1 FSW conditions

Base material	Stainless Clad plate 316L stainless + API 5L X65 180mmB × 400mmL × 18mmt (316L thickness 2mm)
Tool material	PCBN(probe length: 4mm)
Downward force	40kN
Tool rotating speed	200～600rpm
Travel speed	10～40cm/min
Welding method	Stir in plate welding

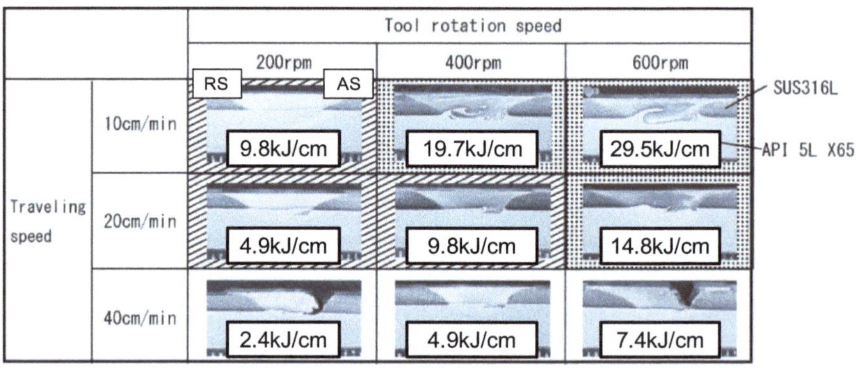

Fig.4 Macroscopic overviews of FSW joint

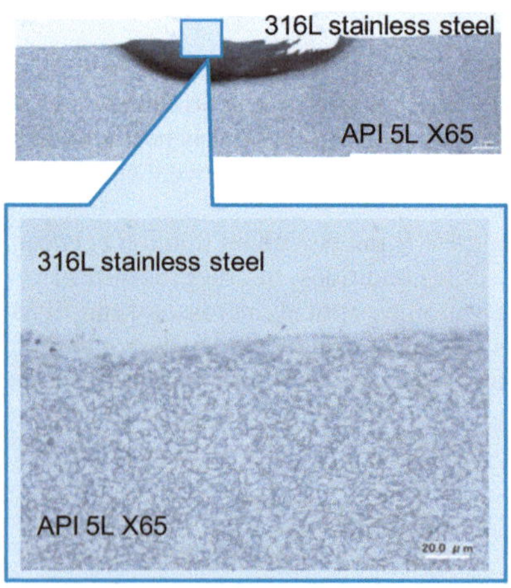

Fig. 5 Microscopic section (x 400)

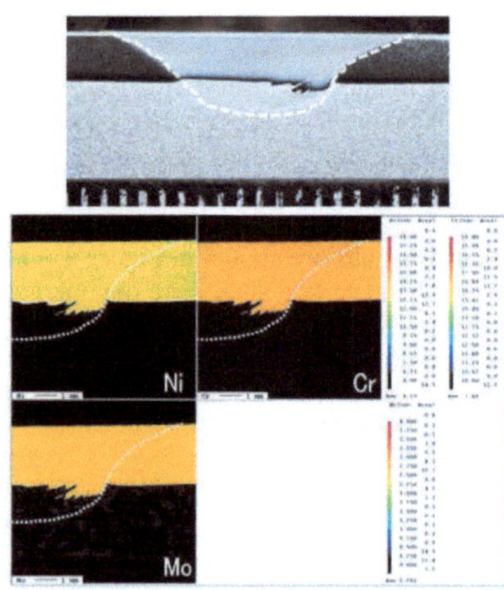

Fig. 6 EPMA maps of Ni, Cr, and Mo

It is expected that those phenomena depend on heat input during FSW. Heat input can be calculated approximately by equation (1) [5]. In the equation, Q is heat input[kJ/cm], μ is coefficient of friction (presumed = 0.2), N is tool rotation speed[rev/sec], p is pressure of stir zone[N/m^2], and R is radius of the tool shoulder[m].

$$Q=4/3 \cdot \pi^2 \cdot \mu \cdot N \cdot p \cdot R^3 \quad (1)$$

As the results shown in Figure 6, two criteria to receive sound joints without any defects and mixing of both the materials were acquired. When the heat input is less than approximately 10[kJ/cm], the good macroscopic section with no dilution of both materials was produced. Moreover, the travel speed is less than 40[cm/min], resulting in the good joint without welding defects.

3.2 Corrosion resistance

The inner surface of FSW joint should also have corrosion resistance at least as same as the base metal and 316L unaffected zone. Pitting potential measurement test was done to certify the corrosion resistance of bead surface after FSW. Test specimens with the size of 10 x 10[mm] were collected every three pieces from stir zone (SZ), advancing side (AS) and retreating side (RS). They were dipped in 3.5% NaCl solution, and were applied voltage increasing by the rate of 20[mV/min]. The voltage was continuously measured when the current density was reached to 100[μ A/cm^2]. Figure 7 shows the test results. Comparing with general data of 316L stainless steel, it was confirmed that bead surface of 316L stainless steel welded by FSW has sufficient corrosion resistance.

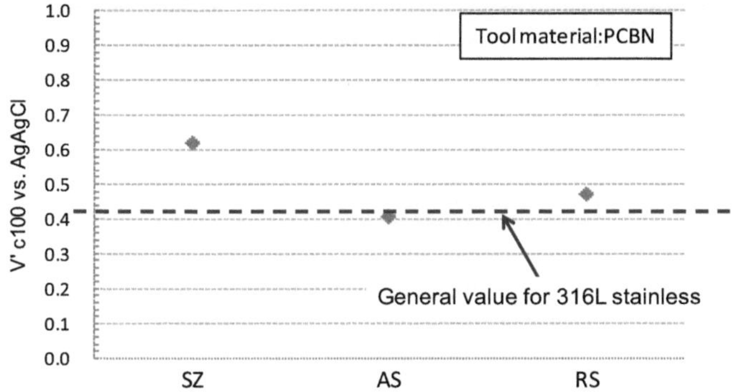

Fig. 7 Pitting potential measurement results

3.3 Mechanical properties

Charpy impact test and Vickers hardness test were performed to confirm mechanical properties of "FSW + MAG" joint. Notch position for Charpy impact test and measurement position for Vickers hardness test are shown in Figure 8. The test temperature for Charpy impact test was 0 centigrade, and test load for Vickers hardness test was 48[kN]. The results are shown in Figure 9.

In Figure 9 (a), it shows that all specimens have adequate Charpy absorbed energy around from 110 to 130[J], which is higher than minimum requirements of each sample (27 [J]) and average (32 [J]). Besides, in Figure 9 (b), the maximum hardness was found in SZ in both case of 316L and X65. The maximum values were 243[Hv5] and 265[Hv5] in 316L and X65, respectively. Those values satisfy specifications of the general project. Therefore, the results of tests indicate that the developing process would be able to apply for a girth weld of actual pipelines.

	Notch position : BOND	Notch position : HAZ(BOND+1mm)
Advancing Side	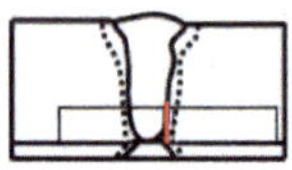	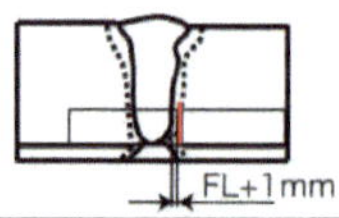
Retreating Side	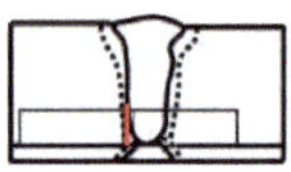	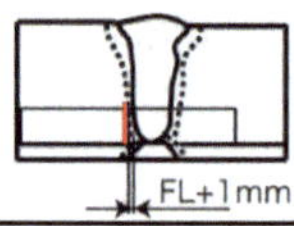

(a) Charpy impact test

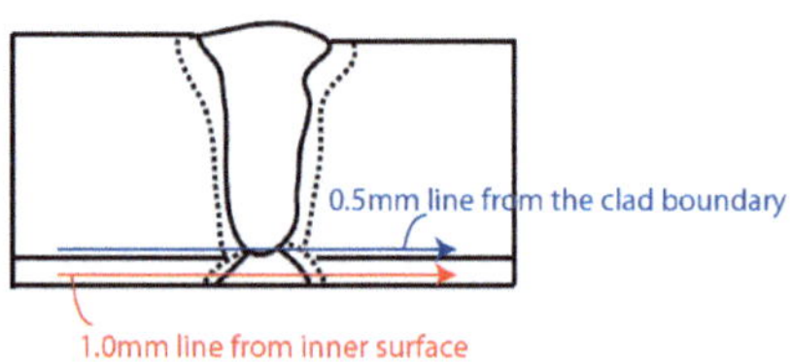

(b) Vickers hardness test

Fig. 8 Measurement positions of mechanical tests

Notch position	Charpy impact energy [J]	
	Individual	Average of the 3pcs.
AS BOND	114,123,128	122
AS BOND + 1mm	112,112,112	112
RS BOND	123,128,131	127
RS BOND + 1mm	117,112,114	114

(a) Charpy impact test

	1.0mm line from inner surface	0.5mm line from the clad boundary
Hardness distribution	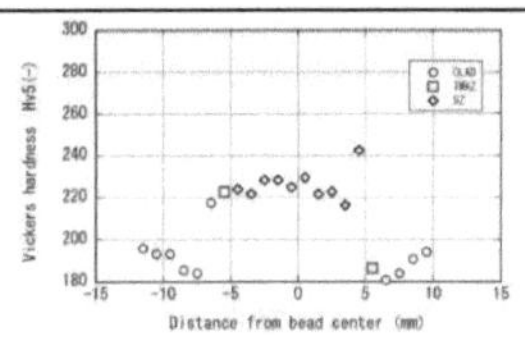	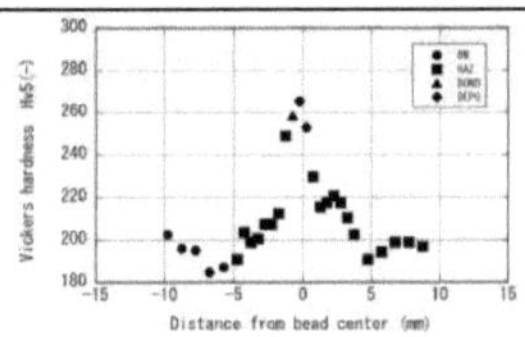
Maximum hardness [Hv5]	243 (Stir zone)	265 (Weld metal)

(b) Vickers hardness test

Fig. 9 Mechanical test results

4. Conclusions

New process, "FSW + MAG" welding process to achieve high productivity for construction of clad pipelines was suggested and practicability for the process was shown by performing some experiments and considerations.
Primary conclusions are shown as below.

- It was confirmed that FSW can be done without dilution between stainless steel and carbon steel. The necessary requirements were not only travel speed less than 40[cm/min] to obtain sound FSW joint without welding defects but also heat input less than 10[kJ/cm] to prevent dilution between stainless and carbon steel.
- FSW bead surface on 316L stainless steel exhibited roughly the same pitting corrosion as the unaffected area.
- As the results of Charpy impact test and Vickers hardness test, it was revealed that the "FSW + MAG" joint had enough mechanical properties for actual pipelines.

5. References

[1] S. Sato, R. Kayano, Y. Nitta, M. Sakuraba, W. Kawakami, T. Maruya, Japan Steel Works, LTD. Technical Report, No.60, 2009

[2] H. Hosoda, Y. Ikuno, T. Hakoda, F. Kimura, Proceedings of "Pipeline Technology Now and Then" at the 8th International Welding Symposium, Japan Welding Society, 2008

[3] T. Hakoda, K. Yanaka, T. Ikezaki, H. Hosoda, N. Sogabe, T. Torii, Nippon Steel Engineering Co., LTD. Technical Report, Vol.1, 2010
[4] Y. Kisaka, F. Kimura, T. Hakoda, H. Hosoda, T. Torii, CanWeld Conference, 2013
[5] Frigaard et al., Proc.1st Int. Symp. FSW, Thousand Oaks ,USA,14-16 June 1999

FATIGUE ASSESSMENT OF FRICTION STIR WELDED DH36 STEEL

Athanasios Toumpis[1], Alexander Galloway[1], Helena Polezhayeva[2], Lars Molter[3]

[1]University of Strathclyde; James Weir Building, 75 Montrose Street; Glasgow G1 1XJ, UK
[2]Lloyd's Register EMEA; 71 Fenchurch Street; London EC3M 4BS, UK
[3]Center of Maritime Technologies e.V.; Bramfelder Str. 164; D-22305, Hamburg, Germany

Keywords: Friction stir welding, Low alloy steel, Fatigue testing, Fracture surface

Abstract

A fatigue performance assessment of 6 mm thick friction stir welded DH36 steel has been undertaken, filling a significant knowledge gap in the process for steel. A comprehensive set of experimental procedures has been proposed; the consequent study extensively examined the weld microstructure, hardness, geometry and misalignments of the samples in support of the tensile and fatigue testing. The effect of varying weld parameters was also investigated.

The typical fatigue performance of friction stir welded DH36 steel plates has been established, exhibiting considerably extended fatigue lives, well above 10^5 cycles at a stress range of 90% of yield strength, irrespective of minor instances of small surface breaking flaws which have been identified. An understanding of the way in which these flaws impact on the fatigue performance has been developed, concluding that surface breaking defects emanating from the friction stir tool's shoulder marks on the weld top surface can act as the dominant factor for crack initiation under fatigue loading.

Introduction

Recently, there has been a fair amount of progress in the development of the fundamental knowledge on friction stir welding (FSW) of steel. A prior publication [1] for instance has established an understanding of the link between the complex metallurgical system that FSW of DH36 steel produces and the resultant mechanical properties through microstructural characterisation and mechanical property testing, also expanding on the commonly applied welding speeds. However, one important mechanical property of steel friction stir welds, fatigue, requires to be investigated in more detail. Fatigue is considered to be the most important failure mechanism for steels; in particular, it is commonly quoted that fatigue accounts for almost 90% of the recorded mechanical service failures [2]. The fatigue life of welded components, where the weld itself contains process related flaws from which cracks can quickly initiate, even in the best quality welds, is commonly much reduced when compared to components that are unwelded. As one example, lack of weld penetration is widely reported as a highly detrimental feature in terms of fatigue life [3].

There is a growing number of publications for FSW of aluminium and other metals examining the materials' fatigue strength. Indicatively for aluminium, Ericsson and Sandstrom [4] investigate the effect of varying welding speed on the fatigue performance of friction stir butt welded high strength Al6082. Although this aluminium focussed research employs only two welding speeds for FSW, thus limiting the value of the analysis, the fatigue strength of FSW is found to be practically unaffected by speed increasing within the industrially acceptable range [4]. A thorough study [5] on the FSW of stainless steel examines the fatigue behaviour of welded AISI 409M ferritic stainless steel with regard to the parent material (PM) properties. The original coarse PM grains are seen to be transformed by FSW into a refined ferrite / martensite banded structure of significantly higher hardness. The resultant dual phase microstructure is responsible for an improvement to fatigue life with regard to the PM; this is attributed to the superior tensile properties and an advantageous post-weld residual stress distribution [5].

Although material property data are gradually being generated for FSW of low alloy steel, the relevant publications evaluating its behaviour under fatigue loading are very limited; one noteworthy study [6] is evaluating the technical potential of FSW as a shipbuilding welding process and how it compares to submerged arc welding (SAW) of DH36 steel. An acicular shaped ferrite microstructure is observed in the thermo-mechanically affected zone (TMAZ), consistent over the mid-thickness of all FSW samples, whereas SAW samples present a typical acicular ferrite microstructure defined around proeutectoid ferrite grains [6]. The SAW plates present substantially more distortion than the FSW plates of the same thickness, and impact toughness levels for FSW and SAW samples are similar and within classification society impact requirements. A relatively limited fatigue testing programme demonstrates that FSW samples exhibit better fatigue performance than the SAW samples of equivalent thickness [6].

Due to the significance of a solid understanding of the fatigue behaviour in the wider acceptance of the process on steel and the lack of pertinent studies on low alloy steel, a detailed and extensive fatigue testing programme of FSW of steel grade DH36 was undertaken and is reported herein. This novel programme assesses the fatigue behaviour of FSW by testing a statistically broad number of samples in constant amplitude uniaxial tensile loading and generating the S-N (stress-life) curve, also characterising the weld microstructure and analysing the fatigue samples' fracture surfaces. Since fatigue performance is a critical design requirement in most cyclically loaded structures as in shipbuilding and other transportation systems, the findings are expected to further support the case for FSW of steel in a wider industrial environment.

Experimental procedures

Steel grade DH36 plates with original dimensions of 2000 mm x 200 mm and of 6 mm thickness were butt welded together at three traverse speeds (100, 250 & 500 mm/min) by The Welding Institute (TWI Yorkshire) using a WRe-pcBN FSW tool. These welding speeds were selected as representative of the welding speeds explored in a previous study [1]. The specific preparation stages, i.e. sectioning, machining and polishing of the fatigue and tensile test samples adhered strictly to BS 7270 [7], and their basic dimensions are illustrated in Figure 1. The samples' sides were polished longitudinally to diminish the contribution of any transverse machining marks to the fatigue performance; still, the samples' top and bottom surfaces were tested in the "as-welded" condition. Since there are no internationally accepted standards for the testing and assessment of welded components under fatigue, a comprehensive and detailed experimental procedure for the fatigue assessment of FSW of steel was employed and is outlined below:

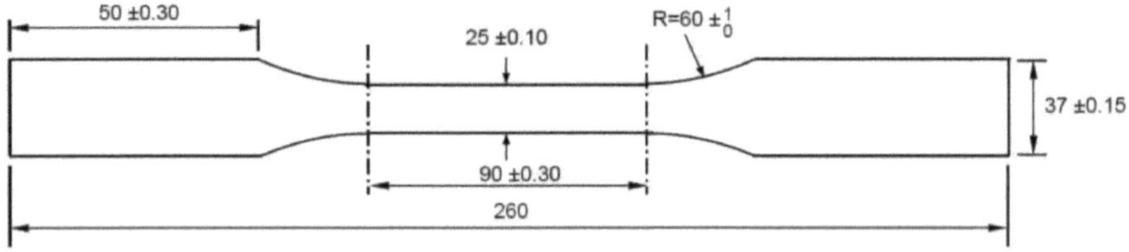

Figure 1. Transverse fatigue and tensile test sample of rectangular cross section (6 mm thick).

- Metallographic examination, to assess the quality of each weld and correlate the observed microstructure to its fatigue performance.
- Hardness measurements, recorded for several positions which were deemed representative of the weld zone, consistently for all three welds.
- Geometry and misalignment measurements; clamping on the fatigue testing machine and consequent axial loading of a sample incorporating such irregularities can induce tensile or compressive stresses on any surface breaking flaws, thus accelerate or hinder any cracks which may initiate from these. Hence, possible irregularities on the samples were measured using strain gauges installed on both sides of three fatigue samples for each stress range by employing a precise and detailed procedure, and a coordinate measuring machine (CMM) to determine the samples' top surface geometry and likely weld misalignment.
- Tensile testing; three samples per weld were subjected to transverse tensile testing in order to identify the yield strength (YS) of the weldment. The trend reported in the previous work [1] was confirmed in this study; all slow and intermediate weld samples fractured in the PM, i.e. the weld exhibits higher tensile strength than the PM, while the fast weld samples fractured in the advancing (AD) side of the weld. The average YS value from the three intermediate weld samples, which was used for calculating all welds' fatigue testing parameters is 382 MPa.
- Fatigue testing was carried out on an Instron 8802 fatigue testing system. A large number of samples were tested for three welding speeds, with the emphasis placed on the intermediate welding speed (samples tested for three stress ranges, i.e. 70%, 80% and 90% of YS). The selection of appropriate stress ranges was informed by trial tests which were initially performed, commencing with stress range of 80% of YS. The effect of varying welding parameters was established by testing samples from the slow and fast speed welds at one stress range (80% of YS) and comparing these results with the basic S-N curve of the intermediate weld. The main variables of each stress range that was tested are summarised in Table I. The stress ratio was maintained equal to 0.1 and the stress frequency was kept constant at 10 Hz during the testing programme. The actual stresses attained by the testing machine vary insignificantly from the calculated values (no more than 0.1%).

Table I. Summary of calculated values of the fatigue testing main variables

Weld speed	Stress range		Maximum stress	Minimum stress	Mean stress	Amplitude
	% of YS	$\Delta\sigma$ (MPa)	σ_{max} (MPa)	σ_{min} (MPa)	σ_m (MPa)	σ_a (MPa)
Intermediate	70	240.7	267.4	26.74	147.1	120.3
	80	275.0	305.6	30.56	168.1	137.5
	90	309.4	343.8	34.38	189.1	154.7
Slow	80	275.0	305.6	30.56	168.1	137.5
Fast	80	275.0	305.6	30.56	168.1	137.5

Microstructural evaluation

The following nomenclature for the various weld zones is identical to the one presented in the earlier study [1]. The metallographic preparation of all samples was performed in a way that the AD side is seen on the left side of the images. A heterogeneous microstructure is exhibited by the weld at 250 mm/min – 300 rpm (intermediate traverse speed); this consists of acicular shaped bainitic ferrite rich regions and ferrite predominant regions of either acicular shape or of random geometry (Figure 2a). The analysis in the prior publication [1] had reported identical phases and concluded that the heterogeneity of the microstructure does not affect the weld's mechanical properties. This has again been verified in the fatigue behaviour of the weld (see later). The microstructure differs towards the bottom and outer sides of the weld, shifting to predominantly refined ferrite grains of random geometry (Figure 2b). This image also features a non-metallic inclusion which appears to have created a discontinuity in the surrounding phase, i.e. a minor cavity. The top surface of the intermediate weld seems mildly uneven, indented by the shoulder's threads. There are a number of incomplete fusion paths, or laps [8] observed particularly on the top outer retreating (RT) side (Figure 2c), with entrapped and interconnected non-metallic inclusions in various stages of oxidation (seen in different shades of grey). These laps could provide crack initiation sites during fatigue testing. The weld root has been fully fused, and the microstructure of this region comprises recrystallized ferrite and pearlite.

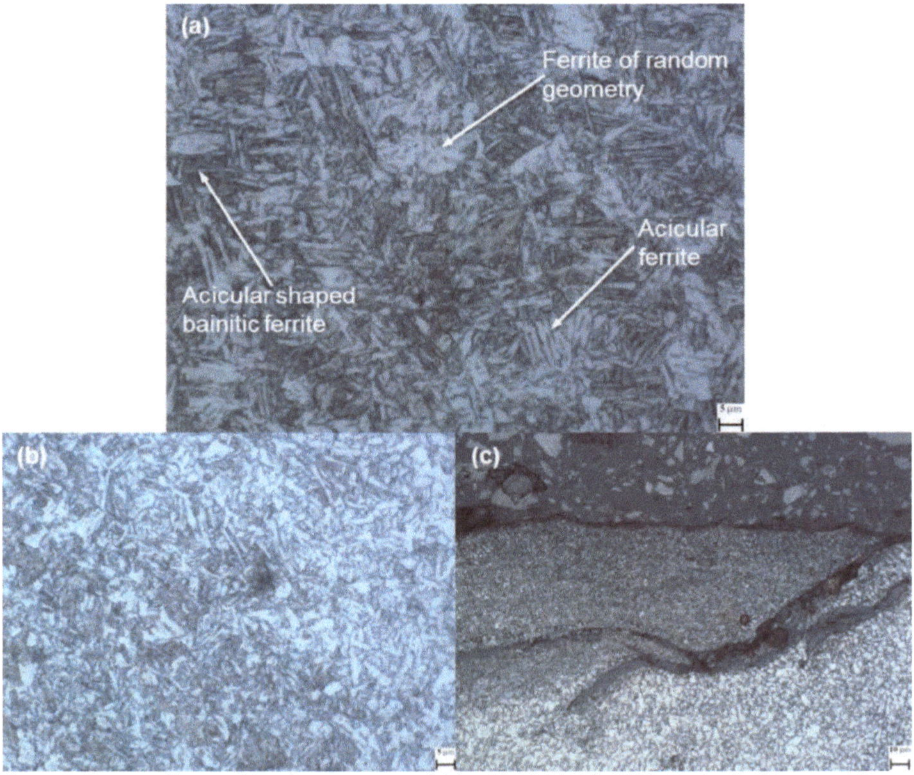

Figure 2. Intermediate weld, microstructure of (a) mid-TMAZ [x1000, Etched], (b) outer AD TMAZ [x1000, Etched] and (c) RT side top surface [x500, Etched].

The slow traverse speed weld (100 mm/min – 200 rpm) presents a ferrite predominant homogeneous microstructure with significant grain refinement in comparison to the PM. The ferrite grains appear to be of random geometry, with minor traces of small acicular shaped grains (Figure 3). As expected for this mild set of welding parameters, no flaws are visible in the bulk of the TMAZ, the transition from the heat affected zone (HAZ) to the TMAZ is smooth, and the non-metallic inclusions introduced in the weld from the plates' surfaces are not completely mixed but remain interconnected.

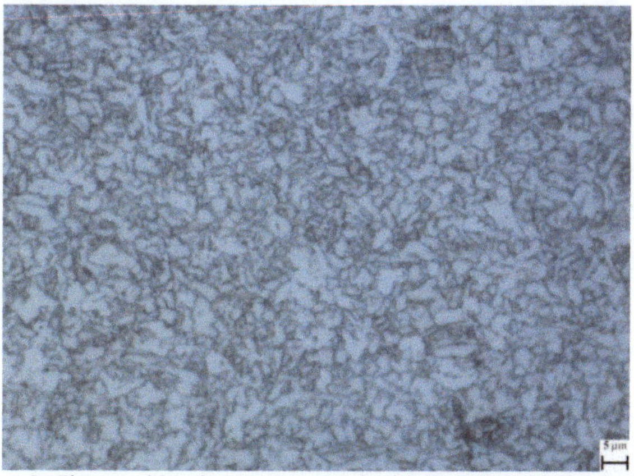

Figure 3. Slow weld, microstructure of mid-TMAZ [x1000, Etched].

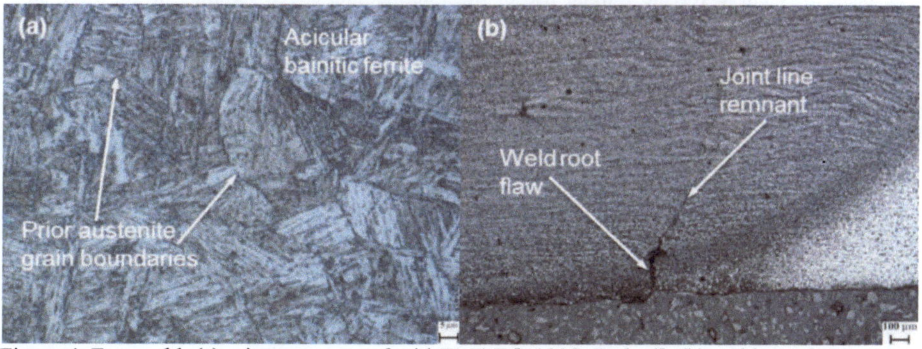

Figure 4. Fast weld, (a) microstructure of mid-TMAZ [x1000, Etched], (b) weld root [x50, Etched].

The fast traverse speed weld (500 mm/min – 700 rpm) features what seems to be a predominantly acicular shaped bainitic ferrite microstructure with small regions of acicula ferrite; this is a fairly heterogeneous structure which should contain stress concentration regions (Figure 4a). The increased bainitic content is a direct consequence of the higher cooling rate due to the higher traverse speed of this weld. More, prior austenite grain boundaries are clearly detected (Figure 4a); acicular shaped grains appear to nucleate perpendicular to these boundaries. The weld presents an uneven top surface with marks on both sides corresponding to the tool shoulder's features. An apparently intermittent insufficient fusion at the weld root is seen to develop (Figure 4b). Since there is almost no stirring action of the tool's pin on the steel in this region, the thin film of non-metallic inclusions on the surface of the two plates being welded is not fully dispersed, thus forming a joint line remnant as an extension of the weld root flaw.

Again, these are stress concentration regions which could provide crack initiation sites hence influence the weld's fatigue performance.

Hardness distribution

The micro-hardness distribution for the three welds is presented in Table II, where the values are supplied as an average of two measurements per position. The hardness values follow the anticipated order; the hardness of the weld is seen to increase as the welding speed is increased. This is attributed to the increasing cooling rate that develops harder phases such as bainite. The microstructural examination above has noted the rise in the bainite content with each speed increment. Broadly, all welds appear harder than the PM but not at levels that can cause concern.

Table II. Micro-hardness (Vickers) measurements for the three weld speeds

Weld	AD top	Mid-top	RT top	Mid-AD	Mid-TMAZ	Mid-RT	Weld root	PM
Slow	254	247	244	230	226	222	225	189
Inter.	254	257	247	266	265	247	240	169
Fast	280	303	318	306	355	356	250	184

Fatigue assessment

The basic S-N curve for the fatigue life of the intermediate speed samples in the three stress ranges is presented in Figure 5. The ultimate fracture position for 19 out of the 20 tests performed on the intermediate weld was the weld's RT side. The fracture initiation sites are found to be the lap defects observed on this side's top surface. Still, all transverse tensile samples from the intermediate weld fractured in the PM; thus, excellent tensile properties do not necessarily predict the fatigue behaviour of a weld and certainly not the fracture position. The original aim of this study was to record fatigue lives within the range of 10^5 to $2*10^6$ cycles.

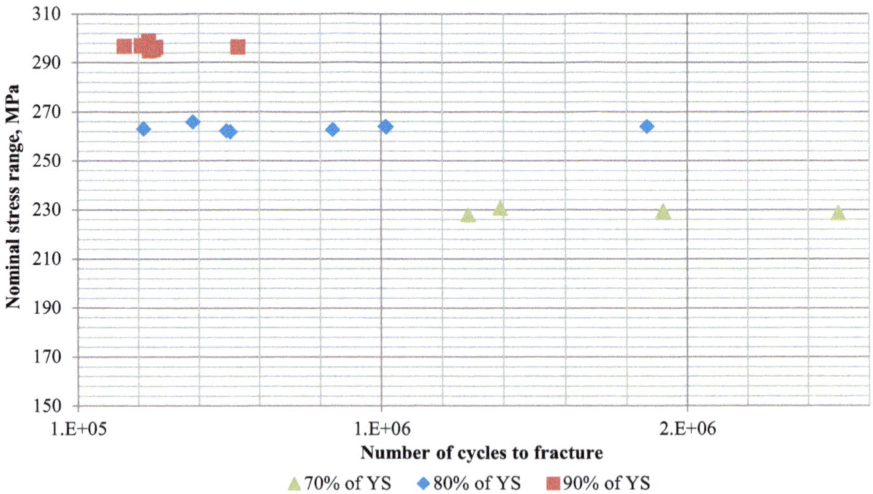

Figure 5. S-N curve for the intermediate weld.

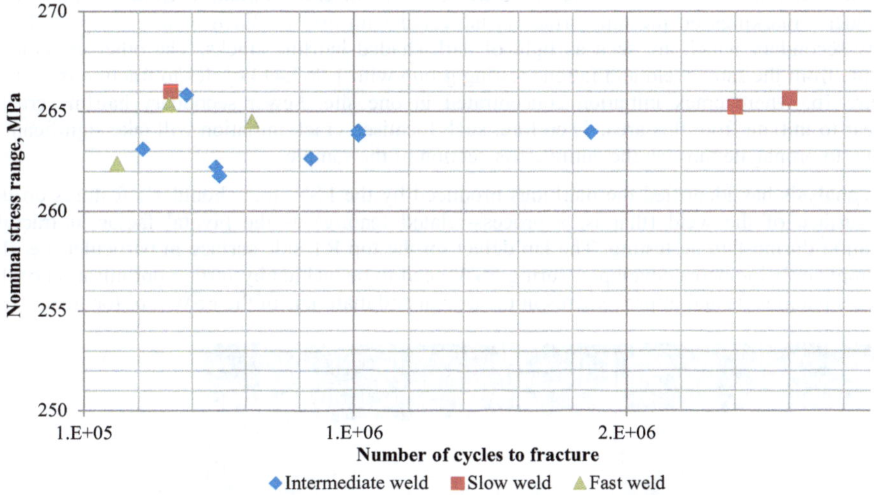

Figure 6. Fatigue test results for 3 welding speeds at 80% of YS.

The results present an excellent picture of the process on steel, with all tests reaching above 10^5 cycles. One intermediate weld test was terminated before fracture since it reached $2.6*10^6$ cycles, far beyond the predetermined objective. The surface breaking flaws appeared to be much less pronounced in this region of the weld, thus did not initiate a crack of significant length within the duration of this test. The fatigue strength of this weld at $2*10^6$ cycles is 235 MPa (Figure 5).

The fatigue test results for the slow and fast weld at 80% of YS are plotted in Figure 6 together with the intermediate weld results of the same stress range. Two samples of the slow weld exhibited excellent fatigue behaviour; the tests were terminated at $2.5*10^6$ cycles and above, and the samples displayed no evidence of crack initiation after testing. One sample fractured apparently prematurely at $4.2*10^5$ cycles (although higher than the lower limit of cycles that was set) on the AD side of the weld, and this is attributed to incomplete fusion paths observed on the weld top surface. The fast weld samples demonstrated rather mixed fatigue behaviour; all three samples recorded fatigue lives higher than 10^5 cycles but with fairly scattered results, from $2.2*10^5$ to $7.2*10^5$, and contrasting fracture regions. Fracture initiated for two samples from the weld root flaw (Figure 4b) whilst the fracture initiation region for the third sample appears to correspond to the boundary between HAZ and TMAZ on the top AD side of the weld.

Fracture surface analysis

A systematic examination of the post-failure fracture surfaces has been performed. A recurrent pattern of fracture is observed in the intermediate weld samples; principally brittle fracture occurred on the outer RT side, with cracks initiating from this side's lap defects (discussed previously). For reporting purposes, two samples' typical fracture surfaces are presented herein as all the surfaces appear almost identical. Figure 7 features the two extreme cases in terms of number of cycles to fracture (lowest and highest) for the intermediate group tested at 80% of YS in order to enhance the fine differences from one sample to another. Figure 7a presents the sample that reached 317,472 cycles to fracture; uniform crack initiation from the FSW tool shoulder's markings on the top surface is indicated with arrows. Each of these sites includes many secondary which are seen as light of dark shaded hairline cracks. The other extreme, a sample from the same weld and fatigue testing group with 1,967,444 cycles to fracture is seen in Figure 7b, where crack initiation concentrated in one site. Fewer secondary hairline cracks appear to initiate from this area. Therefore, such localised crack initiation will take significantly longer to propagate through the entire cross-section of the sample.

This analysis has identified the markings produced by the FSW tool shoulder's features on the top surface of the weld (that is, a process-related feature) as the pivotal factor in fracture initiation during fatigue testing. The lap defect on the top RT side surface in particular, i.e. the incomplete fusion paths developed during FSW need to be tackled by further optimisation of the process parameters and improvements in the design and materials of the FSW tool for steel.

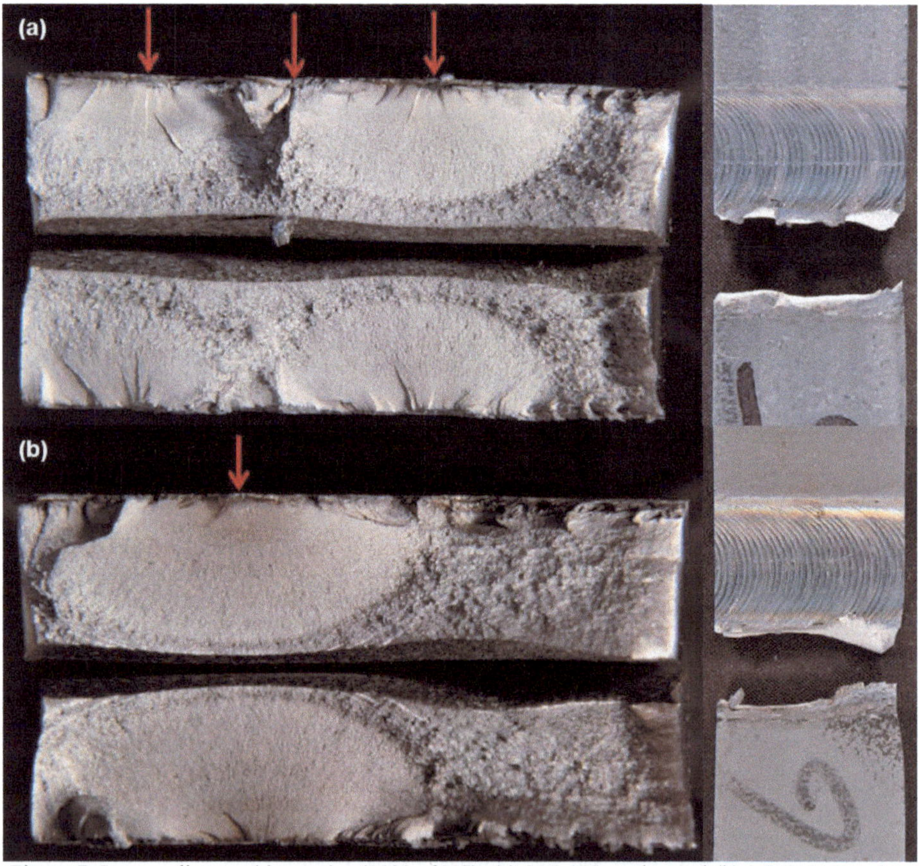

Figure 7. Intermediate weld samples (80% of YS), (a) 317,472 cycles and (b) 1,967,444 cycles.

Conclusions

- A fully developed programme of fatigue assessment of 6 mm thick DH36 steel friction stir welds has been undertaken following a comprehensive set of experimental procedures, also accounting for the effect of varying welding speed.
- Steel friction stir welds have exhibited exceptional fatigue lives, well above 10^5 cycles to fracture, even at a stress range of 90% of yield strength.
- An understanding of the relation between minor surface breaking weld flaws and the weld's fatigue performance has been established. The consistent fracture position has demonstrated that these almost negligible flaws can influence the fatigue life, justifying the need for full penetration welds and for addressing the tool shoulder's markings on the top surface.
- Tensile testing samples fracturing in the parent material and away from the weld proves that except for the yield strength, static loading does not carry important information on the steel's performance in real environments, where cyclic loading is the dominant stress mechanism.

Acknowledgements

The authors gratefully acknowledge the financial support of the European Union which has funded this work as part of the Collaborative Research Project HILDA (High Integrity Low Distortion Assembly) through the Seventh Framework Programme (SCP2-GA-2012-314534-HILDA).

References

1. A. Toumpis, A. Galloway, S. Cater, and N. McPherson, "Development of a process envelope for friction stir welding of DH36 steel - A step change", *Materials and Design*, 62 (2014), 64-75.

2. Flake C. Campbell, ed., *Elements of Metallurgy and Engineering Alloys* (Materials Park, OH: ASM International, 2008).

3. Stephen J. Maddox, *Fatigue Strength of Welded Structures*, 2nd ed. (Cambridge, U.K.: Woodhead Publishing, 2002).

4. M. Ericsson and R. Sandstrom, "Influence of welding speed on the fatigue of friction stir welds", *International Journal of Fatigue*, 25 (2003), 1379-1387.

5. A.K. Lakshminarayanan and V. Balasubramanian, "Assessment of fatigue life and crack growth resistance of friction stir welded AISI 409M ferritic stainless steel joints", *Materials Science and Engineering: A*, 539 (2012), 143-153.

6. N.A. McPherson, A.M. Galloway, S.R. Cater, and S.J. Hambling, "Friction stir welding of thin DH36 steel plate", *Science and Technology of Welding and Joining*, 18 (5) (2013), 441-450.

7. British Standards Institution, "BS 7270:2006. Metallic materials - Constant amplitude strain controlled axial fatigue – Method of test" (London, U.K. 2006).

8. William T. Becker and Roch J. Shipley, eds., ASM Handbook, vol. 11 (Materials Park, OH: ASM International, 2002).

Friction Stir Welding and Processing VIII
Edited by: Rajiv S. Mishra, Murray W. Mahoney, Yutaka Sato, and Yuri Hovanski
TMS (The Minerals, Metals & Materials Society), 2015

USE OF HIGH-POWER DIODE LASER ARRAYS FOR PRE- AND POST-WELD HEATING DURING FRICTION STIR WELDING OF STEELS

B. Baker[1], T. McNelley[2], M. Matthews[3], M. Rotter[3], A. Rubenchik[3], and S. Wu[3]

[1]United States Naval Academy, Annapolis, MD
[2]Naval Postgraduate School, Monterey, CA
[3]Lawrence Livermore National Laboratory, Livermore, CA

Keywords: Friction Stir Welding, High-Power Diode Laser Array, Laser Assisted Friction Stir Welding, Hybrid Friction Stir Welding

Abstract

In this research a high-power diode laser array was used to preheat HY-80 steel to determine the efficacy of using a diode laser array for preheating prior to friction stir welding in order to reduce frictional forces thereby reducing tool wear and increasing welding speeds. Using instrumented plates the temperature profile using diode heating alone was determined in order to validate theoretical models. A high-power diode laser is proposed as a more cost effective and efficient means of preheating compared to other hybrid friction stir welding techniques. Parameters of the diode array were easily controllable and resulted in a preheated area that very closely matches the typical stir zone observed in friction stir welds. A proposed diode laser assisted friction stir welding system is presented, and it is hypothesized that the addition of diode laser preheating will improve tool life and/or increase welding speeds on steels.

Introduction

Wear of tools used during friction stir welding (FSW) is a critical issue and is discussed in detail by Mishra and Ma as well as several other authors [1-4]. For softer metals such as aluminum, frictional forces between the tool and workpiece are relatively low leading to less tool wear thereby allowing the use of simpler tool designs that can have a long useful lifetime; however, FSW of harder materials such as steel alloys results in significantly larger frictional forces causing reduced tool life and requiring stronger, more expensive tools to compensate [5]. The combination of increased tool cost, tool wear, and reduced welding speeds is one of the primary industrial limitations of FSW on steels.

In order to reduce the heat generation due to tool action that is required to achieve FSW, various methods of workpiece preheating have been researched with varying degrees of success and practicality. These methods are often referred to as hybrid FSW techniques and include: electrically assisted FSW [6], vibration assisted FSW [7], conduction assisted FSW [8], gas tungsten arc assisted FSW [9-11], ultrasonic assisted FSW [12], high frequency induction assisted FSW [13], plasma assisted FSW [14], and laser assisted friction stir welding (LAFSW) [15-21]. Several of these methods (electrically assisted, vibration assisted, conduction assisted, and ultrasonic assisted) require an additional physical contact with the workpiece and rely on transfer of energy through the workpiece to the weld area. This limitation is highly application and material dependent and therefore may not be viable for a broad range of applications. The remaining methods (gas tungsten arc assisted, high frequency induction assisted, plasma assisted, and laser assisted) all transmit their energy directly to the workpiece without physical contact; however, they differ widely in their precision and controllability of energy transfer to the weld area alone. Because of the precision and controllability possible with laser material processing LAFSW is considered to be a viable method to reduce the frictional forces experienced during traditional FSW and thereby a practical method to improve tool life and/or welding speed.

Several research groups have demonstrated the ability to combine traditional FSW capability with various laser systems to create functional LAFSW systems. One of the earliest of these systems was developed by Kohn who used a 700 Watt Nd:YAG laser during FSW to join two magnesium alloy plates in a butt weld configuration [21]. In this early demonstration, the efficacy of using a comparatively small laser system to preheat a very small area ahead of the FSW tool (1 cm diameter) to approximately 320°C without detrimental effect to the base workpiece was shown. The plunge and translational frictional forces were determined to be negligible due to the preheating provided by the laser. The preheating used by Kohn raised the relatively soft AZ91D Mg alloy to over one half of its melting temperature leading to the significant reduction in frictional forces during FSW. Since that early demonstration, different LAFSW setups have been used on aluminum and magnesium alloys in traditional butt weld configurations [16, 17, 22, 23] and lap weld configurations [15], as well as nickel based alloys [19], and steels [20, 24] although the total number of LAFSW demonstrations remains limited. In a series of works on 5754H111 aluminum, Campanelli and Casalino evaluated the changes in microstructure, micro-hardness, residual stress, and tensile properties following LAFSW at typical FSW parameters of 500 revolutions per minute (RPM) and 20 millimeters per minute (MMPM) with preheating provided by a 4 kilowatt (kW) Ytterbium fiber laser. It was concluded that LAFSW was capable of improving material conditions of the weld nugget [16, 22, 23]. Adding to the complexity of temperature control during LAFSW, Sun developed a LAFSW setup that used a YAG laser with a conventional FSW machine and observed the formation of a brittle martensite phase within a carbon steel workpiece that could be prevented by judicious selection of welding parameters, most notably tool traverse speed, in order to control the cooling rate following welding [20, 24]. The formation of martensite in carbon steels during rapid cooling from elevated temperatures requires additional control of temperature during preheating, welding, and post-weld cooling that does not exist in other alloy systems. This complexity prevents the indiscriminate heating of steels to high temperatures solely to reduce frictional forces during FSW.

Unlike the large and generally complex Nd:YAG or CO_2 lasers used in current proposed LAFSW systems, high-power diode lasers (HPDL) are smaller, more efficient, and comparatively easier to operate. High-power fiber lasers, which are relatively new, are more compact than other fiber lasers but are more expensive than the diode arrays which pump them. The advantages and potential applications of HPDL systems are discussed in detail in reviews by Li [25], Chen [26], and Bachmann [27]. The programmatic advantages of HPDL systems compared to other laser systems include: lower running cost, higher energy efficiency, smaller size, simpler alignment, and longer service life [25]. Additionally HPDLs can have adjustable wavelengths within a wide range allowing for increased absorption and efficiency. The changes in absorptivity of metals due to variance in irradiance, wavelength, polarization, and angle of incidence as well as due to variance in material properties such as composition, surface roughness, oxide layers, and contamination are beyond the scope of this review, but these aspects as well as the performance of HPDLs on different metals including aluminum, titanium, and steels are discussed in other work by the authors [28]. For the previous reasons HPDL systems have been used for laser assisted machining and laser surface heat treatment for a variety of steels [29-32]; however, the majority of these applications involve rapid heating of steels into the austenite region followed by rapid cooling to form martensite resulting in improved hardness of selected regions of the workpiece. In the proposed application, a HPDL would be used to heat the workpiece (ideally below the austenite region) immediately prior to FSW such that the stir zone (SZ) is preheated and softened resulting in reduced tool wear and overall less heat input from the FSW process. Subsequent cooling could also be assisted by a laser heating process to minimize the cooling rate to prevent the formation of martensite. Given the flexibility of HPDL systems, preheating and post FSW cooling could potentially be controlled; however, the current study focuses primarily on HPDL preheating.

The objectives of this research are to: (1) validate theoretical models of temperature profiles of typical steels using HPDL heating alone, (2) use these theoretical models to establish operating parameters of a HPDL system on a typical hardenable steel, and (3) propose a diode laser assisted FSW (DLAFSW) setup that utilizes laser diode arrays for preheating of the workpiece. These objectives were accomplished by initial laser heating tests on a moving sample workpiece with embedded thermocouples. Subsequent research objectives will determine the effect on FSW tool life with the addition of laser preheating, and determine the material response due to the combined effect of FSW and HPDL preheating. Based on the current research, it is hypothesized that the addition of preheat using laser diode arrays will significantly lower frictional forces, lessen tool wear, and allow for higher tool traverse speeds.

Experimental Procedure

Based on previous FSW experience and its similarity to other hardenable steels, the low carbon alloy steel HY-80 was selected for this study. Plates of HY-80 (MIL-S-16216) were 6.4 mm (0.25 inches) thick and of varying widths and lengths. HY-80 (0.12-0.18 wt% carbon) is used in Navy applications as well as some pressure vessels and acquires its strength and toughness through quenching and tempering treatments. Additional data on HY-80, its use in Naval applications, and response due to FSW alone can be found elsewhere [33-35]. Previous research on FSW of HY-80 has shown that peak temperatures in the SZ are sufficient to create austenite which upon cooling will become tempered or untempered martensite based on the cooling rates. Depending on the intended application the formation of martensite in the weld nugget can be beneficial or detrimental, but in most applications requiring toughness, the formation of a brittle untempered martensite phase is undesirable. Therefore consideration must be given to cooling rates following joining in order to establish a tempered martensitic microstructure. Although considered and established as a future research goal, no specific post welding heat treatment was applied in this research.

To determine temperature distribution during laser heating experiments alone, plates of HY-80 were instrumented with Pt/Re thermocouples at depths varying from 0.635 mm (0.025 inches) to 3.175 mm (0.125 inches) at intervals of 25.4 mm (1 inch) in the direction corresponding to the welding direction (Figure 1). In order to maximize material usage, up to three laser heating passes were accomplished on each plate (Figure 1b). In these cases the preheating runs that passed directly over the line of thermocouples (e.g. the 25A pass in Figure 1b) was used for comparison to theoretical data. Temperature measurements were used for comparison to theoretical models to determine laser operating parameters and verify temperature modelling.

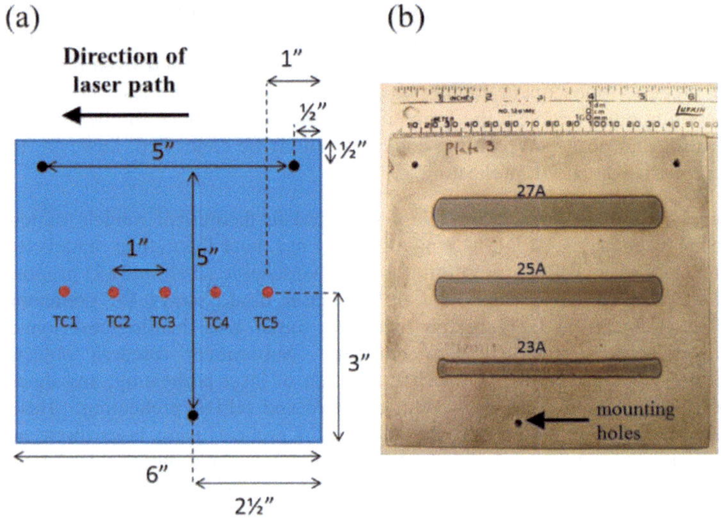

Figure 1: (a) Dimensions and thermocouple (TC) locations in HY-80 plate, (b) top view of actual HY-80 plate with three different laser preheating runs showing the size and shape of the heated area. In (a) thermocouples are inserted from the bottom of the plate at varying depths to determine the temperature profile through the thickness of the plate. In (b) up to three laser preheating passes were accomplished on each plate for material conservation.

The HY-80 workpiece was placed on a motion control stage with an air gap between the workpiece and the stage primarily to allow for thermocouple insertion into the workpiece and to protect the stage from heating damage. This is different than would exist in an actual DLAFSW setup since the workpiece would be mounted to an anvil that would further conduct heat from the workpiece. The moving stage moved the workpiece underneath the laser path with the HPDL stationary at desired speeds which were chosen to correspond to typical FSW traverse speeds. A cartoon summary of the setup is shown in Figure 2 and an actual image of the setup is shown in Figure 3. The experimental setup included the following major components: HPDL array with integrated polarization rotator, HY-80 workpiece, moving stage, CCD camera for video recording, thermocouples embedded in the workpiece, and IR camera to measure surface temperature. The HPDL used was a 5 kW diode array that emits incoherent light at a wavelength of 795 nm. Full details of the specification of the HPDL will be published separately.

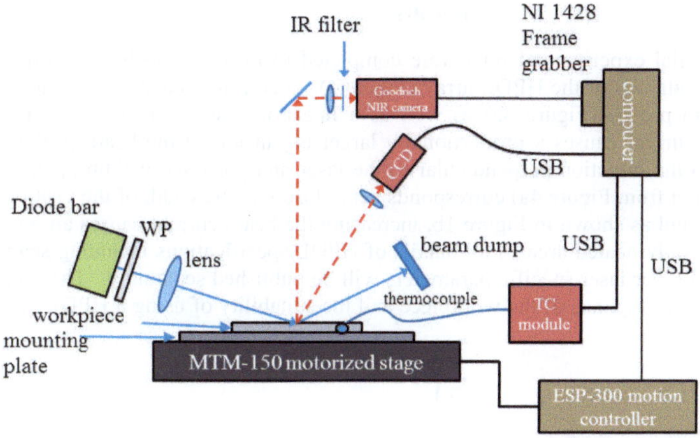

Figure 2: Schematic of the experimental setup used. In this setup the workpiece is moved underneath the HPDL array vice the HPDL moving in front of a FSW tool as would exist in a practical DLAFSW setup.

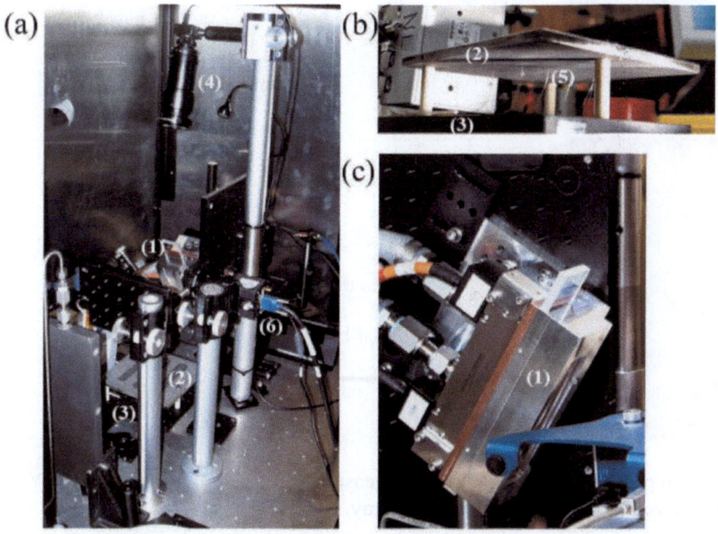

Figure 3: Views of the experimental setup: (a) full view showing all components, (b) close up view of workpiece, thermocouples, and stage, and (c) close up view of the HPDL array. Component numbering: (1) HDPL array, (2) HY-80 workpiece, (3) moving stage, (4) CCD camera, (5) thermocouples, and (6) IR camera.

Results

A series of initial experimental runs were completed to measure the beam profile and establish operating conditions of the HPDL array (Figure 4). Increasing the laser current causes a change in the beam profile (Figure 4a) as well as a linear increase in power (Figure 4b). Increasing the beam current causes a proportionally larger expansion of the beam in the slow axis corresponding to the direction perpendicular to the laser and proposed welding path. The beam width (y-direction from Figure 4a) corresponds very closely to the width of the visibly heat regions in Figure 1b, and as shown in Figure 1b, increasing the beam current causes an increase in the width of the visibly heated area. Full details of HPDL specifications including scattered light, optics used, and other laser specific parameters will be published separately. The focus of this paper is temperature response of the workpiece and the suitability of using a HPDL array in combination with FSW.

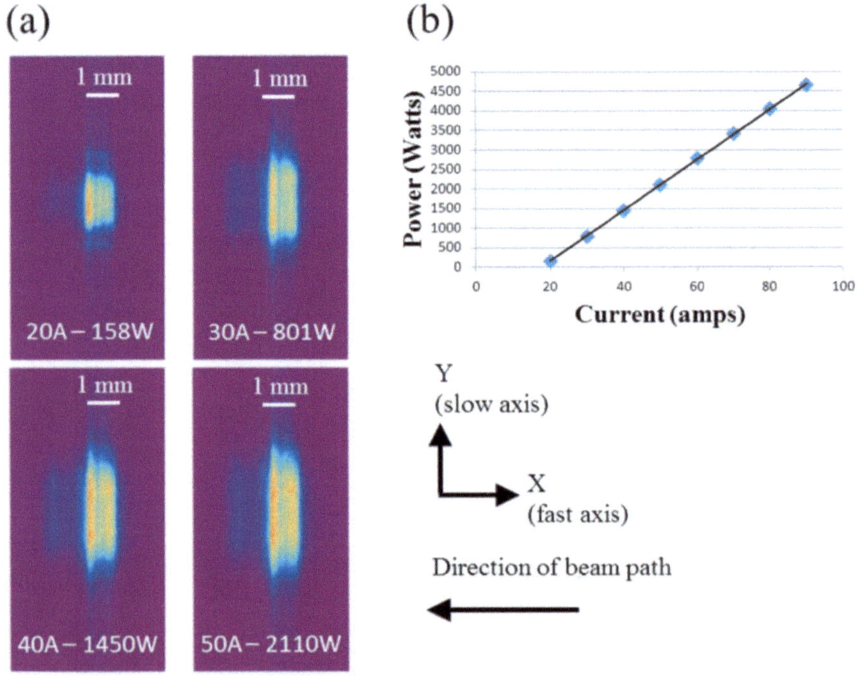

Figure 4: (a) HPDL beam profile at varying currents (measured on a beam dump not the HY-80 workpiece) and (b) power versus current of the HPDL array.

An example of proposed optimal conditions is shown in Figure 5. In this case the workpiece was preheated by the HPDL initially operating at 25A with a 10 sec warmup time followed by a traverse speed of 1.67 mm/sec. This traverse speed correlates to 100 MMPM and was based on previous FSW research on HY-80 that suggests this is a reasonable traverse speed for reasonable tool rotational speeds [35]. After 3 seconds of motion the laser current was increased to 30A (note the increase in heated area of Figure 5 going from right to left). Near the end of the proposed weld path the laser current was reduced to 25A (note the reduction in heated area of Figure 5 going from right to left) prior to securing laser in the vicinity of thermocouple 1 (TC1). The thermocouples in Figure 5 are positioned progressively further from the surface in the line of the laser path (from TC5 to TC1) and as such the subsequent thermocouples each reach a lower peak temperature with the exception of TC5 since the laser began moving away from TC5 before it has fully heated.

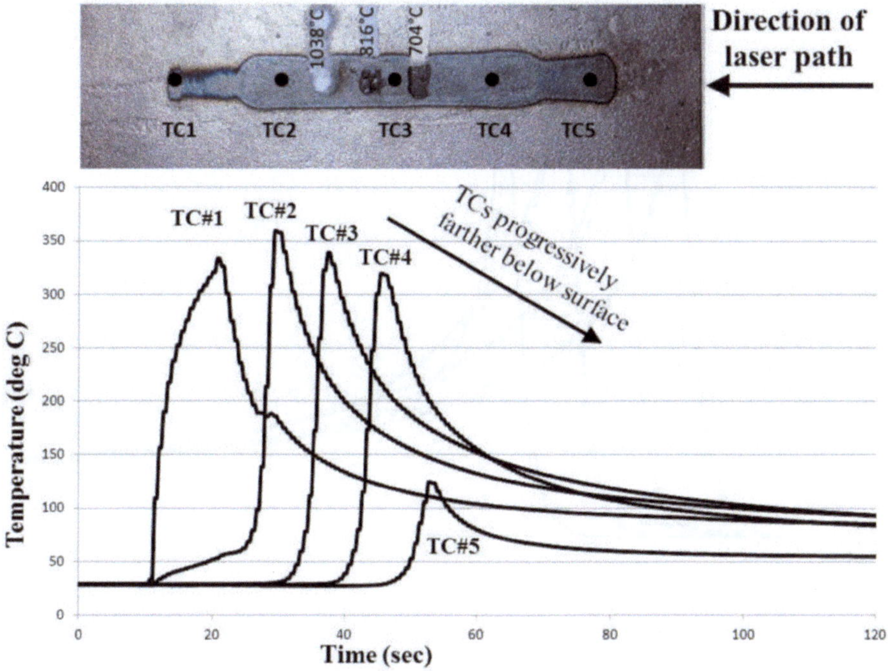

Figure 5: Temperature response of HY-80 workpiece for proposed optimal conditions. In this run, the HPDL is operated at 25A for 10 seconds prior to traversing, traverses at a rate of 1.67 mm/sec, increases to 30A after 3 seconds of motion, and lowers to 25A near the end of the weld path. Thermal lacquers were utilized on the surface of the workpiece and show the approximate surface temperature of the workpiece.

Temperature response of the HY-80 workpiece was modeled using FlexPDE. The heat input from the HPDL was modeled using the actual beam profiles (Figure 4a) and beam power from beam current (Figure 4b). Experimental to theoretical results for a 25A run at 1.67 mm/sec (100 MMPM) are shown in Figure 6. In this case the laser is operating continuously at 25A with a 10 second warmup time prior to traversing followed by traversing at 1.67 mm/sec (100 MMPM). Unlike the case shown in Figure 5, only 4 thermocouples were used (due to a failure of one thermocouple) and the thermocouples are in increasing proximity to the surface (i.e. TC2 is closer to the surface than TC4) resulting in higher peak temperatures progressing down the thermocouple line. This arrangement of thermocouple depths is opposite to the depths listed in Figure 5 explaining the differences in peak trends between Figure 4 and Figure 5.

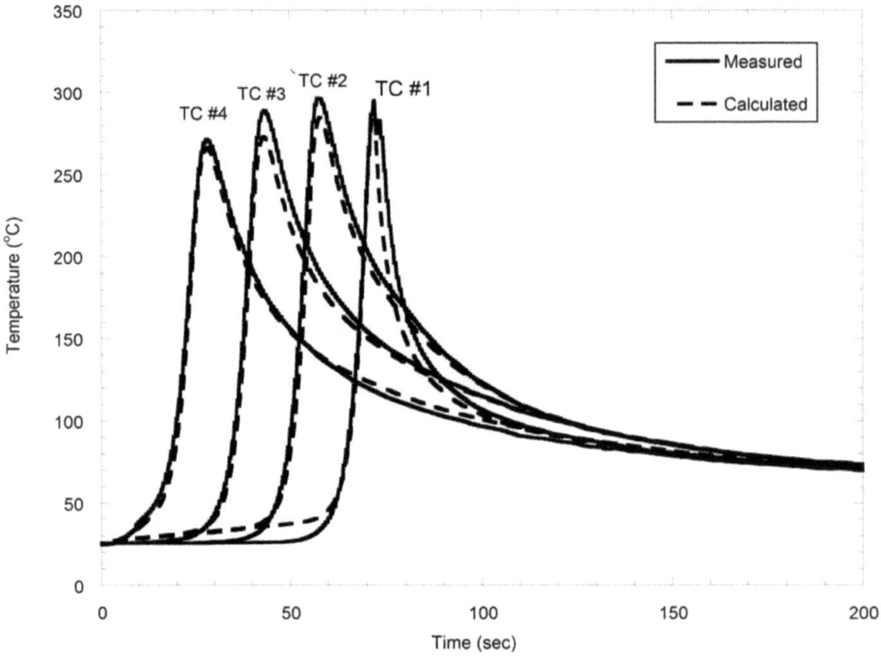

Figure 6: Theoretical to experimental correlation of temperature versus time for a proposed laser preheating run consisting of a 10 second warmup time at 25A followed by 1.67 mm/sec traverse speed at a steady current of 25A.

Theoretical results from FlexPDE in planar view (Figure 7a) and transverse view (Figure 7b) show the theoretical temperature field due to laser preheating alone. In addition to temperature data from thermocouples, thermal lacquers were used on the surface of the workpiece to establish some experimental bounds on the surface temperature of the workpiece.

Temperature sensitive thermal lacquers (Omega Engineering Inc., CT, USA) have been used previously for direct measurement of surface temperatures in other laser assisted processing by the authors and results compared closely with other temperature monitoring devices [36]. An example of the comparison between the theoretical and experimental temperature bounds determined by the thermal lacquers is shown in Figure 8a and 8b. Figure 8a is a magnification of theoretical data from Figure 7 and depicts the temperature field for a 25A pass at 1.67 mm/sec. The experimental surface temperature response of this run is shown in Figure 8b. As the thermal lacquers change color corresponding to their temperature value, the width of the changed area can be compared to the theoretical surface temperature. Comparing these methods, an area of approximately 2 cm in width (i.e. perpendicular to the laser and proposed welding path) is heated to approximately 300°C on the surface. For comparison, a typical friction stir weld and the FSW tool used to make that weld are shown in Figure 8c and 8d respectively.

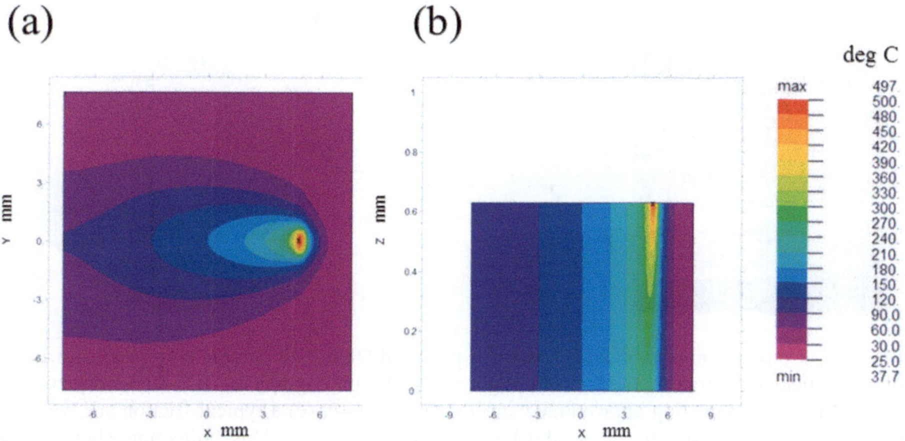

Figure 7: (a) Planar (x-y) view and (b) transverse (x-z) view of the theoretical temperature field during a steady state portion of a 25A run at 1.67 mm/sec.

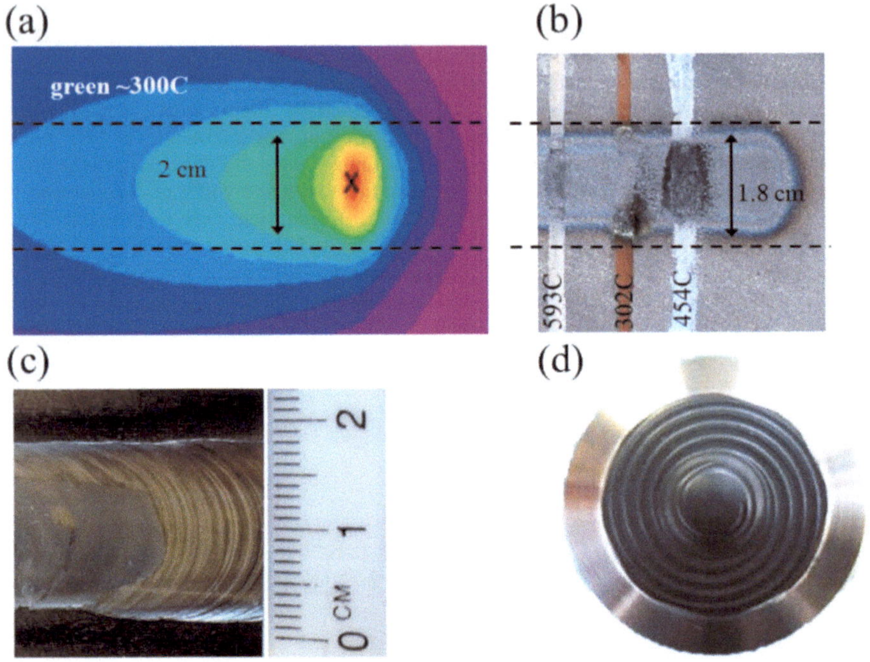

Figure 8: Comparison of preheated areas and potential DLAFSW parameters: (a) magnification of theoretical temperature field (from Figure 7a), (b) photograph of HY-80 workpiece with thermal lacquers depicting approximate surface temperature, (c) a typical friction stir weld, and (d) proposed FSW tool to be used (the tool used for (c)). This collection shows strong correlation of theoretical and experimental heated area to the proposed weld path.

Discussion

Temperature control of the preheated zone in hybrid FSW systems is critical and other hybrid methods have shown that preheating the workpiece may lead to reduced frictional forces and improved tool life. Where many of these systems have come short is practical application of the systems either through complicated connections to the workpiece or expensive and inefficient energy transfer to the weld area alone. This research has shown that a relatively small and inexpensive HPDL system can preheat a workpiece in a very controllable manner that closely approximate theoretical predictions.

In contrast to previous HPDL applications, where only the surface of the part was under consideration, in DLAFSW the entire depth of the workpiece needs to be brought up to the desired temperature (as seen in Figure 7(b)). This can be an issue for low-conductivity materials such as steel. Consequently, in order to accurately model the experiment, temperature-dependent properties for the HY-80 steel must be known. Unfortunately, sufficient temperature-dependent data for HY-80 was not available. As a plausible substitute, the properties for HY-130 [37] were used and were scaled to the curves of the room-temperature data for HY-80 [38]. Figures 9 (a), (b), and (c) show the results for thermal conductivity, density, and specific heat, respectively. Also shown in Figures 9 (a) and (b) are analytic fits used in the FlexPDE model. The specific heat data was imported into FlexPDE as a table, so no fits were used in this case. The good agreement between experiment and model suggests that the temperature-dependent properties derived outside these experiments provide a good description of the experiment. As a result, the model provides a useful tool in determining the optimum parameters needed to produce the desired temperature distribution near the tool.

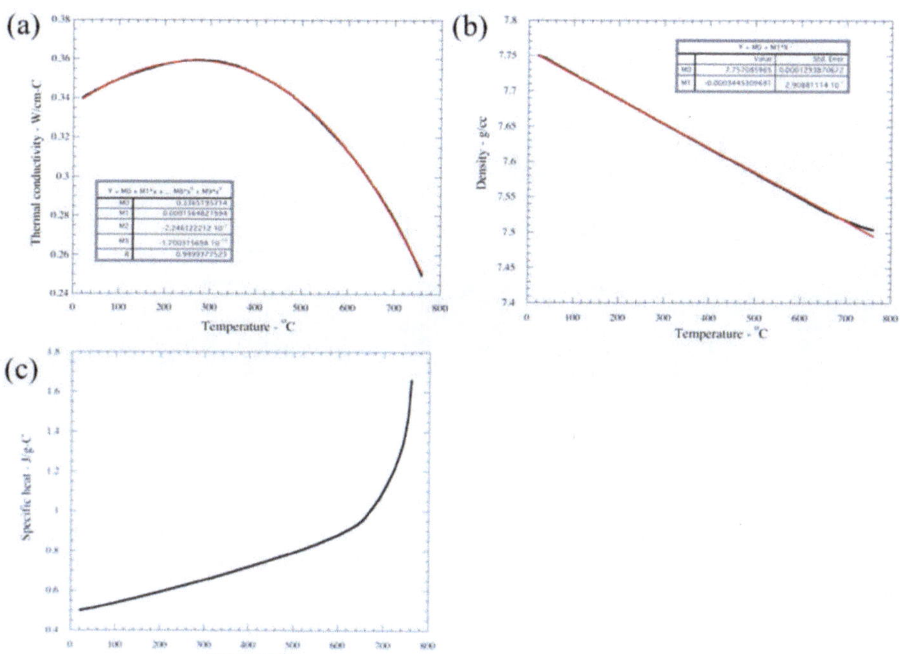

Figure 9: Temperature-dependent properties of HY-80: (a) thermal conductivity, (b) density, and (c) specific heat. Red curves in Figures (a) and (b) show polynomial fits to the curves, with fit coefficients listed in the tables. The specific heat data was imported into FlexPDE as an interpolant table, so no fits were used in this case.

One of the primary advantages of FSW is joining without melting. Although a HPDL can easily melt the HY-80 workpiece, it may also be used at varying combinations of power output and welding speeds to heat the workpiece to a desired temperature regime. This research shows that HY-80 may be heated to below the austenitic transformation region (nominally 700°C) while also heating only the weld area. Figure 10 shows the strong similarity of FSW SZs and HPDL heated areas for two different steels. This simple comparison shows the primary advantage of proposed DLAFSW system in that the parameters of the HPDL can be optimized such that the heated area corresponds only to the proposed welding area. Similarly the temperature of the workpiece immediately prior to FSW can be controlled to optimize the combined DLAFSW process. This research also shows that theoretical modeling of the laser heating can accurately estimate the temperature field within the workpiece.

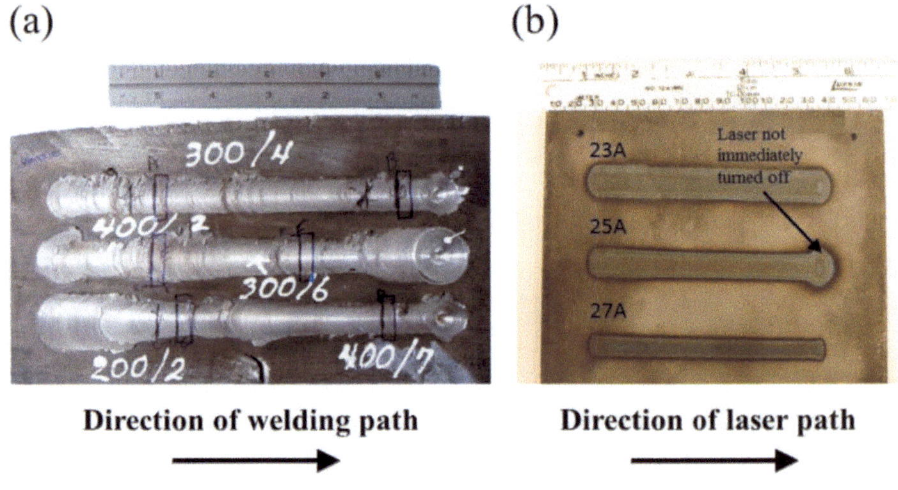

Figure 10: Comparison of (a) actual FSW runs on a steel workpiece (numbers indicate tool rotational speed in revolutions per minute to tool traverse speed in inches per minute) and (b) HPDL preheated areas on HY-80 steel with varying current.

Full material characterization of the preheated area remains to be completed and an initial proposed DLAFSW has been designed (Figure 11) that uses the exact same HPDL setup as used in this research (Figure 3c). Once the HPDL is combined with a traditional FSW system, full analysis of the reduction in frictional forces and improvement in tool life will be completed. The goal of the proposed DLAFSW setup will be to preheat the workpiece to approximately 300°C (in all cases less than the transformation temperature to austenite of 727°C) in order to reduce frictional forces and thereby reduce tool wear while also preventing the formation of untampered martensite in the workpiece.

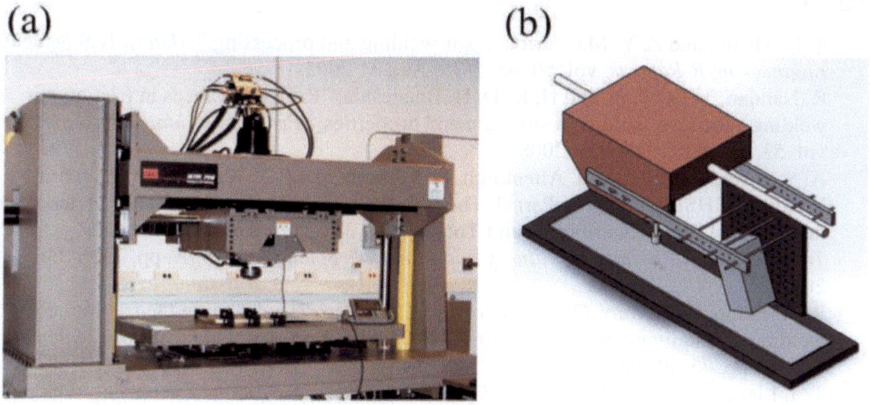

Figure 11: Proposed DLAFSW system consisting of (a) traditional FSW setup and (b) combined FSW setup with the HPDL setup used in this research. The HPDL used in this setup (including the mounting components used) have been designed to mount to the traversing head of the already existing FSW setup shown in (a).

Conclusions

The improved efficiency, compact size, and simplicity of operation of HPDLs make them good candidates for use during machining and surface treatment of steels. Rather than using the heat input from a HPDL to either machine a component or to increase its surface hardening by rapid heating and cooling, a HPDL is recommended for use in conjunction with a traditional FSW system to make a DLAFSW system. Using the experimental and theoretical data developed in this research, DLAFSW parameters may be optimized to preheat the weld area alone with limited impact on the remainder of the workpiece. It is hypothesized that preheating will cause a reduction in tool wear and/or allow for an increase in tool traverse speed for steels.

Acknowledgements

This work was completed by a collaborative research agreement between the United States Naval Academy, Naval Postgraduate School, and Lawrence Livermore National Laboratory. This work is dedicated to the memory of Howard Lowdermilk at Lawrence Livermore National Laboratory for his support of this work.

References

[1] R. S. Mishra and Z. Y. Ma, "Friction stir welding and processing," *Materials Science & Engineering R-Reports,* vol. 50, pp. 1-78, Aug 31 2005.

[2] R. Nandan, T. DebRoy, and H. K. D. H. Bhadeshia, "Recent advances in friction-stir welding - Process, weldment structure and properties," *Progress in Materials Science,* vol. 53, pp. 980-1023, Aug 2008.

[3] A. Steuwer, S. J. Barnes, J. Altenkirch, R. Johnson, and P. J. Withers, "Friction Stir Welding of HSLA-65 Steel: Part II. The Influence of Weld Speed and Tool Material on the Residual Stress Distribution and Tool Wear," *Metallurgical and Materials Transactions a-Physical Metallurgy and Materials Science,* vol. 43A, pp. 2356-2365, Jul 2012.

[4] R. A. Prado, L. E. Murr, D. J. Shindo, and K. F. Soto, "Tool wear in the friction-stir welding of aluminum alloy 6061+20% Al2O3: a preliminary study," *Scripta Materialia,* vol. 45, pp. 75-80, Jul 2001.

[5] T. J. Lienert, W. L. Stellwag, B. B. Grimmett, and R. W. Warke, "Friction stir welding studies on mild steel - Process results, microstructures, and mechanical properties are reported," *Welding Journal,* vol. 82, pp. 1S-9S, Jan 2003.

[6] W. A. Ferrando, "The Concept of Electrically Assisted Friction Stir Welding (EAFSW) and Application to the Processing of Various Metals," N. S. W. C. C. Division, Ed., ed. West Bethesda, MD, 2008.

[7] H. Montazerolghaem, M. Badrossamay, and A. F. Tehrani, "Investigation of Vibration Assisted Friction Stir Welding," in *Material Forming - Esaform 2012, Pts 1 & 2.* vol. 504-506, M. Merklein and H. Hagenah, Eds., 2012, pp. 741-746.

[8] P. C. Sinclair, W. R. Longhurst, C. D. Cox, D. H. Lammlein, A. M. Strauss, and G. E. Cook, "Heated Friction Stir Welding: An Experimental and Theoretical Investigation into How Preheating Influences Process Forces," *Materials and Manufacturing Processes,* vol. 25, pp. 1283-1291, 2010.

[9] H. Bang, H. Bang, G. Jeon, I. Oh, and C. Ro, "Gas tungsten arc welding assisted hybrid friction stir welding of dissimilar materials Al6061-T6 aluminum alloy and STS304 stainless steel," *Materials & Design,* vol. 37, pp. 48-55, May 2012.

[10] B. HanSur, B. HeeSeon, S. HyunJong, and J. SungMin, "Joint properties of dissimilar Al6061-T6 aluminum alloy/Ti-6%Al-4%V titanium alloy by gas tungsten arc welding assisted hybrid friction stir welding," *Materials & Design,* vol. 51, pp. 544-51, Oct. 2013.

[11] E. Scutelnicu, D. Birsan, and R. Cojocaru, "Research on Friction Stir Welding and Tungsten Inert Gas assisted Friction Stir Welding of Copper," *Proceedings of the 4th International Conference on Manufacturing Engineering, Quality and Production Systems (MEQAPS 2011). Recent Advances in Manufacturing Engineering,* pp. 97-102, 2011.

[12] K. Park, B. Kim, and J. Ni, *NUMERICAL SIMULATION OF PLUNGE FORCE DURING THE PLUNGE PHASE OF FRICTION STIR WELDING AND ULTRASONIC ASSISTED FSW*, Proceedings of the ASME International Mechanical Engineering Congress and Exposition, 2009.

[13] B. M. Tweedy, W. Arbegast, and C. Allen, "Friction stir welding of ferrous alloys using induction preheating," *Friction Stir Welding Processing III,* pp. 97-104, 2005.

[14] D. K. Yaduwanshi, S. Bag, and S. Pal, "Effect of Preheating in Hybrid Friction Stir Welding of Aluminum Alloy," *Journal of Materials Engineering and Performance,* vol. 23, pp. 3794-3803, Oct 2014.
[15] N. Able, F. Pfefferkorn, and Asme, *Laser-assisted friction stir lap welding of aluminum,* 2005.
[16] S. L. Campanelli, G. Casalino, C. Casavola, and V. Moramarco, "Analysis and Comparison of Friction Stir Welding and Laser Assisted Friction Stir Welding of Aluminum Alloy," *Materials,* vol. 6, pp. 5923-5941, Dec 2013.
[17] W.-S. Chang, S. R. Rajesh, C.-K. Chun, and H.-J. Kim, "Microstructure and Mechanical Properties of Hybrid Laser-Friction Stir Welding between AA6061-T6 Al Alloy and AZ31 Mg Alloy," *Journal of Materials Science & Technology,* vol. 27, pp. 199-204, Mar 2011.
[18] M. Merklein and A. Giera, "Laser assisted Friction Stir Welding of drawable steel-aluminium tailored hybrids," *International Journal of Material Forming,* vol. 1, pp. 1299-1302, Apr 2008.
[19] K. H. Song, T. Tsumura, and K. Nakata, "Development of Microstructure and Mechanical Properties in Laser-FSW Hybrid Welded Inconel 600," *Materials Transactions,* vol. 50, pp. 1832-1837, Jul 2009.
[20] Y. F. Sun, Y. Konishi, M. Kamai, and H. Fujii, "Microstructure and mechanical properties of S45C steel prepared by laser-assisted friction stir welding," *Materials & Design,* vol. 47, pp. 842-849, May 2013.
[21] G. Kohn, Y. Greenberg, I. Makover, and A. Munitz, "Laser-assisted friction stir welding," *Welding Journal,* vol. 81, pp. 46-48, Feb 2002.
[22] G. Casalino, S. Campanelli, A. D. Ludovico, N. Contuzzi, and A. Angelastro, "Study of a fiber laser assisted friction stir welding process," *High Power Laser Materials Processing: Lasers, Beam Delivery, Diagnostics, and Applications,* vol. 8239, 2012.
[23] G. Casalino, S. L. Campanelli, N. Contuzzi, A. Angelastro, and A. D. Ludovico, "Laser Assisted Friction Stir Welding of aluminum alloy lap joints: microstructural and microhardness characterizations," *Proceedings of the SPIE - The International Society for Optical Engineering,* vol. 8963, pp. 896316 (8 pp.)-896316 (8 pp.), 2014 2014.
[24] Y. F. Sun, J. M. Shen, Y. Morisada, and H. Fujii, "Spot friction stir welding of low carbon steel plates preheated by high frequency induction," *Materials & Design,* vol. 54, pp. 450-457, Feb 2014.
[25] L. Li, "The advances and characteristics of high-power diode laser materials processing," *Optics and Lasers in Engineering,* vol. 34, pp. 231-253, Oct-Dec 2000.
[26] W. Q. Chen, C. S. Roychoudhuri, and C. M. Banas, "DESIGN APPROACHES FOR LASER-DIODE MATERIAL-PROCESSING SYSTEMS USING FIBERS AND MICROOPTICS," *Optical Engineering,* vol. 33, pp. 3662-3669, Nov 1994.
[27] F. Bachmann, "Industrial applications of high power diode lasers in materials processing," *Applied Surface Science,* vol. 208, pp. 125-136, Mar 2003.
[28] A. Rubenchik, S. Wu, V. K. Kanz, M. Leblanc, W. H. Lowdermilk, M. Rotter, *et al.,* "Temperature-Dependent 780-nm Laser Absorption by Engineering Grade Aluminum, Titanium, and Steel Alloy Surfaces," *Optical Engineering,* vol. 53, December 2014.
[29] B. Shi and H. Attia, "Integrated Process of Laser-Assisted Machining and Laser Surface Heat Treatment," *Journal of Manufacturing Science and Engineering-Transactions of the Asme,* vol. 135, Dec 2013.

[30] F. Lusquinos, J. C. Conde, S. Bonss, A. Riveiro, F. Quintero, R. Cornesana, *et al.*, "Theoretical and experimental analysis of high power diode laser (HPDL) hardening of AISI 1045 steel," *Applied Surface Science,* vol. 254, pp. 948-954, Dec 2007.
[31] S. K. Panda, V. H. B. Hernandez, M. L. Kuntz, and Y. Zhou, "Formability Analysis of Diode-Laser-Welded Tailored Blanks of Advanced High-Strength Steel Sheets," *Metallurgical and Materials Transactions a-Physical Metallurgy and Materials Science,* vol. 40A, pp. 1955-1967, Aug 2009.
[32] H. Hugel, "New solid-state lasers and their application potentials," *Optics and Lasers in Engineering,* vol. 34, pp. 213-229, Oct-Dec 2000.
[33] S. R. Heller, "An Evaluation of HY-80 Steel as a Structural Material for Submarines," *Naval Engineers Journal,* 1965.
[34] W. C. Stewart, "Feasibility of Underwater Friction Stir Welding of HY-80 Steel," M.S., Mechanical and Aerospace Engineering, Naval Postgraduate School, Monterey, CA, 2011.
[35] G. W. Young, "Evaluation of Friction Stir Processing ofHY-80 Steel Under Wet and Dry Conditions," M.S., Mechanical and Aerospace Engineering, Naval Postgraduate School, Monterey, CA, 2012.
[36] S. T. Yang, M. J. Matthews, S. Elhadj, V. G. Draggoo, and S. E. Bisson, "Thermal transport in CO2 laser irradiated fused silica: In situ measurements and analysis," *Journal of Applied Physics,* vol. 106, Nov 2009.
[37] J. P. Mabry, "Prediction and Control of Residual Stresses and Distortionin HY-130 Thick Pipe Weldments," PhD Dissertation, Massachusettes Institute of Technology, 1979.
[38] T. J. Holmquist, "Strength and Fracture Characteristics of HY-80, HY-100, and HY-130 Steels Subjected to Various Strains, Strain Rates, Temperatures, and Pressures," DTIC Document AD-A233061, 1987.

FRICTION STIR WELDING AND PROCESSING VIII

High Temperature Materials II

Friction Stir Welding and Processing VIII
Edited by: Rajiv S. Mishra, Murray W. Mahoney, Yutaka Sato, and Yuri Hovanski
TMS (The Minerals, Metals & Materials Society), 2015

PERFORMANCE ENHANCEMENT OF Co-BASED ALLOY TOOL FOR FRICTION STIR WELDING OF FERRITIC STEEL

Yutaka S. Sato[1], Masahiro Miyake[1], Shinichi Susukida[1], Hiroyuki Kokawa[1], Toshihiro Omori[1], Kiyohito Ishida[1], Shinya Imano[2], Seung Hwan C. Park[2], Itto Sugimoto[2], Satoshi Hirano[2]

[1]Graduate School of Engineering, Tohoku University;
6-6-02 Aramaki-aza-Aoba, Aoba-ku; Sendai 980-8579, Japan
[2]Hitachi Research Laboratory, Hitachi, Ltd.;
7-1-1 Omika-cho; Hitachi 319-1292, Japan

Keywords: Friction stir welding, Co-based alloy, Welding tool, Ferritic steel

Abstract

Co-based alloy tools having the γ/γ' microstructure with distribution of hard phases hardly experience large deformation and severe wear during friction stir welding (FSW) of ferritic steels at the welding temperature lower than the γ'-solvus. Further examinations implied that the tool wear were suppressed more effectively on Co-based alloy tool with larger volume fraction of hard phases with average size of 2 to 5 μm. In this study, Co-based alloy tools with different volume fractions of hard phases were designed, and effect of hard phases on tool wear was examined in FSW of 0.45% carbon steel. FSW trial showed that the tool wear decreased with increasing volume fraction of hard phases. At the volume fraction of 36%, the wear volume was successfully reduced to two third of that of the tool exhibiting the good wear resistance in a previous study. This study showed a possibility that the tool performance is effectively enhanced by further microstructure optimization of the tool material.

Introduction

The tool damage is a critical issue in friction stir welding (FSW) of steels and Ti alloys [1-3]. The tool should maintain sufficient strength to constrain the material to be welded at the softening temperature, and also be resistant to fatigue, mechanical wear, cracking, chipping, and chemical reactions with both the atmosphere and the weld material [1]. To date, refractory metals and superabrasive materials have been used as the tool materials for FSW of steels [1,3,4]. These materials are available to FSW of steels, but their usage is still limited in the practical applications because of the low cost-effectiveness.

As a cost-effective tool material for FSW of steels, the authors have developed a Co-based alloy strengthened by precipitation of intermetallics, $Co_3(Al,W)$ γ'-phase, with an $L1_2$ structure [5]. The alloy exhibits yield strength higher than 500 MPa at 1000°C, which is higher than minimum requirement of yield strength of FSW tool material for steels (400 MPa at 1000°C) [6]. This alloy hardly contains expensive rare elements, and it can be manufactured through a simple production method consisting of casting, heat-treatment and subsequent machining. In a preliminary study, the authors designed some Co-based alloys having the γ/γ' microstructure with distribution of hard phases, such as carbides and/or intermetallics, and examined feasibility of the Co-based alloy tool for FSW of some steels and titanium alloys [7-10]. Co-based alloy tool exhibited high feasibility to FSW of some steels and titanium alloys, but the wear of the tool shoulder was sometimes significant especially at the FSW temperature higher than the γ'-solvus temperature, resulting from dissolution of γ' precipitates in the vicinity of the tool shoulder surface [11]. On the other hand, the large deformation and severe wear of the tool were remarkably suppressed at the FSW temperature lower than the γ'-solvus. In this case, change in microstructure arising proximal to the tool shoulder surface was negligible, and suppression of the tool wear were more effective in the Co-based alloy tool with larger volume fraction of hard phases with average size of 2 to 5 μm [12]. According to the previous result, the performance and life of Co-based alloy tool might be enhanced by an increase in volume fraction of hard phases during FSW of steels. In this study, Co-based alloy tools with different volume fractions of hard phases were designed, and effect of volume fraction of hard phases on wear of Co-based alloy tool was examined in FSW of 0.45% carbon steel.

Experimental Procedures

4 kinds of γ/γ'-type Co-based alloys with the different volume fractions of hard phases were designed in this study. A Co-based alloy used in this study was representative of Co-based alloys exhibiting the good wear resistance in a previous study [12], expressed as "alloy B" throughout this paper, whose volume fraction of hard phases was about 9%. The ingots of Co-based alloys were produced by induction melting followed by heat treatment to obtain the γ' precipitates in the γ matrix.

Microstructures of these alloy themselves were examined by scanning electron microscopy (SEM) and electron probe micro-analyzer (EPMA). Samples for SEM were prepared by mechanical polishing and subsequent etching in an H_2SO_4 + HNO_3 solution. The γ'-solvus temperatures were quantified by differential scanning calorimetry (DSC). As the mechanical properties, the high-temperature strength, ball-on-disc wear resistance, and room-temperature hardness were individually measured. The high-temperature strength was examined by compression test with cylindrical specimen of 8 mm in diameter and 12 mm in height at a constant strain rate of 2.0×10^{-4} s^{-1} at temperatures from 1073 K (800°C) to 1373 (1100°C). The ball-on-disc wear test was performed with steel ball and Co-based alloy disc at an axial load of 19.6 N at room temperature for 7.2 ks (2 h). The relative sliding speed between the ball and the disc was 150 mm/s. After the ball-on-disc wear test, wear volume of Co-based alloy disc was quantified, and then wear rate (mm^3/m), defined as wear volume of Co-based alloy disc per unit sliding length, was calculated. The room-temperature hardness was measured at a load of 9.8 N by Vickers hardness test.

FSW tools were machined from the heat-treated ingots of Co-based alloys. Dimension and appearance of the tool probe are shown in Figure 1. The tool design included a 15 mm diameter shoulder and an unthreaded probe with a length of 1.7 mm. The probe was tapered from 6 mm at the shoulder and 3.5 mm at the probe tip. FSW trials were applied to 0.45wt%C steel at a tool rotational speed of 150 rpm and a travel speed of 1 mm/s with argon gas cooling for further reduction of the tool temperature. A weld length of an FSW pass was 450 mm, which was repeated to increase the total weld length.

The tool shape was captured after each FSW pass using KEYENCE VHX-1000 digital microscope, and then the tool wear volume was quantified. Since the shoulder experienced severer wear than the probe, "shoulder wear rate (mm^3/m)," which was defined by an equation (shoulder wear rate = wear volume of shoulder periphery region / sliding length of the shoulder periphery during FSW), was used in this study. In this equation, the shoulder periphery region corresponded to a region between 6 mm and 7 mm from the tool center, and the sliding length at 6.5 mm from the tool center was used as the sliding length of the shoulder periphery.

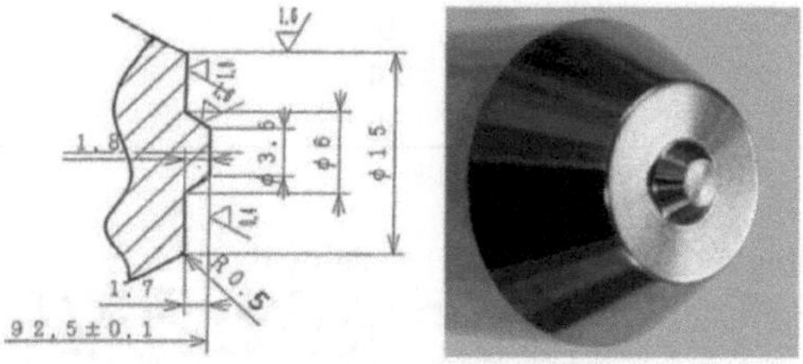

Figure 1. Dimension and appearance of tool tip used in this study.

Results and Discussion

Microstructures and properties of Co-based alloy themselves

SEM images of the Co-based alloys designed in this study are shown in Figure 2. All alloys consist of the γ/γ' microstructure (gray region) with distribution of irregularly shaped hard-phases (white regions). From chemical composition analyses by EPMA, the hard phases in alloy A were mainly MC-type carbides, while two kinds of hard phases, i.e., M_6C-type carbide and μ phase (Co_7W_6), were found in the other alloys. Volume fractions of hard phases, quantified by point counting to SEM images, were 2%, 9%, 24% and 36% in alloys A, B, C and D, respectively. Average sizes of hard phases were also quantified using the areas of hard phases by assuming the spherical hard-phases. Volume fractions and average sizes of hard phases are summarized in Table 1. The γ/γ' microstructures of the Co-based alloys used are presented in Figure 3. All alloys have spherical γ' phases in the γ matrix, although the size and volume fraction of γ' phases are different among the alloys.

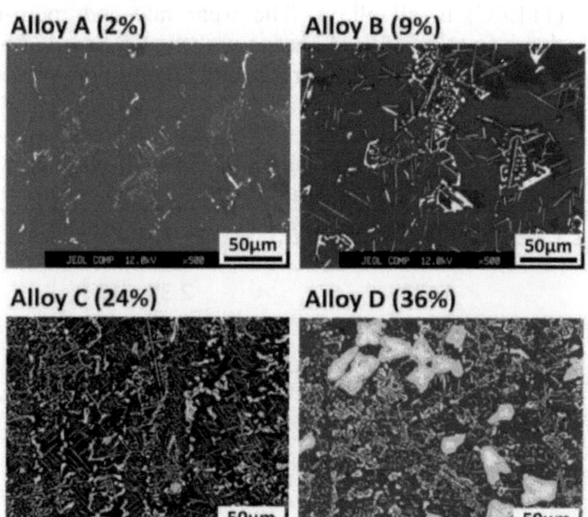

Figure 2. SEM images of the Co-based alloys designed in this study.

Table 1. Volume fractions and average sizes of hard phases in the Co-based alloys used.

Alloy ID	Volume fraction (%)	Average size (μm)
Alloy A	2	1.0
Alloy B	9	2.5
Alloy C	24	1.9
Alloy D	36	2.7

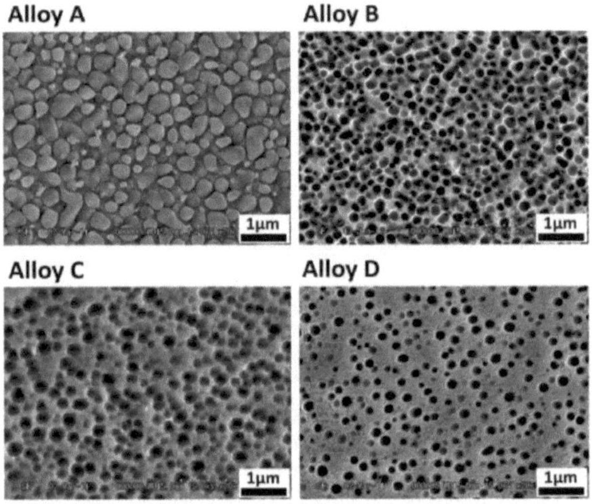

Figure 3. SEM images of the γ/γ' microstructures of the Co-based alloys used.

The γ'-solvus temperature, wear rate obtained by ball-on-disk wear test, and room-temperature hardness of the Co-based alloy themselves are given in Table 2. The γ'-solvus temperatures are higher than 1392 K (1119°C) in all alloys. The wear rate and room-temperature hardness probably depend on volume fraction of hard phases, i.e., the Co-based alloy with higher volume fraction of hard phases exhibited the lower wear rate and higher room-temperature hardness.

Table 2. γ'-solvus temperature, wear rate obtained by ball-on-disk wear test, and room-temperature hardness of the Co-based alloys used.

Alloy ID	γ'-solvus temperature (K)	Wear rate (mm^3/m)	Hardness (Hv)
Alloy A	1451	5.38×10^{-3}	402
Alloy B	1423	2.39×10^{-3}	502
Alloy C	1409	9.47×10^{-4}	546
Alloy D	1392	5.49×10^{-4}	562

Evaluation of Co-based alloy tools

Shape change in the tool tip of alloy D during FSW is shown in Figure 4. The macroscopic wear was hardly observed from the appearances, but the slight wear of the shoulder periphery was visible in the tool-shape profiles. The shoulder wear rates of the tools are presented in Figure 5. The wear during FSW of 0.45wt%C steel is more suppressed in the Co-based alloy tool with the higher volume fraction of hard phases. At the volume fraction of 36%, the shoulder wear rate was successfully reduced to two third of that of the tool developed in a previous study (alloy B) [12]. SEM image near the shoulder surface of the tool made of alloy D after FSW at a total weld length of 3.6 m is shown in Figure 6. The region in the vicinity of the shoulder surface subjected to FSW has roughly the same microstructure as the solidified and subsequently heat-treated ingot (alloy D in Figure 2), but it seems that the γ/γ' microstructure is more preferentially worn than the hard phases at the tool shoulder surface. This result implies that the tool wear depends on an area of the γ/γ' microstructure contacting the weld material on the shoulder surface.

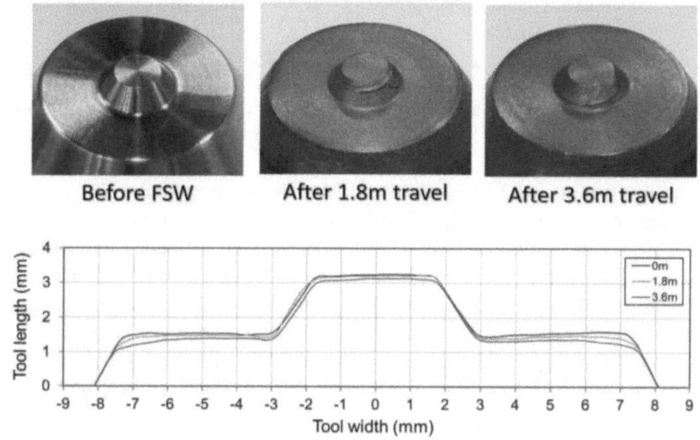

Figure 4. Changes in appearances and shape-profiles of the tool tip made of alloy D during FSW.

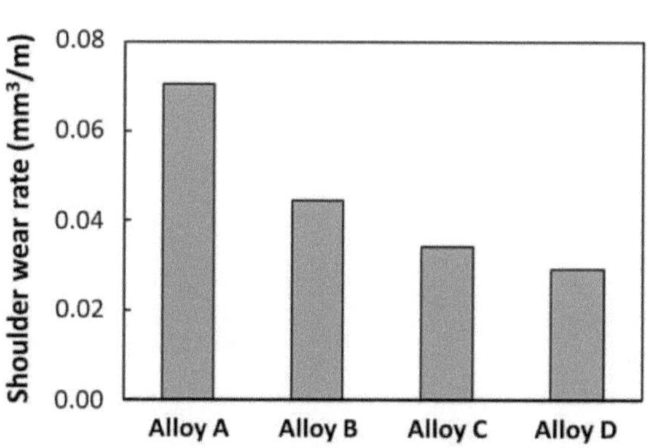

Figure 5. The shoulder wear rates of the tools.

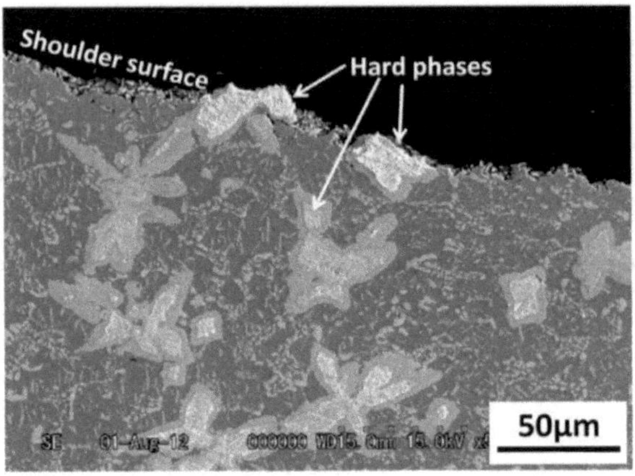

Figure 6. SEM image of near the shoulder surface on a cross section, cut parallel to the axial direction, of the alloy D tool after FSW at a total weld length of 3.6 m.

Effect of volume fraction of hard phases on tool wear is given in Figure 7. As above-mentioned, the tool wear decreases with increasing volume fraction of hard phases, but the tendency is found to be deviated from a linear relationship. The deviation from a linear relationship would be due to difference in size and/or type of hard phases. As shown in a previous study [12], the shoulder wear rate depended on average size of hard phases as well as the volume fraction, i.e., the high volume fraction of hard phases with size of 2 to 5 μm effectively suppressed the tool wear. Moreover, it is likely that type of hard phases should affect the tool wear because the hardness (wear resistance) of the hard phase themselves is different. Furthermore, heterogeneities of side and distribution of hard phases might be microstructural factors governing the tool wear in the Co-based alloy tool. Fundamental examination to clarify these effects on tool wear is currently ongoing. There would be a possibility that the microstructure optimization of the tool material based on such knowledge results in further enhancement of the performance in the Co-based alloy tool.

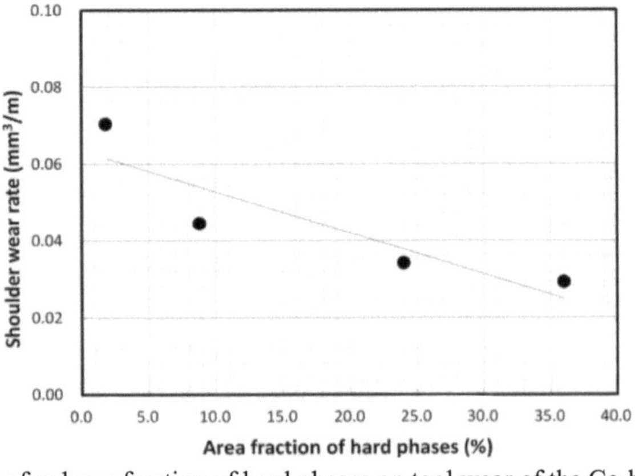

Figure 7. Effect of volume fraction of hard phases on tool wear of the Co-based alloy tool.

Summary

In this study, Co-based alloy tools having the γ/γ' microstructure with different volume fractions of hard phases were designed, and then effect of volume fraction of hard phases on the tool wear was examined. The tool wear decreased with increasing volume fraction of hard phases. At the volume fraction of 36%, the wear volume was successfully reduced to two third of that of the tool exhibiting the good tool performance in a previous study. This study showed a possibility that the tool performance is effectively enhanced by further microstructure optimization of the tool material.

Acknowledgements

The authors are grateful to Mr. A. Honda, Dr. K. Kobayashi and Dr. Y. Takaku of Tohoku University for the preparation of specimens and testing. They also wish to thank Prof. K. Adachi, Prof. K. Maruyama and Dr. S. Mironov for their valuable discussions. This work was supported by a Grant-in-Aid for Scientific Research (B) and partially a program named "Next Generation Automobiles in Miyagi" entrusted by Ministry of Education, Culture, Sports, Science and Technology, Japan.

References

1. C.D. Sorensen and T.W. Nelson, "Friction Stir Welding of Ferrous and Nickel Alloys," *Friction Stir Welding and Processing*, Edited by R.S. Mishra and M.W. Mahoney, ASM International, Materials Park, OH, (2007), 111-121.

2. H. Bhadeshia and T. DebRoy, "Critical Assessment: Friction Stir Welding of Steels," *Science and Technology of Welding and Joining*, **14** (2009), 193-196.

3. B. Thompson, S.S. Babu, "Tool Degradation Characterization in the Friction Stir Welding of Hard Metals," *Welding Journal*, **89** (2010), 256-261.

4. T. Miyazawa, Y. Iwamoto, T. Maruko, and H. Fujii, "Development of Ir based tool for friction stir welding of high temperature materials," Science and Technology of Welding and Joining, 16 (2011), 188-192.

5. J. Sato, T. Omori, K. Oikawa, I. Ohnuma, R. Kaimuna, and K. Ishida, "Cobalt-Base High-Temperature Alloys," *Science*, **312** (2006), 90-91.

6. W. Gan, Z.T. Li, and S. Khurana, "Tool materials selection for friction stir welding of L80 steel," *Science and Technology of Welding and Joining*, **12** (2007), 610-613.

7. Y.S. Sato, M. Miyake, H. Kokawa, T. Omori, K. Ishida, S. Imano, S.H.C. Park, and S. Hirano, "Development of a cobalt-based alloy FSW tool for high-softening-temperature materials," *Friction Stir Welding and Processing VI*, A John Wiley & Sons, Inc., and TMS, (2011), 3-9.

8. S.H.C. Park, T. Nagahama, S. Hirano, S. Imano, Y.S. Sato, H. Kokawa, T. Omori, and K. Ishida, "Friction Stir Welding of Titanium Alloy using a Cobalt-based Alloy Tool," *Proceedings of the 12th World Conference on Titanium*, (2011), 1616-1620.

9. S.H.C. Park, S. Hirano, S. Imano, Y.S. Sato, H. Kokawa, T. Omori, and K. Ishida, "Friction-stir welding of high-softening-temperature materials using cobalt-based alloy tool," *Materials Science Forum*, **706-709** (2012), 996-1001.

10. I. Sugimoto, A. Sato, S.H.C. Park, S. Hirano, S. Imano, Y.S. Sato, H. Kokawa, T. Omori, and K. Ishida, "Friction-stir-welding of thick carbon steels using Co-based alloy tool," *Friction Stir Welding and Processing VII*, A John Wiley & Sons, Inc., and TMS, (2013), 101-106.

11. M. Miyake, Y.S. Sato, H. Kokawa, Y. Takaku, T. Omori, K. Ishida, S. Imano, S.H.C. Park, and S. Hirano, "Material properties governing wear of Co-based alloy tool during friction stir welding of steels," *Proceedings of 9th International Friction Stir Welding Symposium*, Huntsville, AL, (2012), CD-ROM.

12. Y.S. Sato, M. Miyake, H. Kokawa, T. Omori, K. Ishida, S. Imano, I. Sugimoto, S.H.C. Park, and S. Hirano, "Tool material factors for suppression of wear in Co-based alloy tool during friction stir welding of 0.45C steel," *Proceedings of 10th International Friction Stir Welding Symposium*, Beijing, China, (2014), CD-ROM.

Friction Stir Welding and Processing VIII
Edited by: Rajiv S. Mishra, Murray W. Mahoney, Yutaka Sato, and Yuri Hovanski
TMS (The Minerals, Metals & Materials Society), 2015

STABILIZATION OF THE RETAINED AUSTENITE IN STEEL BY FRICTION STIR WELDING

Takuya Miura[1], Rintaro Ueji[1], Hidetoshi Fujii[1]

[1]Joining and Welding Research Institute, Osaka University
11-1 Mihogaoka, Ibaraki, Osaka, 567-0047, Japan

Keywords: Friction stir welding, Fe-Ni-C steel, Retained austenite, TRIP

Abstract

The phase transformation in steel during friction stir welding (FSW) was studied using the Fe–24wt%Ni–0.1%C alloy. When the FSW condition was adapted in order that the maximum temperature was sufficiently reached for austenitization, the volume fraction of the retained austenite in the stir zones was significantly increased. A crystallographic measurement revealed that the martensite transformation was inhibited after the stirring during which the austenite was plastically deformed. Additional experiments clarified that the dislocation and grain boundary introduced by the plastic deformation of austenite may provide a barrier against the martensite transformation. The stabilized austenite enhances the work hardening rate and maximum strength at room temperature due to the transformation induced plasticity (TRIP).

Introduction

Friction stir welding (FSW) is a solid-state welding technique patented in 1991 at The Welding Institute (TWI) in the United Kingdom [1]. The development of the rotation tools and studies of the appropriate process conditions have facilitated the application of FSW to various steels [2-8]. The FSW process consists of plastic flow and frictional heat; FSW can therefore be regarded as a thermomechanical process. This is similar to the hot rolling of steel in which phase transformations occur during the thermomechanical process. Consequently, the phase transformation should also be controlled by the FSW conditions. Actually, some researchers have clarified that the microstructure in the steel joints can be controlled by the FSW conditions, such as the tool rotation speed and the welding speed as follows. Fujii et al. [2] revealed the variation in the microstructure of carbon steel associated with a change in the process temperature controlled by the welding speed. Cui et al. [3] clarified that several different microstructures can be obtained in high carbon steel through control of both the tool rotation speed and the welding speed. Sato et al. [4] explained the relation between the cooling rate of the FSW and microstructural evolution in high-carbon steel. Saeid et al. [5] achieved a sound weld by FSW without changing the ratio between phases in duplex stainless steel by preventing the transformation during FSW at a relatively low processing temperature. Ghosh et al. [6] estimated the effect of temperature and strain rate on the microstructure during FSW by considering the processing parameters.

We are attempting to more aggressively control the phase transformation by FSW. As the authors clarified [7], the volume fraction of retained austenite in a Cr–Mo steel joint is increased by the FSW. Moreover, the joint with the retained austenite exhibits an excellent strength and elongation during tensile tests at room temperature. Fujii et al. [8] reported similar phenomena in ferritic stainless steels. However, the mechanisms of the stabilization of the retained austenite during FSW have not been sufficiently clarified. To discuss the mechanism, we need to determine details of the microstructural changes induced by systematic changes in the process parameters, such as the rotation speed. The limited volume fraction of the retained austenite and the occurrence of diffusional transformation conceal the structure of the prior austenite. High-Ni steels were used in the study of the martensite transformation because the martensite start temperature (Ms) is relatively low (on the order of a few tens °C); therefore, the parent conditions and the resultant martensite crystal can be easily compared. Moreover, at high cooling rates, the high-Ni steels hardly undergo a diffusional transformation. The aim of this study was to clarify the microstructural change during FSW and its effects on the microstructure formation and mechanical properties of the Fe–Ni–C steel.

Experimental procedure

The chemical composition of the steel used in this study was 24wt%Ni–0.1%C–bal. Fe; it was prepared by vacuum induction melting and was shaped into a sheet of 1.8 mm thickness by being hot rolled at 1073 K. After the surface oxides were removed, the sheets were stir-in-plate friction stir welded at tool rotation speeds of 200 or 400 rpm. The tool traveling speed was fixed at 400 mm/min. The plugging load during the processes was maintained in the range of 2500–3700 kg to achieve a constant plugging depth. A tool made of tungsten carbide, with a 12-mm shoulder diameter, a 4-mm probe diameter, and a 1.4-mm probe length was used with an advancing tilt angle of 3° from the normal direction. The microstructure was observed on the welding-direction (WD) plane, which is parallel to both the transversal direction (TD) and the normal direction (ND). To prevent oxidation of the sample and tool, Ar gas was used as the shielding gas at the flow rate of 20 L/min.

An optical microscope was used to examine the specimens prepared by being cut, mechanically polished, and then etched in a 3% Nital solution at room temperature. The microstructures and fracture surfaces were observed via field-emission scanning electron microscopy (FE-SEM) on an instrument equipped with an electron backscatter diffraction (EBSD) system. The section for the EBSD observations was electronically polished in a solution of 900 mL CH_3COOH + 100 mL $HClO_4$ at approximately 290 K and 20 V. Areas with dimensions of 80 × 80 μm^2 were scanned using an EBSD step size of 0.1 μm. The data points with lower CI values than 0.1 were excluded from analysis. The transverse tensile specimens with a gauge length of 2 mm were used at a constant crosshead speed of 0.01 mm/min at room temperature or 373K. To avoid being affected by an inhomogeneous thickness distribution, the specimens were polished to the thickness of 1.3 mm. The specimens for the microstructural analysis and tensile tests were obtained from each bead, except for the first 10 mm, which represented the transitory stage.

Results and discussion

<u>Base metal</u>

Figure 1 shows the phase distribution map of the base metal (BM) of the 24%Ni-0.1%C steel. The microstructure of the BM was composed of martensite and retained austenite. The martensite variants with a length and width of approximately 10 μm and 2–3 μm, respectively, were observed in the prior austenite grains with a grain size of approximately 30–60 μm. A similar microstructure was reported in a similar chemical composition [9].

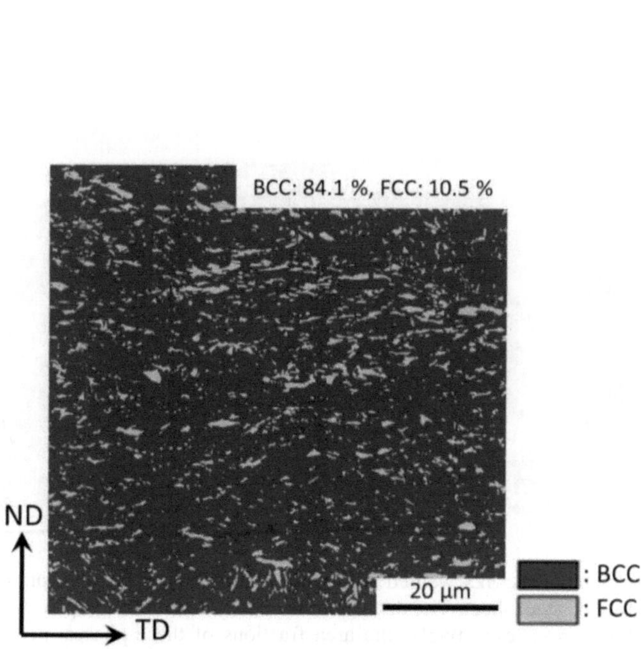

Figure 1. Phase map of the BM of 24%Ni–0.1%C steel sheet. The area fractions of these phases are denoted on the image.

Microstructure changes

Optical macroscope images in the cross-sectional planes perpendicular to the welding direction of the samples welded at the tool rotation speeds of 200 rpm (a) and 400 rpm (b) are shown in Fig. 2. The advancing side (AS) was arranged on the right side. No defect is observed in both welds. The HAZ and TMAZ are difficult to identify under all of the investigated FSW conditions, probably because both the recrystallization and phase transformation took place with some distribution. The distributions of the microstructure in the SZs were inhomogeneous. The optical microscope images of the SZs mainly show brightly etched areas with numerous black bands whose morphology resembles the "growth rings" in wood. In this study, to simplify the comparison between the welds, we performed further research of the SZs with the most characteristic microstructures.

Figure 3 shows the phase distribution maps in the regions located at a depth of 0.7 mm from the top surface in the center of the SZs of the 24%Ni-0.1%C steel welded at 200 rpm (a) and 400 rpm (b). After the FSW, the area fraction of the retained austenite significantly increases.

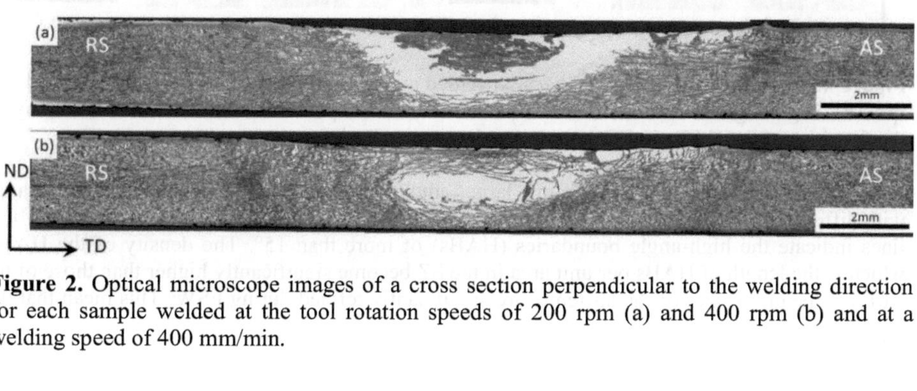

Figure 2. Optical microscope images of a cross section perpendicular to the welding direction for each sample welded at the tool rotation speeds of 200 rpm (a) and 400 rpm (b) and at a welding speed of 400 mm/min.

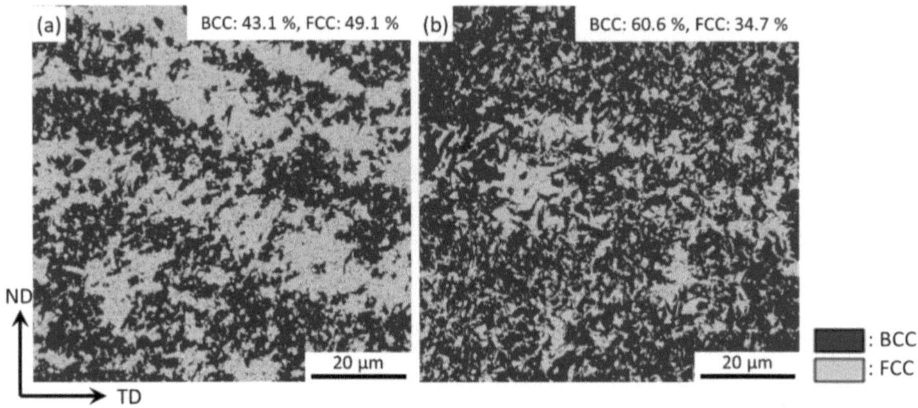

Figure 3. Phase maps of the SZs welded at the tool rotation speeds of 200 rpm (a) and 400 rpm (b) for 24%Ni–0.1%C steel sheet. The distributions of the BCC and FCC phases are displayed as dark gray and light gray, respectively; the area fractions of these phases are denoted on each image.

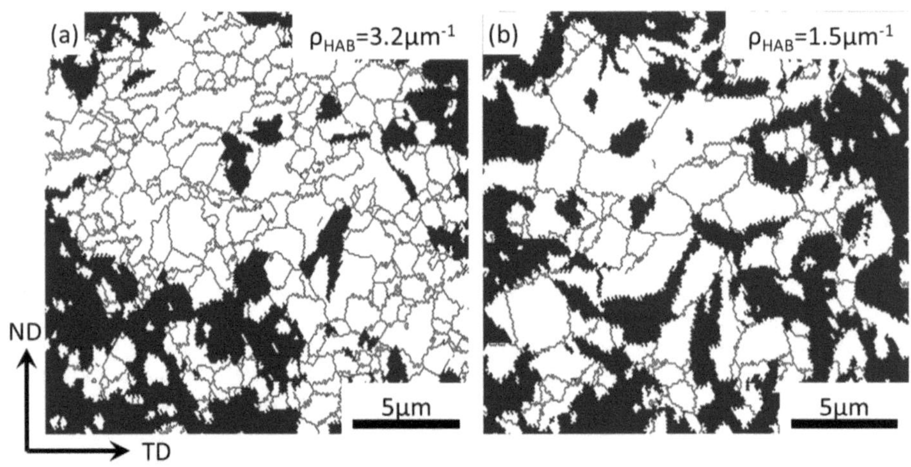

Figure 4. Grain boundary maps of the SZs welded at tool rotation speeds of 200 rpm (a) and 400 rpm (b). High-angle boundaries (HABs) (>15°) in the FCC phase of each material are displayed; the densities of the HABs in FCC phase are denoted on each image.

A lower tool rotation speed resulted in a larger amount of the retained austenite. The boundaries with different crystallographic misorientations in the FCC phase are displayed in Fig. 4. The bold lines indicate the high-angle boundaries (HABs) of more than 15°. The density of the HABs, which is the length of HABs per unit area in the SZ become significantly higher than those of the BM, i.e. the grain sizes of the austenite are significantly refined during FSW. This mean that the retained austenite was significantly strengthened. Moreover, the grain sizes become smaller with a decreasing tool rotation speed.

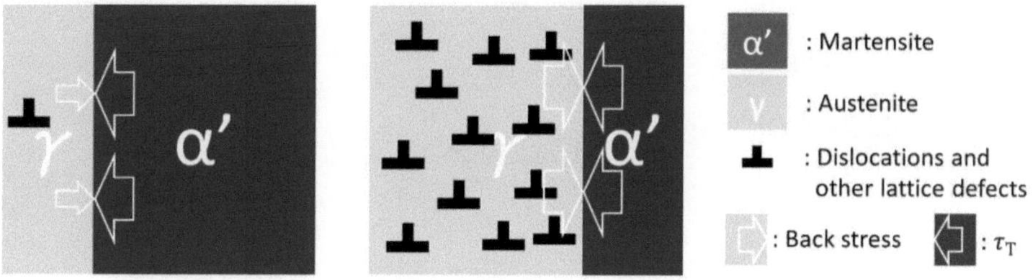

Figure 5. Schematics of mechanical stabilization of the retained austenite.

The mechanisms of the stabilization of retained austenite during the FSW are discussed from a view point of microstructural evolution in austenite. The grain refinement can be achieved through recrystallization and can be accelerated via plastic deformation at a higher strain rate during the FSW. The SZs were reheated to above the A_1 or A_3 temperature and the prior austenite grains were refined by the severe plastic deformation during the FSW. Chatterjee et al. [10] provided an estimation of the mechanical stabilization of austenite based on the following eq. (1). The stress τ_T driving the motion of the interface originates from the chemical free energy of transformation; this stress is expressed as

$$\tau_T = \phi \Delta G, \tag{1}$$

where ϕ is a constant assumed to be equal to unity and ΔG is the magnitude of the driving force for the transformation from austenite to martensite (Fig. 5 left). When the back stress provided by the accumulated lattice defects overcomes τ_T, the transformation can be inhibited (Fig. 5 right). In addition, Kajiwara [11] indicated that the increase in difficulty for the plastic accommodation with the decreasing grain size is the main factor that decreases the Ms in the fine-grained austenite. Consequently, the amount of retained austenite was significantly increased in the SZ.

Tensile properties of SZ

Figure 6 shows the nominal stress–displacement curves at room temperature or 373 K of the SZ welded at the tool rotation speed of 200 rpm and at a welding speed of 400 mm/min. The SZ exhibits a higher work hardening rate and higher maximum strength at room temperature than at 373 K. Figure 7 shows phase distribution maps of the SZs after the tensile test at each temperature. The area fraction of the austenite significantly decreased during the tensile test at room temperature while the 373 K fractured specimen maintained the same amount of retained austenite. This significant dependence of the temperature on the plastic behaviors is very similar to that reported in the study of TRIP [12]. TRIP with an appropriate stability of the austenite can enhance both the strength and elongation. Therefore, the austenite stabilized by FSW can make possible the management of both the high strength and large ductility by the TRIP mechanism.

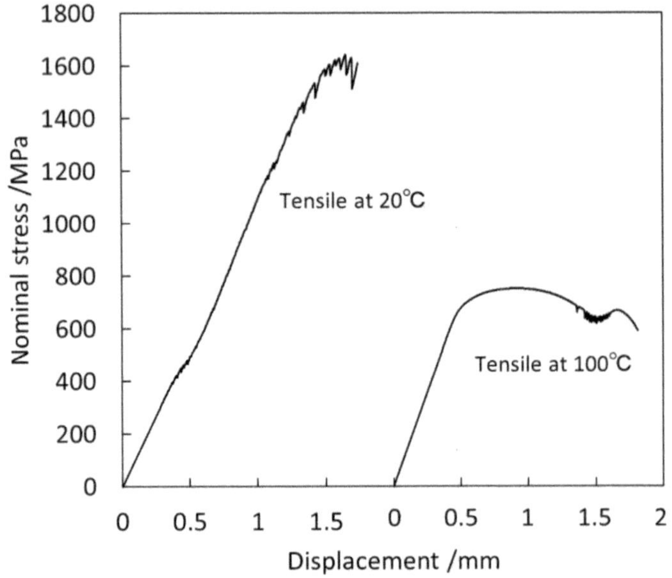

Figure 6. Displacement–nominal stress curves at room temperature (left) or 373 K (right) of the SZ welded at a tool rotation speed and welding speed of 200 rpm and 400 mm/min, respectively.

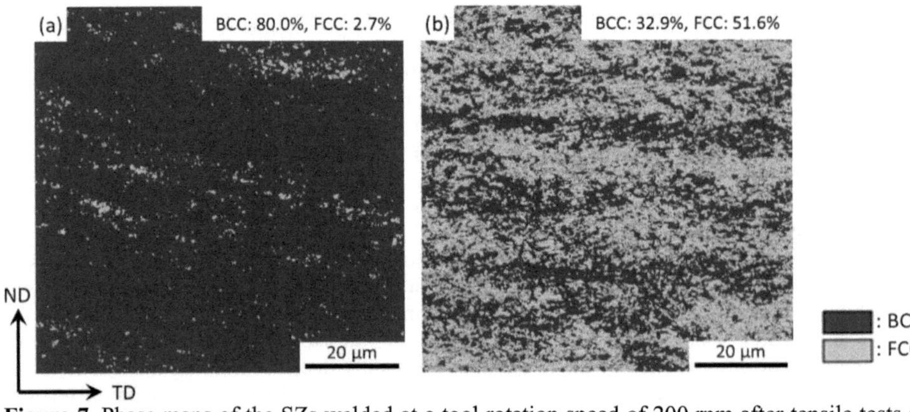

Figure 7. Phase maps of the SZs welded at a tool rotation speed of 200 rpm after tensile tests at room temperature (a) or 100 ℃ (b) for 24%Ni–0.1%C steel sheet. The distributions of BCC and FCC phases displayed as dark gray and light gray, respectively; the area fractions of these phases are denoted on each image.

Effects of the rapid cooling using liquid CO_2

The retained austenite in the SZ can be stabilized by the high density lattice defects such as dislocations and grain boundaries introduced by the severe plastic deformation during the stirring, as shown in Fig. 5. However, the densities of these lattice defects may decrease through recovery or grain growth during the cooling stage after stirring. Therefore, the stabilities of the retained austenite in the SZ may be enhanced with an increasing cooling rate. With an aim to clarify the effects of the cooling rate on the stabilization of the retained austenite in the SZ, the rapid cooling FSW using liquid CO_2 was evaluated.

The 24%Ni–0.1%C steel sheets were stir-in-plate friction stir welded at a tool rotation speed of 200 rpm and tool traveling speed of 200 mm/min. The plugging load during the processes was maintained at 3700 kg. Ar gas was used as the shielding gas at the flow rate of 20 L/min. In order to increase the cooling rate, the liquid CO_2 was sprayed right behind of the welding tool. By the rapid cooling using the liquid CO_2, the microstructure in the SZ is cooled until appropriately 230 K within a few minutes after the stirring.

Figure 8 shows the phase distribution maps in the regions located at a depth of 0.7 mm from the top surface in the center of the SZs welded at the tool rotation speed of 200 rpm and welding speed of 200 mm/min without (a) or with rapid cooling (b). The SZ with rapid cooling is constructed by the larger amount of the retained austenite than that without rapid cooling and BM. This result means that the increased cooling rate after the FSW effectively increase the austenite in the SZ.

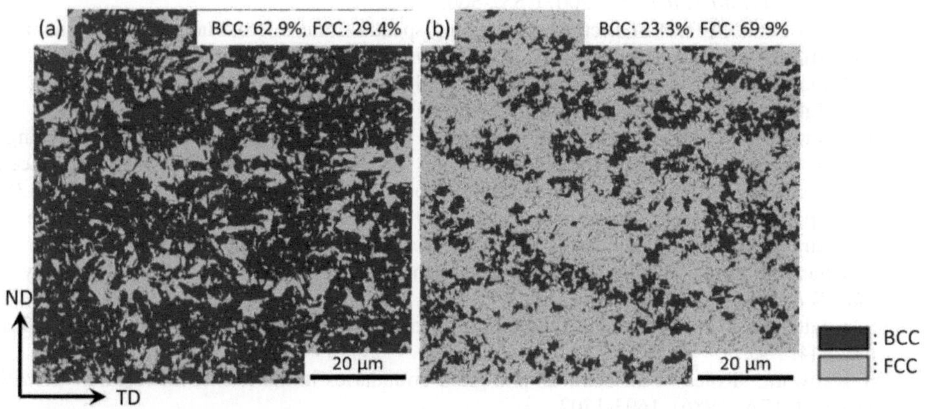

Figure 8. Phase maps of the SZs welded at a tool rotation speed of 200 rpm and at a welding speed of 200 mm/min without (a) or with (b) rapid cooling using liquid CO_2 for 24%Ni–0.1%C steel sheet. The distributions of BCC and FCC phases displayed as dark gray and light gray, respectively; the area fractions of these phases are denoted on each image.

Conclusion

Along with the optimization of the welding parameters, the retained austenite can be stabilized in the SZ due to the high densities of lattice defects introduced by the severe plastic deformation during the FSW. This stabilized retained austenite exhibits TRIP which contributes to the management of both the high strength and large ductility at room temperature. The additional post cooling method using liquid CO_2 is very effective in enhancing the stability of the retained austenite in the SZ.

Acknowledgement

This study was supported by the Japan Science and Technology Agency (JST) under Collaborative Research Based on Industrial Demand "Heterogeneous Structure Control: Towards Innovative Development of Metallic Structural Materials"; a future pioneering program commissioned by the New Energy and Industrial Technology Development Organization (NEDO); the Global COE Programs from the Ministry of Education, Sports, Culture, Science; and a Grant-in-Aid for Science Research from the Japan Society for Promotion of Science and Technology of Japan, ISIJ Research Promotion Grant. The authors would like to acknowledge all these great supports.

References

1. W.M. Thomas et al., International Patent Application, (1991), No. PCT/GB92/02203.
2. H. Fujii, et al., "Friction stir welding of carbon steels," *Mat. Sci. Eng. A*, 429A (2006), 50-57.
3. L. Cui et al., "Friction stir welding of a high carbon steel," *Scripta Mater.*, 56 (2007), 637-640.
4. Y.S. Sato et al., "Microstructural evolution of ultrahigh carbon steel during friction stir welding," *Scripta Mater.*, 57 (2007), 557-560.
5. T. Saeid et al., "Effect of friction stir welding speed on the microstructure and mechanical properties of a duplex stainless steel," *Mat. Sci. Eng. A*, 496A (2008), 252-268.
6. M. Ghosh et al., "Analysis of microstructural evolution during friction stir welding of ultrahigh-strength steel," *Scripta Mater.*, 63 (2010), 851-854.
7. T. Miura et al., "Phase transformation behavior of Cr-Mo steel during FSW," (Proceedings of Int. Sympo. on Visualization in Joining & Welding Science through Advanced Measurements and Simulation, Suita, Osaka, Japan, 28-30 November 2012), Vol.1, 137-138.
8. H. Fujii et al., "High strength and ductility of friction-stir-welded steel joints due to mechanically stabilized metastable austenite," *Scripta Mater.*, 70 (2014), 39-42.
9. M Umemoto et al., "The morphology of martensite in Fe-C, Fe-Ni-C and Fe-Cr C alloys," *Mat. Sci.*, 18 (1983), 2893-2904.
10. S. Chatterjee et al., "Mechanical stabilisation of austenite," *Mater. Sci. Tech.*, 22 (2006), 641-644.
11. S. Kajiwara, "Roles of dislocations and grain boundaries in martensite nucleation," *Metall. Trans. A*, 17A (1986), 1693-1702.
12. W.S. Park et al., 2010. "Strain-rate effects on the mechanical behavior of the AISI 300 series of austenitic stainless steel under cryogenic environments," *Mater. Design*, 31 (2010), 3630-3640.

Friction Stir Welding and Processing VIII
Edited by: Rajiv S. Mishra, Murray W. Mahoney, Yutaka Sato, and Yuri Hovanski
TMS (The Minerals, Metals & Materials Society), 2015

Study of Mechanical Properties and Characterization of Pipe Steel Welded by Hybrid (Friction Stir Weld + Root Arc Weld) Approach

Yong Chae Lim[1], Samuel Sanderson[2], Murray Mahoney[3], Andrew J. Wasson[4], Doug P. Fairchild[4], Yanli Wang[1], and Zhili Feng[1]

[1]Materials Science & Technology Division, Oak Ridge National Laboratory;
One Bethel Valley Road; Oak Ridge, TN 37831, USA
[2]MegaStir Technologies LLC;
333 West 2230 North; Provo, UT 84604, USA
[3]Consultant;
1161 W. Sunburst Lane; Midway, UT 84049, USA
[4]ExxonMobil Upstream Research Company;
3319 Mercer St.; Houston, TX 77027, USA

Keywords: Friction stir welding, pipe steel, hybrid approach, root arc weld, mechanical properties

Abstract

Friction stir welding (FSW) has recently attracted attention as an alternative construction process for gas/oil transportation applications due to advantages compared to fusion welding techniques. A significant advantage is the ability of FSW to weld the entire or nearly the entire wall thickness in a single pass, while fusion welding requires multiple passes. However, when FSW is applied to a pipe or tube geometry, an internal back support anvil is required to resist the plunging forces exerted during FSW. Unfortunately, it may not be convenient or economical to use internal backing support due to limited access for some applications. To overcome this issue, ExxonMobil recently developed a new concept, combining root arc welding and FSW. That is, a root arc weld is made prior to FSW that supports the normal loads associated with FSW. In the present work, mechanical properties of a FSW + root arc welded pipe steel are reported including microstructure and microhardness.

Introduction

Conventional fusion welding typically is used to construct pipelines for transportation of gas and oil. When fusion or arc welding processes are applied to large diameter or thicker wall pipe, welding process time is significantly increased. This is because a larger volume of parent metal is required to be melted and deposited (from filler metals) in the weld area. As a result, multiple arc welding passes are generally required to make a complete weld joint. In addition, distortion and residual stress, due to considerable heat from fusion welding with multiple arc welding passes, are additional concerns for final assembly of pipelines and subsequent mechanical strength.

This manuscript has been authored by UT-Battelle, LLC, under Contract No. DE-AC05-00OR22725 with the U.S. Department of Energy. The United States Government retains and the publisher, by accepting the article for publication, acknowledges that the United States Government retains a non-exclusive, paid-up, irrevocable, world-wide license to publish or reproduce the published form of this manuscript, or allow others to do so, for United States Government purposes.

Continuing research efforts to develop a new/or innovative process is critical to further improve economical aspects for pipeline construction. Recently, friction stir welding (FSW) has been developed and applied to high temperature and/or high strength materials with the development of new tool materials and new tool designs [1,2,3,4]. FSW is a unique solid state joining process with lower temperature generated during welding resulting in benefits compared to fusion welding. Also, FSW provides improved mechanical properties, at times even better than parent metal properties due to the refined grain size in weld zone [5,6,7]. One economic advantage is the ability of FSW to join in a single pass the entire wall thickness, or nearly the entire wall thickness based on the pin length of the tool. Therefore, FSW can be an alternative method that replaces/or combines with conventional fusion arc welding for pipeline construction. Also, FSW does not require the use of a filler metal, where conversely, conventional fusion arc welds does need filler metal. Clearly, there are both cost and productivity benefits associated with FSW compared to fusion or arc welding process for on and off-shore pipeline construction [8]. However, when FSW is applied to a pipe geometry, an internal backing is required to support the downward force from FSW. Based on pipe dimensions, an internal backing support requires additional equipment that adds to the cost of fabrication.

In order to overcome the internal support issue, ExxonMobil recently invented a unique concept [9]. That is, both fusion arc welding and FSW are used together. Prior to FSW, a root arc weld with a butt joint configuration is made on the inner diameter of the pipe wall to support the normal load from FSW. As a continuing research effort, this paper describes an investigation of the mechanical properties of pipe steel (API 5L X70 grade) welded by this hybrid approach. A digital image correlation (DIC) technique was employed during the transverse tensile tests to monitor local strain fields and to correlate failure initiation locations. Microhardness in the welded area was measured and correlated to microstructures observed in the weld zone.

Experimental Approaches

Materials

The material used in this study was line pipe X70 with dimensions of 762 mm outer diameter (OD) and 15.8 mm wall thickness. Based on the mill certification, the average yield and tensile strength of the base metal is 619.7 MPa and 677 MPa, respectively. Additionally, elongation is 38.6%. Short sections of the pipe were welded by gas metal arc welding (GMAW) in the 1GR position on the internal diameter (ID) prior to FSW on the OD. These short sections were assembled and tack welded by gas tungsten arc welding (GTAW) on the OD using an external line-up clamp prior to GMAW the ID.

Root Arc Welding and Friction Stir Welding (ExxonMobil Concept)

Figure 1 illustrates a schematic of the butt joint configuration for FSW and root arc welding. The V-groove dimension, including bevel angle and depth for the root arc weld is shown. The friction stir welds on the OD of the 762 mm pipe (API 5L X70) were completed with the MegaStir orbital FSW system. The concept of utilizing both a friction stir weld on the OD and an arc weld on the ID were applied previously to 320 mm (12.7 inch) diameter and 12 mm (0.5 inch) wall thickness pipe and the larger diameter pipe evaluated herein [10]. The internal arc weld in that study adequately supported the friction stir weld on the OD surface.

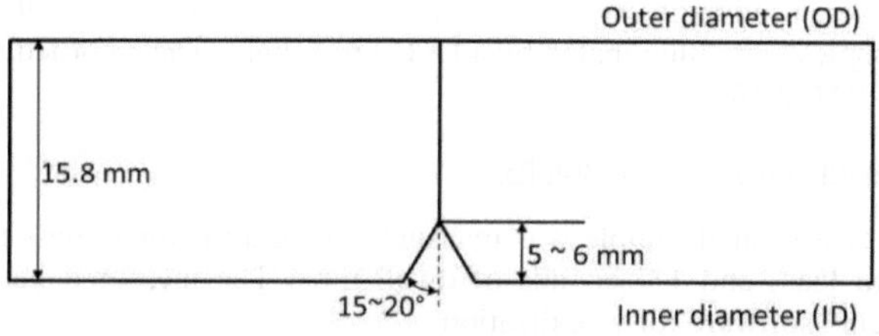

Figure 1. A schematic of the butt joint configuration for root arc welding prior to FSW.

In the current study, the depth of penetration of the GMAW arc weld was between 5 mm and 6 mm. A summary of the GMAW process parameters are given at Table 1. Various process parameters of travel speed and spindle speed were explored, ranging from 0.635 mm/s to 1.693 mm/s and 150 RPM to 250 RPM. The combination of parameters that produced defect free welds included a traverse speed of 1.27 mm/s and spindle speed of 175 RPM. Force control was used throughout, and the range of spindle-axis loads was 77.84 kN to 80 kN. The tool design and material selected for the present work was a standard convex scroll should step spiral (CS4) design made with MS80 consisting of 80% cBN with 20% second phase binder. The corresponding FSW tool pin length was 11 mm with a shoulder diameter of 38 mm and pin base diameter of 16 mm. However, based on measurements of the weld on the crown surface, the full shoulder diameter was not engaged.

Table 1 Summary of welding process parameters used for root arc welding.

Welding process	Gas Metal Arc Weld (GMAW)
Weld position	1G
Wire class	ER70S-4
Wire size (mm)	0.9 (0.035 inch)
Shielding gas	75%Ar/25%CO2
Shielding gas flow rate (m^3/h)	0.99 (=35 CFH)
Electrical polarity	DCEP
Welding currents (Amps)	240-270
Arc voltage (Volts)	24
Wire feed rate (mm/s)	190.5 (=450 inch/min)
Travel speed (mm/s)	9.99 (23.6 inch/min)
Number of passes	Single

Mechanical Tensile Testing with DIC Measurements

To assess mechanical strength of the welded pipe, steel plate was sectioned into rectangular bars transverse to the weld direction for tensile testing. Mechanical tensile tests were conducted on an MTS 22 kip servo-hydraulic machine driven by an MTS model 407 controller at a cross head rate of 1.02×10^{-3} mm/s at room temperature. Additionally, DIC technique (Correlated Solutions, VIC-3D), which is an optical method to measure deformation and strain, was used in order to study the deformation and local strain of the material during tensile testing. Test samples were cleaned with acetone and then random speckle patterns were produced on the transverse section of the tensile specimens by sequentially spraying white and black paint. Two digital cameras of the stereo system were used to observe and record a cross sectional view of the tensile specimens during testing. Samples were illuminated by a LED green light. The recorded data were analyzed by software (VIC-3D® 2010).

Microhardness Mapping and Metallography

Vickers hardness of a welded sample was measured using a microhardness tester with 300 µm spacing, a 500 gm load, and 13 seconds of dwell time. The measured hardness values were transferred to Origin® software for visualization.

In order to reveal the microstructure at the weld zone, samples were cut and mounted for metallography. Initially, the mounted samples were ground using different grades of silicon carbide papers and finally diamond paste solutions (6, 3, and 1µm) were used for final polishing. A 5% Nital solution was used to etch the samples. Optical and field emission scanning electron microscopy were used to characterize microstructures of the welded samples.

Results & Discussion

FSW+ Root Arc Weld Sample

Figure 2 shows OD and ID images of a section of girth weld produced by the hybrid approach (FSW and root arc weld). Figure 2a shows a top view of the pipe steel with the friction stir weld zone along the circumferential direction. A spiral arc weld (red dot lines) is shown at about a 45 degree angle to the pipe axis. This spiral arc weld intersects the friction stir weld and root arc weld. Also, the root arc weld (white dot box) on the ID is shown in Figure 2b. The weld widths for the friction stir weld and arc weld were measured to be ~25.8 mm and ~12.7 mm, respectively.

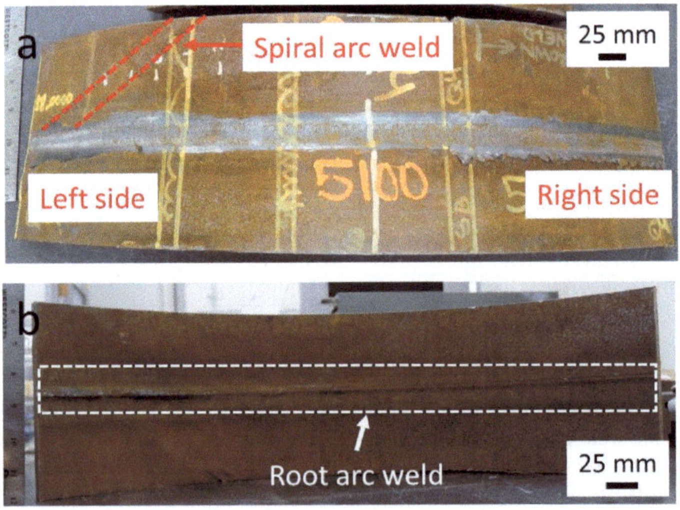

Figure 2. (a) Top view image showing FSW on line pipe steel. Spiral arc weld (red dot lines) is shown. (b) Inner surface view showing the root arc weld on line pipe steel (white color dotted box).

Transverse Tensile Test Results

In order to study the overall mechanical strength of the line pipe steel joined by this hybrid concept, transverse tensile tests were performed. As shown in Figure 3, rectangular tensile bars were prepared by cutting the welded steel pipe from both left and right sides. The overall length of the tensile bar was approximately 163.6 mm and the plate thickness 15.9 mm. From Figure 3a and 3b, the spiral arc weld (red color box) is observed close to the friction stir weld. Also, a tensile specimen cut from the right side of welded pipe steel is shown in Figure 3c and 3d, displaying misalignment of the two pipe steels. The root arc weld is indicated by a red dotted circle for each specimen. One reason the left and right sides are differentiated is that unevenness of the welded pipe steels was found on the right side of the welded line pipe steels. In addition, no spiral arc weld was found on the right side. For this reason, the term "left" and "right" is only used to define the different cutting locations for tensile specimens. Therefore, transverse tensile tests were performed using specimens cut from both sides to study overall mechanical strength in the "as-is" condition.

Figure 3. (a) Top view image of a transverse tensile specimen cut at the left side of the welded line pipe steel plate with the spiral arc welding shown (red color box). (b) Cross section view of Figure 3a showing the spiral arc weld. (c) Top view image of the tensile specimen cut at the right side. (d) Cross section view of tensile specimen, already shown in Figure 3c, showing unevenness or misaligned two pipes. Note that the root arc weld is marked by a red dot circle.

A summary of transverse tensile test data is given in Table 2. A total of five samples were used for the tensile tests and the average tensile strength from the left and right side samples is found to be 601.1 MPa and 652.5 MPa, respectively. For comparison, the tensile strength of the BM is 677 MPa based on the mill certification. The average tensile strength from the left side sample is somewhat (~ 76 MPa) lower than the tensile strength of the BM. This discrepancy could result from the smaller cross sectional area for the left side samples. That is, the FSW process produced a depression at the cap, resulting in smaller cross section area, Figure 4a. Therefore, the smaller cross section caused a more severe state of stress at an earlier point in the loading as compared to the right side tensile sample that performed better. On the other hand, the average tensile strength from the right side samples is almost comparable to the tensile strength of the BM, without significant loss of material at the cap, Figure 4b.

Table 2. Summary of transverse tensile tests on welded X70 line pipe (testing samples were cut from the left and right side of welded line pipe steels).

Left side		Right side	
Sample ID	Tensile strength (MPa)	Sample ID	Tensile strength (MPa)
#1_L	617.8	#1_R	639.1
#2_L	595	#2_R	664
L_#2	606.7	R_#2	655.7
L_#3	594.3	R_#3	643.3
L_#4	592.2	R_#4	660.5
Average	601.2	Average	652.5

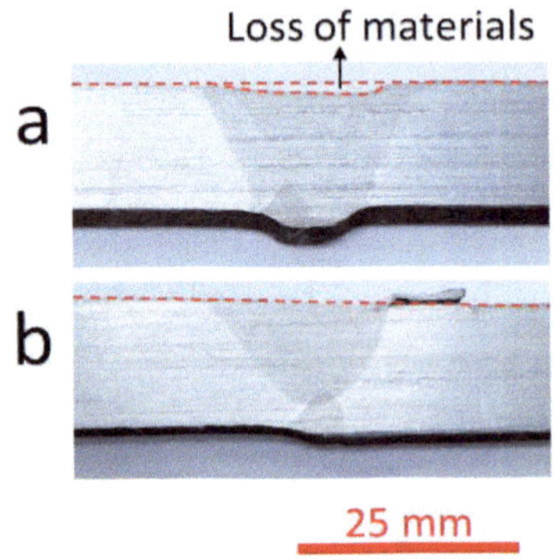

Figure 4. (a) Cross section view for left side sample, showing depression at the cap. (b) Cross section view for right side sample, showing no depression at the cap.

Figure 5a and 5b show top and cross sectional views of a fractured tensile sample following tensile testing. Furthermore, Figure 5c shows a cross sectional view of a fractured tensile specimen after etching with a Nital solution. Failure locations were found in both the SZ and root arc weld. Additional samples also showed the same failure locations. Assessment of the failure location is presented below, followed by a discussion of microhardness and microstructure.

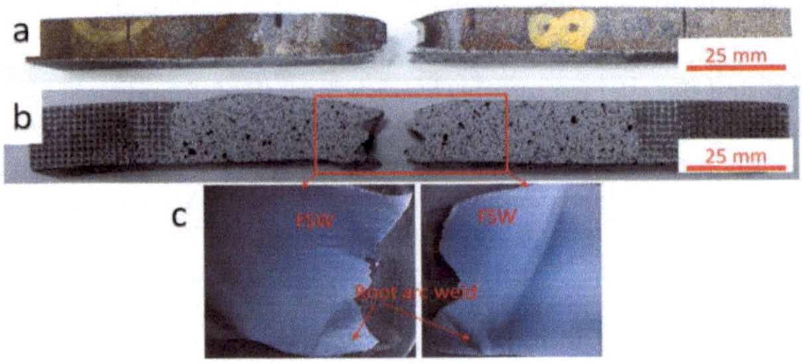

Figure 5. (a) Top view image of a fractured tensile specimen (left side sample) after tensile testing. The failure location was found to be in the SZ. (b) Cross sectional image of a broken

tensile bar with random speckles (white and black) for DIC measurements. (c) Cross sectional image of the broken sample after etching. Both the SZ and root arc weld are seen as fracture locations.

To assess the failure initiation location and local strain fields, a series of DIC images were taken during tensile testing and are shown in Figure 6. A fracture initiation location was found between the SZ and root arc weld (white dot circle), Figure 6a. After fracture initiation, the failure propagated into both the SZ and root arc weld finally resulting in fracture of the tensile sample. The local strain, ranging from 0.26 to 0.3 near the fracture initiation location, is much higher than the surrounding area (ranging from 0.08 to 0.13). As previously stated, a smaller cross section area, due to depression at the cap during FSW, can induce a more severe state of stress at an earlier point in the loading leading to higher local strains. For this reason, the failure location was initiated from a higher local strain area. This is consistent with the tensile test result. A joint area with higher local strains will be further characterized by microhardness measurement and microstructural analysis in a later section.

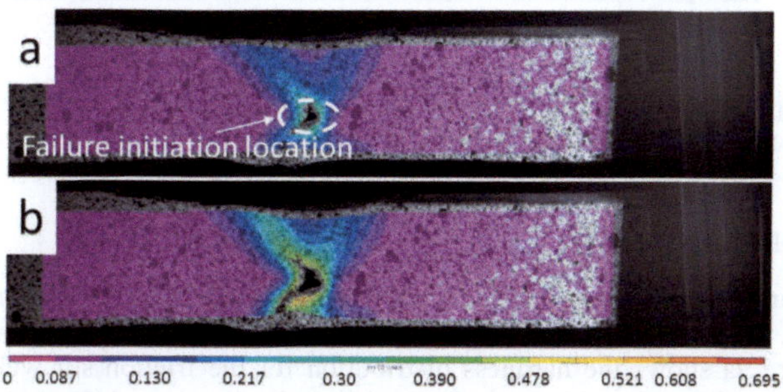

Figure 6. Sequential images of local strain fields during tensile test measurements made by DIC. (a) Fracture initiation location is seen (white dot circle). (b) Subsequently, the fracture propagates through both the friction stir weld and root arc weld, resulting in failure.

Similarly, fractography of the tensile specimen from the right side of welded line pipe steel is shown in Figure 7. Unlike the fractured tensile specimens from the left side of pipe steel, all failure locations were consistently found at the upper part of the SZ, but away from the root arc weld. This may be correlated to unevenness or misaligned of the two line pipe steels, causing a stress concentration where the geometry sharply changed. According to DIC measurements (Figure 8), the fracture initiation location was near the curvature where the geometry was suddenly changed. As a result, higher stress was concentrated in this area during the tensile test, leading to failure initiation.

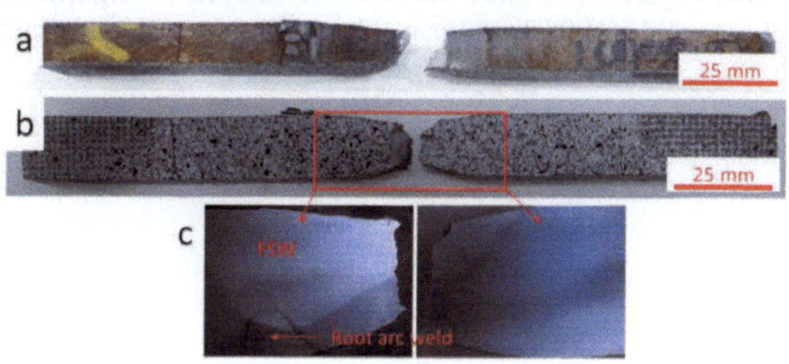

Figure 7. (a) Top view image of a broken tensile specimen (right side sample) after tensile testing. The failure location was found to be in the SZ, but away from root arc weld. (b) Cross sectional image of a broken tensile bar with random speckles (white and black) for DIC measurements. (c) Cross sectional image of the fractured sample after etching. Both the SZ and root arc weld are seen.

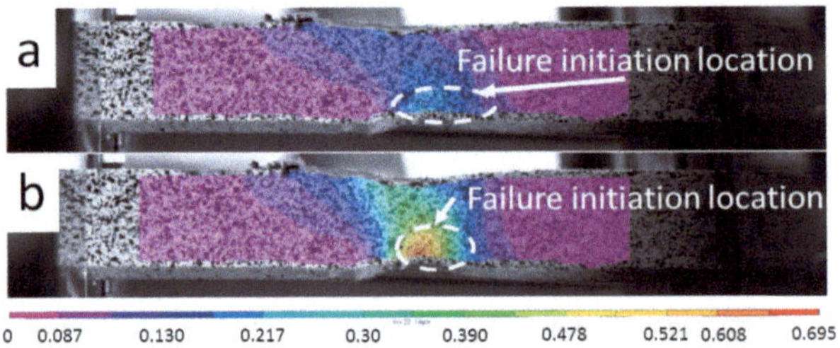

Figure 8. Sequential images of local strain fields during tensile tests measured by DIC. (a) Fracture initiation location is seen (white dot circle). (b) The fracture initiation propagates, resulting in final failure.

Microhardness and Metallography

Vickers microhardness was measured to correlate the strength of each weld, including the spiral arc, friction stir, and root arc welds. This sample was cut from the left side of the welded line pipe steel. Figure 9a shows the hardness distribution for the friction stir weld, spiral arc weld (adjacent to the friction stir weld, marked with black dot lines), and the root arc weld regions, where the x axis of the hardness graph is transverse to the weld direction. The hardness profiles along the three dashed lines (red color) in Figure 9a are further plotted in Figure 9b.

For the BM, the average microhardness was measured to be 225 HV. In the SZ, the average microhardness is similar to the hardness of the BM. The average microhardness in the HAZ of the friction stir weld ranged from 190 to 200 HV. However, as shown, the microhardness at the intersection between the bottom of the friction stir weld and the top of the root arc weld is reduced, varying between 170 and 180 HV. The local hardness drop in this region contributed to the fracture initiation site because of relative softness, and thus lower strength, compared to other weld zone regions. This observation corresponds to fractography and DIC measurements found in Figure 5 and 6. The reason why this zone is locally softer than other areas is discussed further with the microstructure analysis. Similarly, Vickers microhardness was measured for the sample cut from the right side, Figure 10. Unlike the left side sample, no significant hardness drop was observed at the root arc weld and friction stir weld zone for the right side sample.

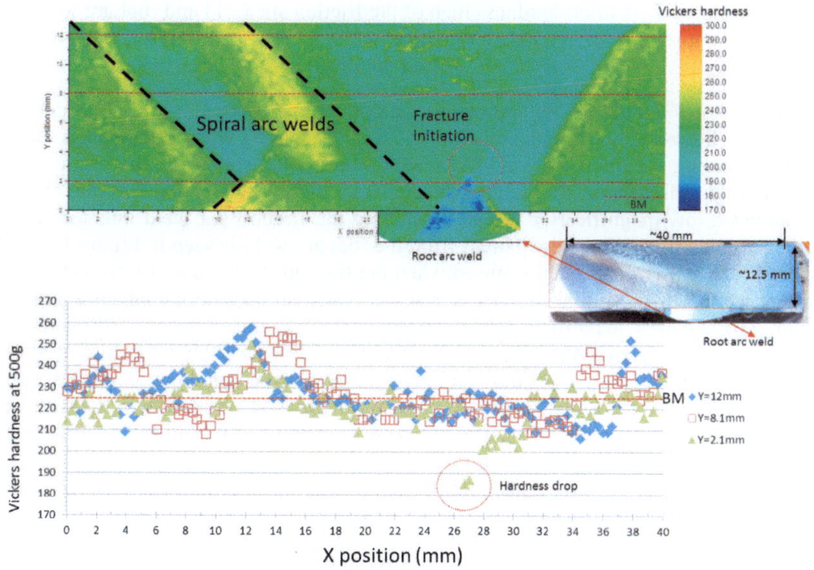

Figure 9. (a) Vickers microhardness map of the spiral arc weld (black dot lines), friction stir weld and root arc weld areas from the left side sample. A hardness drop was found at the intersection of the friction stir weld and root arc weld. (b) Vickers hardness plot along the x direction with different Y locations (red dot lines).

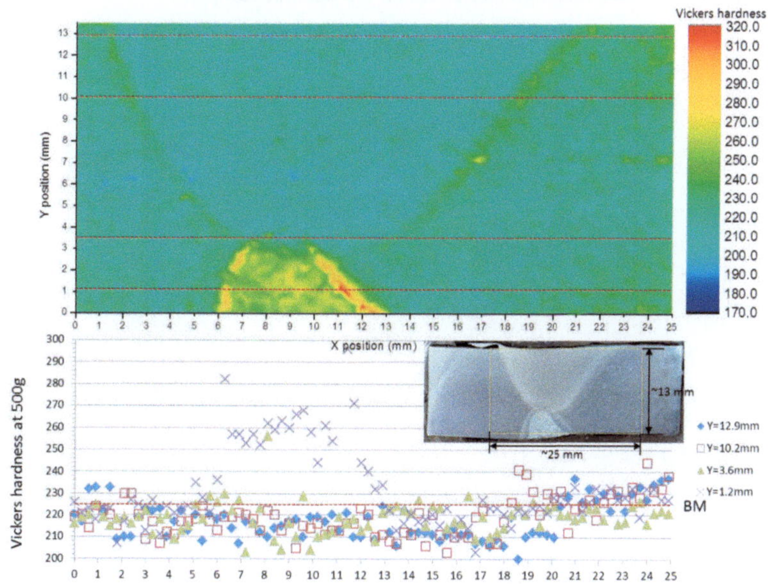

Figure 10. (a) Vickers microhardness map of the friction stir weld and root arc weld areas for the right side sample, and (b) Vickers hardness plot along the x direction with different Y locations (red dot lines).

Figure 11 shows a macrograph and magnified optical microscopic images (red dot boxes) at the intersection between the friction stir weld and root arc weld. Figure 11a displays the macrograph image, including the spiral arc weld (adjacent to the friction stir weld), the friction stir weld, and root arc weld. The friction stir weld intersected both the spiral arc weld and root arc weld. Figure 11b shows a low magnification image illustrating the friction stir weld intersecting the root arc weld. An upward material flow pattern from the root arc weld is seen in Figure 11b. This upward material flow pattern is possibly caused when the friction stir weld tool intersected the top of the root arc weld. This is similar to what is observed when an FSW tool contacts a supporting anvil. Figure 11c shows a magnified optical image in the SZ and soft zone. Similarly, Figure 11d shows the soft zone and BM. In the soft zone, a mostly coarse ferrite (white) microstructure is observed compared with the other regions. The low measured microhardness values in this region (170 to 180 HV) support the existence of ferrite. Also, grain size in the soft zone is larger than the grain size in either the SZ or BM. Ferrite is known to be softer than other common steel microstructures, such as bainite or martensite. The microstructure of the X70 pipe steel is typically a mixture of ferrite and bainite depending on thermal and mechanical processing conditions.

One potential reason for the presence of a soft region with coarse ferrite near the bottom of the SZ for the left side sample can be due to the proximity of the FSW SZ to the root arc weld. The SZ of the friction stir weld appears to have penetrated the root arc weld to a greater degree for the left side sample versus the right side sample. This is evidenced by the fact that the left side sample shows distinct "wisps" of stirred material being pulled from the root arc weld whereas for the right side sample no "wisps" of stirred material appear. This condition means that the FSW thermal cycle would have had a greater effect on the left side sample.

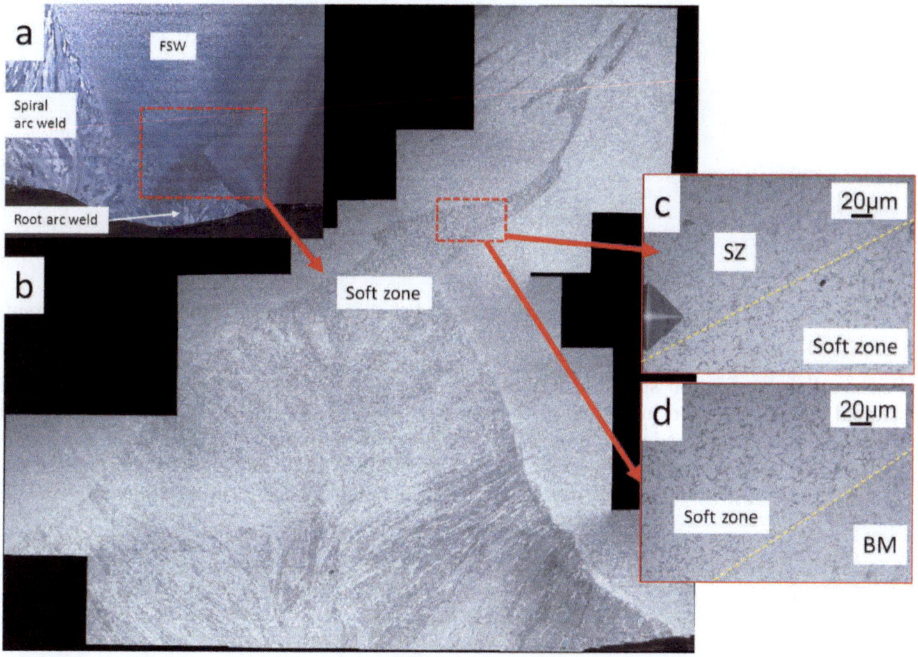

Figure 11. Macrograph and magnified images (red dot boxes) at the intersection between the friction stir weld and root arc weld. (a) Macrograph image showing three different regions (FSW, spiral arc weld, and root arc weld). (b) Low magnification optical image at the interface between the friction stir weld and root arc weld. (c) Magnified optical image between the SZ and soft zone (d) Magnified optical image between the soft zone and BM.

Another consideration is the thermal cycle of the FSW pass. Cooling rates associated with root arc welding are significantly faster than the cooling rate following FSW. That is, the soft zone with significant amounts of ferrite may be due to cooling rates slower than intended for arc weld chemistries. To access this assumption, heat inputs for the root arc weld and friction stir welding processes were estimated based on welding process parameters. First, heat input for the root arc weld was calculated based on the following equation:

$$Q = \eta \frac{VI}{s}$$

where Q is heat input (J/mm), η is arc efficiency, V is arc voltage (volt), I is welding current (amp), and s is a travel speed (mm/s). An arc efficiency (η) of 0.8 was used. This is a typical efficiency for gas metal arc welding (GMAW) [11]. Arc voltage and welding current were 24 volts and 270 amps, respectively. Travel speed was 9.99 mm/s. The calculated heat input for root arc welding is 0.52 kJ/mm. Next, heat input for FSW was estimated based on equations derived from a torque based model [12].

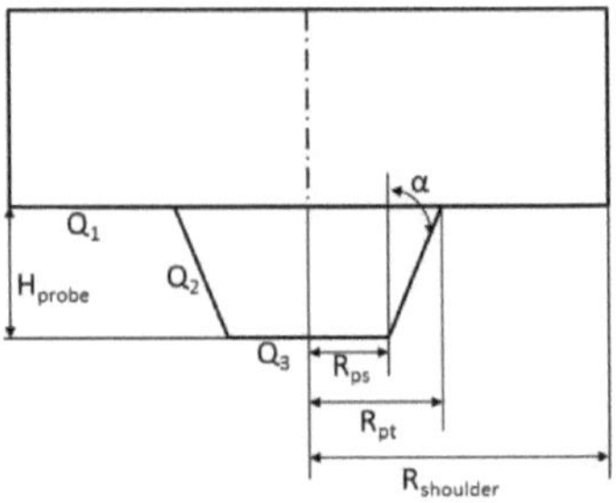

Figure 12. A schematic of a friction stir weld tool showing heat generations at three different locations of the FSW tool.

For simplicity, a friction stir weld tool shape was considered as a taper cylindrical pin profile without threads. As seen in Figure 12, heat can be generated at three different locations of the friction stir weld tool; under the tool shoulder (Q_1), at the tool pin side (Q_2), and at the tool pin tip (Q_3). The total heat generation (Q_{total}) is a summation of three different heat generations (i.e., $Q_{total}= Q_1+ Q_2+Q_3$). Analytical equations for heat generation at each location are as follows [13]:

$$Q_1 = \frac{2}{3}\pi\omega\tau_{contact}(R^3_{shoulder} - R^3_{PT})$$

$$Q_2 = 2\cdot\pi\cdot\omega\cdot\tau_{contact}\cdot\frac{H_{probe}}{cos\alpha}\cdot\left(\frac{R_{PS}+R_{PT}}{2}\right)^2$$

$$Q_3 = \frac{2}{3}\pi\omega\tau_{contact}R^3_{PS}$$

where ω is a tool rotational speed (18.33 rad/s), $\tau_{contact}$ is contact shear stress, H_{pt} is pin length (11 mm), α is the taper angle of the pin (26 °), $R_{shoulder}$ is the tool shoulder radius (12.9 mm), R_{PT} is the radius of the probe at top part (8.02 mm), and R_{PS} is the radius of the probe at the bottom (2.57 mm), respectively. Because the full diameter of the tool shoulder was not engaged, an effective tool shoulder radius of 12.9 mm was based on the measurement of the friction stir weld crown surface. This value was used for the heat input calculation.

Contact shear stress was calculated as follows:

$$\tau_{contact} = \frac{\mu F}{\pi R^2_{shoulder}}$$

where μ is a friction coefficient (0.3), and F is the plunge load during FSW (80 kN). Finally, heat input for FSW is calculated by:

$$Q_{FSW} = \frac{Q_{total}}{TS_{FSW}}$$

where TS_{FSW} is the travel speed for FSW (1.27 mm/s). The calculated total heat input for FSW is 3.7 kJ/mm, that is, about 7 times higher than the heat input from the root arc weld. This large heat input for FSW is due to the high heat generation from the large shoulder of the FSW tool and slower travel speed during FSW. As a result, the cooling rate for FSW is expected to be much slower than the cooling rate for the root arc weld.

Using the estimated heat input and peak temperature, microstructures at the interface between the top of the root arc weld and bottom of the friction stir weld can transform to austenite during the heating cycle and subsequently transform to coarse ferrite with the slow cooling rate experienced following FSW. This can be a potential reason for the soft zone in the joint area between the root arc weld and the friction stir weld.

There is one other potential reason that coarse ferrite may have formed in the region noted. Many arc welding consumables are designed to produce (fine grained) acicular ferrite as nucleated by weld metal oxide inclusions. This key transformation phenomena occurs during weld pool solidification and subsequent cooling of the austenite. This phenomenon of acicular ferrite nucleation may not occur during a reheating event created by subsequent FSW.

Conclusions

In summary, a hybrid (friction stir welding + root arc weld) approach was successfully used to demonstrate the ability to join line pipe (API 5L X70 grade steel) without internal support. Transverse mechanical tensile tests were used to study the overall mechanical strength of the line pipe steel welded by this hybrid concept. Tensile specimens cut in the left side of the assembled line pipe steels showed slightly lower tensile strength (~71 MPa) than the BM due to a smaller cross section resulting from the depression at the cap during FSW. However, tensile strength of specimens cut from the right side of the assembled line pipe steel was almost comparable to the strength of the BM. The local microstructure and hardness changes were measured for the left side sample. These results identified a local soft zone at the intersection of the friction stir weld and root arc weld that created a premature failure initiation site. This local soft zone was apparently affected by the thermal cycle of the FSW pass and the subsequent reheating effect on the root arc weld. Digital image correlation measurements showed good correlation with transverse tensile tests and microhardness measurements. In conclusion, the overall mechanical properties of welded line pipe steels by this hybrid concept are comparable to the base metal but this small localized soft zone merits further evaluation and a remedy.

Acknowledgements

The authors would like to thank the US Department of Energy, MegaStir, and ExxonMobil for financial support. Oak Ridge National Laboratory (ORNL) is managed by UT-Battelle, LLC for the U.S. Department of Energy under Contract DE-AC05-00OR22725.

References

1. Q. Liu, R. Steel, J. Peterson, S. Horman, M. Collier, D. Marshall, J. Davis, and M. Mahoney, “Advances in Friction Stir Welding Tooling Materials Development,” Proceedings of the Twentieth (2010) International Offshore and Polar Engineering Conference, Beijing, China, ISOPE, 2010, vol. 4., pp. 298-302.

2. C. Sorensen and B. Nielsen, “Exploring Geometry Effects for Convex Scrolled Shoulder, Step Spiral Probe FSW Tools,” Friction Stir Welding and Processing V, Edited by R. Mishra, M. Mahoney, and T. Lienert, TMS 2009, San Francisco, California, February 2009, pp. 85-92.

3. A. De, H.K.D.H. Bhadeshia, and T. DebRoy, "Friction stir welding of mild steel: tool durability and steel microstructure," Mater Sci Tenchnol, 30 (9) (2014), 1050-1056.

4. T.J. Lienert, W.L. Stellwag, B.B. Grimmett, and R.W. Warke, "Friction stir welding studies on mild steel - Process results, microstructures, and mechanical properties are reported," Weld J, 82 (2003), 1-9.

5. W.M. Thomas , P.L. Threadgill, and E.D. Nicholas, "Feasibility of friction stir welding steel," Sci Technol Weld Joi, 4 (1999), 365-372.

6. Y.C. Lim, S. Sanderson, M. Mahoney, X. Yu , D. Qiao, Y. Wang , W. Zhang, and Z. Feng, "Characterization of Multilayered-Multipass Friction-Stir Weld on ASTM A572 G50 steel," in press, Welding Journal.

7. J. Fujii , L. Cui, N. Tsuji , M. Maeda , K. Nakata , and K. Nogi, "Friction stir welding of carbon steels," Mater Sci Eng A. 429 (2006), 50-57.

8. A. Kumar, D. P. Fairchild, M. L. Macia, and T. D. Anderson, "Evaluation of Economic Incentives And Weld Properties For Welding Steel Pipelines Using Friction Stir Welding" The Twenty-first International Offshore and Polar Engineering Conference, 19-24 June, Maui, Hawaii, USA, 2011.

9. D.P. Fairchild, S.J. Ford, A. Kumar, N.E. Nissley, N.E. Biery, and M.I. Macia, "Butt Weld and Method of Making Using Fusion and Friction Stir Welding," US Patent, No. 7,874,471 B2, Jan. 25, 2011.

10. M. Mahoney, S. Sanderson, S. Larsen, R. Steel, Z. Feng, A. Wasson, D. Fairchild, S. Packer, and D. Fleck, "Methods to Friction Stir Weld Onshore and Offshore Steel Linepipe with Consistent Full Penetration", Proceedings of the TWI 10th International Friction Stir Welding Symposium, Beijing, China, 19-21May 2014.

11. J.N. DuPont and A.R. Marder, "Thermal efficiency of arc welding processes," Weld J, 74 (1995), 406s-416s.

12. H. Schmidt, J. Hattel, and J. Wert, "An analytical model for the heat generation in friction stir welding," Model Simul Mater SC, 12 (2004), 143-157.

13. V.S. Gadakh and K. Adepu, "Heat generation model for taper cylindrical pin profile in FSW," J Mater Res Technol, 2 (2013), 370-375.

Friction Stir Welding and Processing VIII
Edited by: Rajiv S. Mishra, Murray W. Mahoney, Yutaka Sato, and Yuri Hovanski
TMS (The Minerals, Metals & Materials Society), 2015

IMPROVED TEMPERATURE AND DEPTH CONTROL DURING FSW OF COPPER CANISTERS USING FEEDFORWARD COMPENSATION

Lars Cederqvist[1], Olof Garpinger[2], Anton Cervin[3], Isak Nielsen[4]

[1]SKB, Oskarshamn, Sweden
[2]ALTEN, Gothenburg, Sweden
[3]Lund University, Dept. of Automatic Control, Sweden
[4]IAN Control, Linkoping, Sweden

Keywords: friction stir welding, copper, automatic control, tool depth control

Abstract

The welding procedure to seal copper canisters requires variable power input throughout the 45 minute long weld cycle to keep the probe temperature within the process window. By using a cascaded loop that determines the power input requirement, the controller will not be dependent on repeatability in the necessary power input between weld cycles, and the lag time in the probe temperature measurement will not be critical. Due to fast-changing thermal boundary conditions during the downward sequence, a feedforward to the power input was designed to further improve controller performance.

In addition to the cascade controller adjusting the tool rotation rate, a depth controller is adjusting the axial force to control the shoulder depth. The purpose is to eliminate flash due to excessive shoulder depth and to control the position of the probe tip, which influences the size and shape of the hook defect produced. Controlling depth is challenging for several reasons, including deflection in the welding machine and thermal expansion of the weld material, and also results in cross-coupling between axial force and spindle torque. The cross-coupling was handled by another feedforward compensator that adjusts the tool rotation rate based on the commanded axial force.

The implemented controllers and feedforward compensations have been evaluated over several welds with good results, where the depth is kept within ±0.1mm when the tool reaches the joint line sequence and the probe temperature is held within ±5°C during full circumferential welds compared with a process window of ±60°C.

Introduction to cascade temperature control

The Swedish Nuclear Fuel and Waste Management Company (SKB) will join at least 12,000 lids and bottoms to the extruded 5 cm thick copper tubes containing Sweden's nuclear waste using FSW. Unlike most FSW applications where straight welds are made on flat plates, the welding procedure to seal copper canisters requires variable power input throughout the weld cycle to keep the probe temperature within the process window. This is due to variable thermal boundary conditions throughout the different sequences in the weld cycle (see Figure 1).

In 2007 Cederqvist and Bolmsjö [1] presented the idea that controlling the power input could lead to faster response to process changes than by only controlling the tool temperature (measured by a thermocouple in the probe), and in 2008 Cederqvist et al [2] presented the cascade controller that controls power input and tool temperature by adjusting the tool rotation rate. In 2010 Cederqvist et al [3] first presented results using the cascade controller, and others [4-5] have since started to use the same idea/principle; cascade control of tool temperature by controlling the power input.

Figure 1. Sequences in a weld cycle: 1. Start, 2. Downward, 3. Joint line, 4. Overlap, 5. Parking.

Current cascade controller including feedforward compensation

The current controller structure is illustrated in the block diagram in Figure 2. In the inner loop of the cascade controller, the actual power input (ActPower) is controlled by the power regulator (PowerReg) via adjustment of the tool rotation speed (SetRPM). In the outer loop, the actual tool temperature (ActTemp) is controlled by the temperature regulator (TempReg) via adjustment of the desired power input (SetPower). The cascade controller version 1.0 used PI-controllers for both the power and temperature regulators. In the current implementation (version 2.0), both controllers have been re-tuned to suppress process disturbances as efficiently as possible. Analyzing the trade-off between control signal (RPM) noise injection and error in injected power, it was found that a pure I-controller gave optimum performance for the inner power controller. Using the same tool [6], a PID-controller was synthesized for the outer temperature controller. The derivative action stabilizes the temperature dynamics and allows for more aggressive integral action, and hence betters the disturbance rejection.

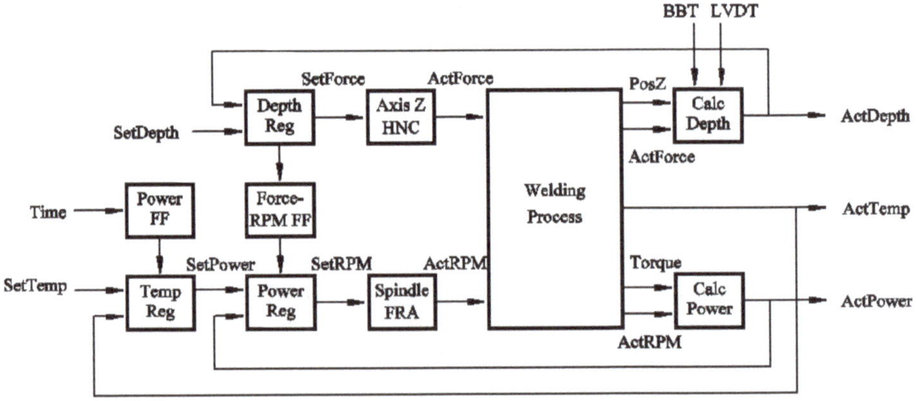

Figure 2. Cascade controller structure including feedforward compensators (PowerFF, Force-RPM FF)

During the downward sequence, the process experiences growing torque disturbances, which cannot be completely eliminated using integral action. Therefore, a time-based power feedforward compensator (PowerFF) was designed based on a simple ramp model of the power loss. The compensator is activated when the downward sequence starts and its output is frozen when the actual power reaches a pre-set threshold value.

In the tool depth loop, the actual tool depth (ActDepth) is controlled by the depth regulator (DepthReg) via adjustment of the axial force (SetForce). When the axial force is adjusted, another feedforward (Force-RPM FF) tries to counter the cross-coupling to the temperature control loop.

Cascade controller results

In Table 1, the results for the 360° joint line sequences from different welding series are presented. As reference, the demonstration series was controlled by manual changes of tool rotation rate (and axial force) by a welding operator. It can be seen that the current controller maintains the temperature to within ±5°C at the joint line sequence, with a standard deviation of less than 1°C.

Table 1. Analysis of probe temperature data during the joint line sequence for full welds.

Name of series	Cycles	T_{avg}-T_r	σ_{avg}	T_{min}-T_r	T_{max}-T_r
Demonstration series (2004)	20	+4	6	-52	+49
Controller version 1.0 (2010)	8	+0.6	1	-7	+8
Controller version 2.0 (2013)	2	+0.4	1	-3	+5

T_{avg}-T_r is the average deviation from the reference probe temperature (T_r), σ_{avg} is the standard deviation, T_{min}-T_r is the lowest temperature relative to T_r, and T_{max}-T_r is the highest temperature relative to T_r.

Introduction to tool depth controller

Historically, the tool depth can vary between cycles although constant probe temperature is achieved, and that the axial force and welding speed are held at same and constant values. The objective of the tool depth controller is therefore to maintain the tool depth at a constant value (resulting in no flash formation) throughout the weld cycle. The reasons that the tool depth needs to be constant are:

1. The position of the probe tip determines if a hook defect is produced (and the size of the defect). The hook defect reduces the corrosion barrier of 5 cm copper.
2. Flash formation during the downward sequence can cause spindle torque disturbances (and fluctuating probe temperature) during the overlap sequence.
3. Different tool depths and flash formations result in different spindle torque, which will make it more difficult for the cascade controller and to evaluate the weld data.

The tool depth controller that is currently implemented is developed over several years. Different approaches have been investigated, and much work effort has been put on modeling the process to obtain a model that can be used for control synthesis. In 2012 Nielsen [7] introduced and evaluated dynamic models that describe the process. These models are used in the synthesis and evaluation of suitable control structures for the tool depth regulator. Furthermore, in 2012 Nielsen [7] and in 2013 Nielsen et al [8] a Non-linear Model Predictive Control (NMPC) scheme is for the first time introduced to control both the tool depth and the tool temperature during FSW. The solution is evaluated using simulations of a detailed dynamical model of the process, and the simulation result indicates that this strategy could successfully be used to achieve high performance control of the process. However, due to implementation issues, this controller has not been tested on the real process.

Tool depth measurements

In order to control the tool depth at a desired value, a reliable measurement of the depth is necessary. Currently there are two different types of sensors available for this purpose; a spindle position sensor (henceforth called Position Z) and a linear displacement transducer that measures the distance from the tool to the canister surface (henceforth called LVDT), see Figure 3. In the controller that is implemented on the welding machine it is possible to choose between the two, but the Position Z measurement has been used as the main sensor. The two sensors measures the depth in slightly different ways, both having their pros and cons.

When the Position Z measurement is used, there are a few major challenges and factors that must be taken into consideration; the deflection of the welding machine due to the high axial forces, thermal expansion of the canister and non-centering of the canister in the welding machine. The LVDT measurement is not affected by these challenges, but its main challenges is that it is positioned approximately 12 cm in front of the welding tool, i.e. it is not measuring at the tool tip which is desired. Attempts have been made to remedy the drawbacks with both sensors, and they will be presented below.

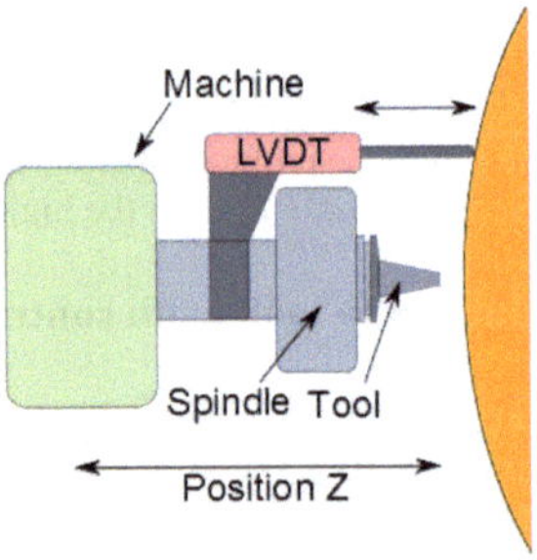

Figure 3. Schematic overview of the welding machine and its sensors. The Position Z sensor measures the spindle position relative to the machine, and the LVDT sensor measures the spindle position relative to the canister surface, but approximately 12 cm ahead of the tool.

By using information from both sensors, the deflection of the machine can be measured as the difference in the Position Z and LVDT measurements. By applying axial forces in the range 10 kN to 100 kN at different positions along the weld, a simple model for the deflection was obtained. It was observed that the deflection could be described by a constant times the axial force. The value of the constant (gain) is 0.032 mm/kN.

The thermal expansion also affects the tool position measurements, since increased heating of the copper canister will expand the material and push the tool further away from the canister's centre. This expansion will be registered by the Position Z measurement, and will be treated as a false change in tool depth if it is not compensated for. Depending on the propagation of the expansion, and due to the placement of the LVDT, it is possible that also the LVDT measurement register effects from the expansion as well, which will indicate a false change in tool depth. In the current implementation, no compensation of the thermal expansion is used. However, the effects are assumed to be repetitive and can be compensated for by adjusting the reference depth.

A similar problem to thermal expansion arises from the fact that the canister cannot be perfectly centred in the welding machine. This non-uniform deviation of the canister's surface (from the centre of the welding machine) also affects the Position Z measurements and must be compensated for. This is done by measuring the eccentricity before the weld and using this information to correct the measurements. However, it is not yet fully known if the position of the canister changes during welds, and this must be further investigated. A thorough description of the welding machine and its sensors is given in [7].

In 2010 Longhurst et al [9] and Lammlein et al [10] argued that the spindle torque is a good indicator of tool depth. However, since the process is not in steady state in this application the spindle torque is not used in this work.

Current tool depth controller including feedforward compensation

The proposed tool depth controller is a PI-controller that uses the measured tool depth to compute the commanded axial force in order to track the desired tool depth reference, and the above mentioned cascaded temperature controller that regulates the tool temperature.

It is observed that if the axial force is kept constant during the joint line sequence, the tool depth will remain fairly constant throughout the weld. Hence, the most important objective of the controller is to repetitively regulate the process such that the desired depth is obtained when the tool enters the joint line sequence. Due to this, the tool depth controller is only active during the start and downward sequences, and the axial force is held constant during the joint line sequence.

Since the axial force can affect the cascaded temperature controller in a negative manner, and since the tool depth during the start and downward sequences are allowed to vary, a smooth tool depth reference is used to minimize the use of large and rapidly changing axial forces. The controller is thus smoothly changing the axial force such that the reference tool depth is met when the tool is entering the joint line. The tool depth reference used is computed by filtering the desired tool depth by a third order low pass filter and can be seen as the dashed line in Figure 4.

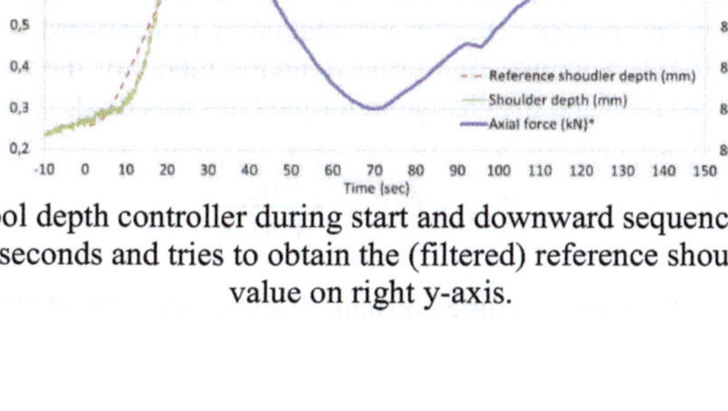

Figure 4. Active tool depth controller during start and downward sequences. The controller is activated at time 0 seconds and tries to obtain the (filtered) reference shoulder depth. *denotes value on right y-axis.

The implemented tool depth controller has been evaluated during several welds using Position Z measurements. The Position Z measurements have been compensated for the eccentricity and the deflection using the methods described in the previous section. The result is seen in Figure 5, where data from nine different welds are presented. The tool reaches the joint line after 135 seconds. The upper part of the figure shows the axial forces that are commanded by the controller. For most welds, the controller manages to keep the tool depth within the desired interval (±0.1mm) when the joint line is reached. For the three welds which are not in the interval, the controller hits the lower bound (82 kN) on the axial force. This means that the welds are too deep although the minimum allowed axial force is commanded. It can be seen that all welds start with a too shallow/low tool depth (positive control error), and the controller commands a larger axial force to plunge the tool deeper into the material.

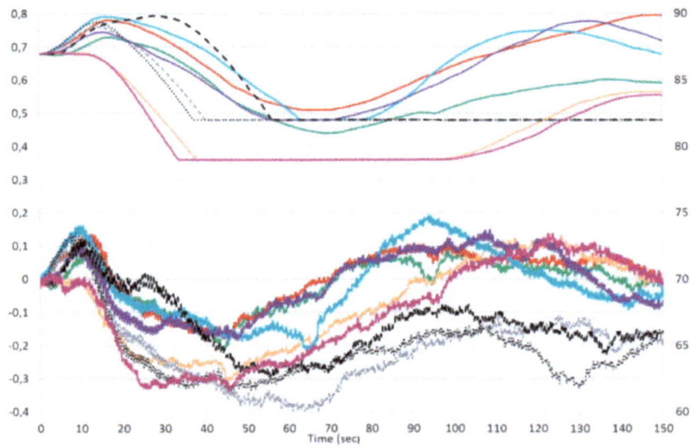

Figure 5. The control error for nine different welds with the tool depth controller active. The tool depth control error (reference minus actual tool depth) is on the left y-axis and the commanded axial force on the right y-axis.

When the tool depth controller adjusts the commanded axial force in order to maintain the desired tool depth also the spindle torque is changed due to varying friction. In Figure 6 the commanded axial force and the measured spindle torque are plotted for five different welds. From the figure it is clear that the torque is highly affected by the changes in the axial force and this change will affect the cascade temperature controller. The fairly rapid changes in torque will be handled as disturbances in the process by the cascade controller.

Since the change in force is known (computed by the tool depth controller) the effects of the change in torque has been counter acted by introducing a feedforward loop from commanded axial force to spindle RPM. A simple first-order dynamic model of the cross-coupling between the change in axial force (ΔF_z) and the change in torque ($\Delta\tau$) was used,

$$G_{\Delta\tau\Delta F_z}(s) = \frac{K_{\tau F_z}}{1 + sT_{\tau F_z}}$$

where $K_{\tau F_z}$ and $T_{\tau F_z}$ are the gain and time constant of the model. The RPM adjustment was then computed as

$$RPM_{FF} = -\frac{SetRPM \cdot \Delta\tau}{\tau}$$

where τ is the current actual spindle torque. Hence, the commanded RPM is partly determined from the current commanded axial force; the results using this feedforward compensation link will be added to the final version of the paper.

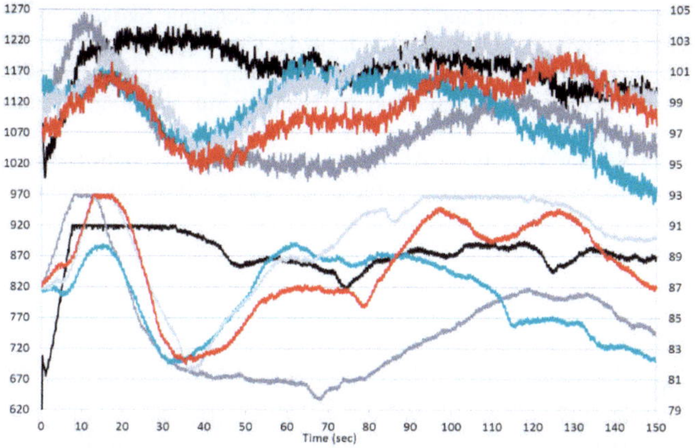

Figure 6. The spindle torque (lower plot) is affected by the axial force (upper plot). The changes in torque due to changes in axial force will affect the cascaded temperature controller. The spindle torque is on the left y-axis and the commanded axial force on the right y-axis.

Although the current implemented depth controller manages to keep the tool depth within the process window of ±0.1 mm for the majority of the welds, an even more robust solution is needed to guarantee that the desired depth reference is met for all welds. Furthermore, it is important to investigate how well the used measurements (Position Z and LVDT) are correlating with the actual depth, i.e. how accurate they measure the tool depth. It is possible that new sensors are needed to get better measurements of the actual tool depth.

Summary

The development of the two controllers and the use of feedforward loops in both have shown the following:

- A cascade controller controls the probe temperature to maintain the desired value of 845°C and can normally within ±5°C at the joint line sequence, with a standard deviation of less than 1°C. Thus, the risk of producing cavities, which occur at probe temperatures below 790°C, or probe fractures, which occur above 910°C, is very small.

- Controlling the duration of the dwell sequence via the power input increases the resulting tool depth repeatability between welds. Future studies (including full welds and analyses of joint line hooking/remnants) will determine the need for and use of the tool depth controller during the reference welding procedure.

References

1. Cederqvist L, Bolmsjö G. Industrialisation of novel welding process to seal copper canisters containing Sweden´s nuclear waste. Proceeding of 2007 Swedish Production Symposium, Gothenburg, Sweden, 2007.
2. Cederqvist L, Bolmsjö G, Sorensen CD. Adaptive control of novel welding process to seal cansiters containing Sweden's nuclear waste using PID algorithms. Proceedings of 18th Intl. conference on Flexible Automation and Intelligent Manufacturing, Skövde, Sweden, 2008.
3. Cederqvist L, Sorensen C D, Reynolds A P, Garpinger O. Reliable FSW of copper canisters using improved process and regulator controlling power input and tool temperature. Proceedings of 8th International Symposium on Friction Stir Welding, Timmendorfer Strand, Germany, 2010.
4. Mayfield D W, Sorensen C D. An improved temperature control algorithm for friction stir processing. Proceedings of 8th International Symposium on Friction Stir Welding, Timmendorfer Strand, Germany, 2010.
5. Ross K, Sorensen C D. Development and Implementation of a Robust Temperature Control Algorithm. Proceedings of 9th International Symposium on Friction Stir Welding, Huntsville, USA, 2012.
6. Garpinger O. Design of Robust PID Controllers with Constrained Control Signal Activity. Licentiate Thesis LUTFD2/TFRT-3245-SE, Department of Automatic Control, Lund University, Sweden, 2009.
7. Nielsen I. Modeling and control of friction stir welding in 5 cm thick copper canisters. Master Thesis LiTH-ISY-EX--12/4567--SE, Department of Automatic Control, Linköping University, Sweden, 2012.
8. Nielsen I, Garpinger O, Cederqvist L. Simulation based evaluation of a nonlinear model predictive controller for friction stir welding of nuclear waste canisters. Proceedings of 2013 European Control Conference (ECC), Zurich, Switzerland, July 17-19, 2013.
9. Longhurst W R, Strauss A M, Cook G E, Fleming P A. Torque control of friction stir welding for manufacturing and automation. The International Journal of Advanced Manufacturing Technology, April 2010.
10. Lammlein D H, Longhurst W R, DeLapp D R, Fleming P A, Strauss A M, Cook G E. The friction stir welding of hemispheres - a technique for manufacturing hollow spheres. International Journal of Pressure Vessels and Piping, August 2010.

Friction Stir Welding and Processing VIII
Edited by: Rajiv S. Mishra, Murray W. Mahoney, Yutaka Sato, and Yuri Hovanski
TMS (The Minerals, Metals & Materials Society), 2015

FRICTION STIR WELDING OF STEELS USING A TOOL MADE OF IRIDIUM-CONTAINING NICKEL BASE SUPERALLOY

Tatsuya Nakazawa[1], Yutaka S. Sato[2], Hiroyuki Kokawa[2], Kiyohito Ishida[2],
Toshihiro Omori[2], Kunihiro Tanaka[1], Koichi Sakairi[1]

[1]Technical Division, Tanaka Kikinzoku Kogyo K.K.
2-73 Shinmachi, Hiratsuka, Kanagawa 254-0076, Japan
[2]Graduate School of Engineering, Tohoku University
6-6-02 Aramaki-aza-Aoba, Aoba-ku, Sendai, Miyagi 980-8579, Japan

Keywords: Friction stir welding; Stainless steel, Ir-containing Ni base superalloy, Microstructure, Mechanical properties

Abstract

An addition of Ir to Ni base super alloy can increase the high temperature strength. An Ir-containing Ni base superalloy produced through a process of rapid solidification casting, solution heat treatment and subsequent aging was found to exhibit a compressive strength greater than 500 MPa at 1000°C, which suggested it would be a suitable material to use in a friction stir welding (FSW) tool for high-softening-temperature materials, such as stainless steels and carbon steels. Precision tilt casting of this Ir-containing Ni base superalloy into a copper mold followed by appropriate heat treatment, resulted in tools with a fine γ/γ' microstructure containing intergranular carbides. These tools were subsequently proven capable of producing welds on AISI 304 stainless steel and ultrahigh carbon steel without large deformation or severe wear.

Introduction

Ni-base superalloys are one of the most widely used materials in high temperature applications such as space ships, atomic power generators and turbine engines. Such alloys are strengthened by the precipitation of a fine γ' (Ni_3Al) phase with a $L1_2$ structure in a γ matrix. The strengthening effect of this γ' phase exhibits a positive correlation with temperature in Ni-base superalloys, but the service temperature of such alloys is generally lower than 1000°C due to the relatively low melting point of Ni (1455°C).
Platinum group metals (PGM) are known to have a high melting point, as well as excellent corrosion resistance at temperatures above 1000°C. Indeed, both Pt and Ir have been already used in high temperature applications, such as crucibles for crystal growth, spark plug electrodes and glass melting equipment, because of their high heat resistance. The Ni-Ir system in particular is quite popular because of its complete solid solubility (Figure 1) [1]. The formation of a γ/γ' Ir_3(Al, W) phase with a $L1_2$ structure has also been found to occur in the case of Ir addition to Ni base superalloy with a suitable chemical composition [2].

The friction stir welding (FSW) of steels has been achieved using the tool made of high melting point materials, such as W-Re alloy [3], W carbide [4], Co-base alloy [5], polycrystalline cubic boron nitride (cBN), and composites of cBN [6]. However, these tools often experience problems in term of excessive wear, oxidation and deformation. Gan *et al.* have also demonstrated that a strength of at least 400 MPa at 1000°C is needed in tool materials used for the FSW of steels to prevent tool deformation [7]. Given that Ir-containing Ni base superalloy exhibits a strength well above 400 MPa at 1000°C, it is possible this could be used as a tool material for the FSW of steels. This study is therefore focused on preliminary trials into the FSW of steels using Ir-containing Ni base superalloy tools.

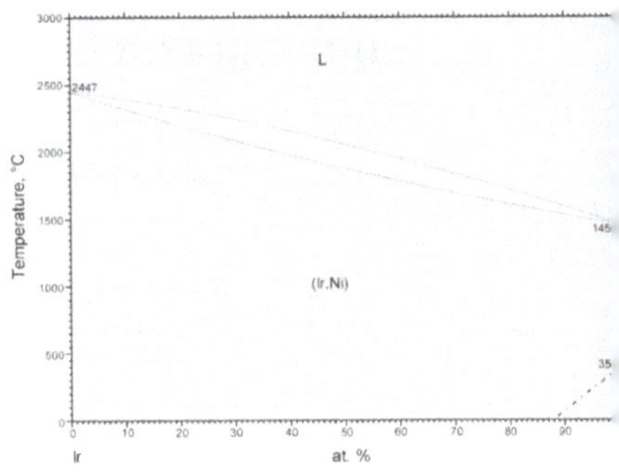

Figure 1. Ni-Ir binary phase diagram.

Experimental Procedure

The chemical composition of the Ir-containing Ni-base superalloy used in this study is shown in Table 1. This master ingot was initially produced via melting, and was subsequently re-melted in an oxide crucible using a high-frequency melting furnace, then rapidly solidified into a Cu mold with an appropriate tool shape placed at the bottom of the furnace chamber. This as-cast alloy was solution-heat-treated under vacuum for a few hours at 1300°C, followed by aging for 1 day at 800°C, and was then lathed to a precise tool shape. Tools with probe lengths of 2 mm and 5 mm were prepared for FSW trials, as shown in Figure 2. Compression testing was also conducted to compare their strength with common as-cast Ni base superalloy (Mar-M247).

Table I. Chemical composition of Ir-containing Ni-base superalloy. (wt. %)

Ni	Ir	Al	W	C	Other
38	25	4	14	0.1	18.9

The plate materials for FSW testing were a AISI 304 stainless steel (AISI304) and an ASTM A295 52100, though only AISI304 was used in FSW trials with the 5 mm probe length tool. A tool welding speed of 1 mm/s and rotation speed of 150 rpm were used in FSW trials with the 2 mm probe, whereas an 80 rpm rotation speed was used for the 5 mm probe in conjunction with a tool travel speed of 20 mm/min (beginning to 5 mm) and 40 mm/min (5 mm to end). In FSW trials using both probe lengths, an Ar shielding gas was used to prevent oxidation of the welded plate.

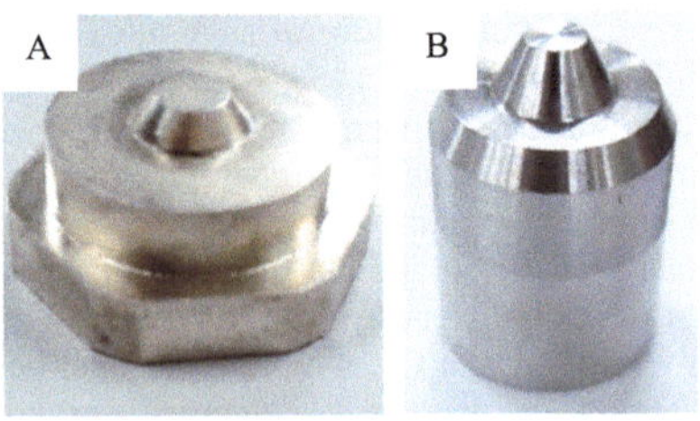

Figure 2. (A) 2 mm (B) 5 mm probe length tool used for FSW.

The appearance of the weld and macroscopic images of its cross section were obtained by optical microscopy. Vickers hardness profiles of a cross section of the weld were measured at a load of 0.3 kgf. Transverse tensile tests were conducted at a rate of 0.5 mm/min at room temperature to measure the mechanical strength of the weld.

Result and Discussion

Alloy characteristics

The rapid solidification induced in the Ir-containing Ni base superalloy by using a Cu mold resulted in a fine dendritic structure, as shown in Figure 3 (A), with a high concentration gradient due to macro segregation subsequently produced by solution heat treatment (Figure 3(B)). With aging by heat treatment, γ' phases with a cuboidal structure were formed (Figure 3(C)) and carbides precipitated at the grain boundaries (Figure 3(D)). The temperature dependence of the 0.2% yield strength of this alloy was found through compression testing to be 595 MPa at 1000 °C, which as shown in Figure 4, is much greater than that of as-cast Ni superalloy (Mar-M247). This would suggest that Ir-containing Ni base superalloy is a more suitable tool material for the FSW of steels.

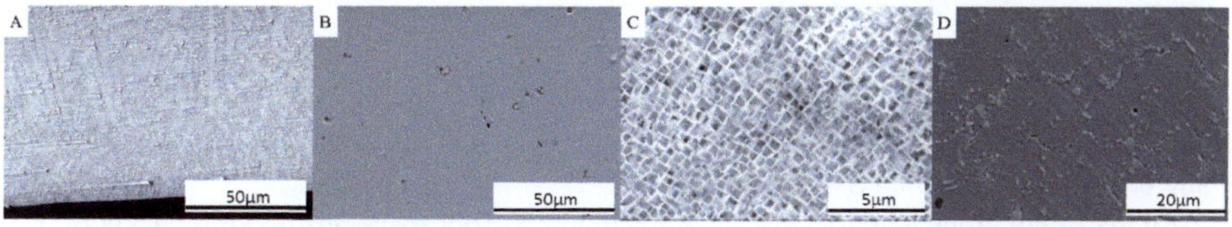

Figure 3. Microstructure of a Ir-containing Ni base superalloy (A) in an as-cast state and (B) after solution heat treatment. (C) Highly magnified image and (D) microstructure after aging.

FSW trials with a 2 mm long probe

From the weld appearance produced on AISI304 (Figure 5), and its cross section (Figure 6), it is clear that no weld defects or surface flash were formed by the use of an Ir-containing Ni base superalloy tool. Furthermore, the profile of the tool before and after FSW (Figure 7) shows that no significant tool wear occurred, even after a FSW of 3150 mm. The Vickers hardness profile of the weld in Figure 8 shows an increase in hardness in the heat-affected zone and stir zone, with the maximum hardness of 220 to 230 Hv in the stir zone being, about 1.5 times greater than that of the base material (Hv180). Similar results were obtained in FSW trials with ASTM A295 52100 at a weld length of 1400 mm (Figure 9 and 10), which clearly shows that Ir-containing Ni base superalloy is suitable for the FSW of both stainless and high carbon steels.

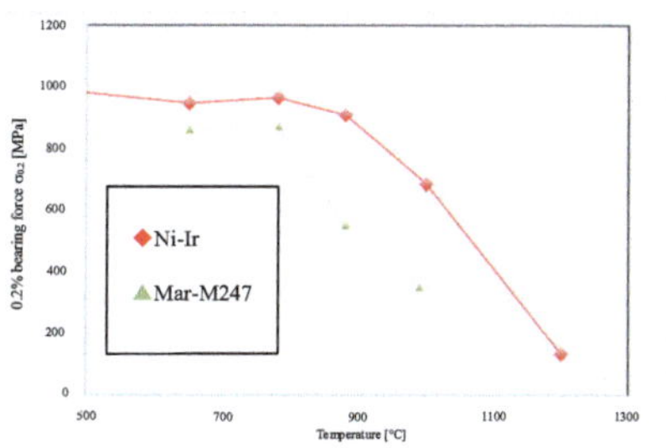

Figure 4. Compression test results for different high-temperature resistant alloys.

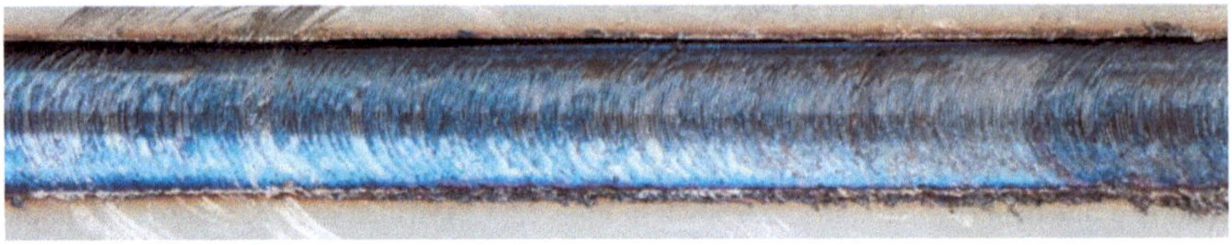

Figure 5. Weld appearance of a FSW on AISI304 (2 mm probe length).

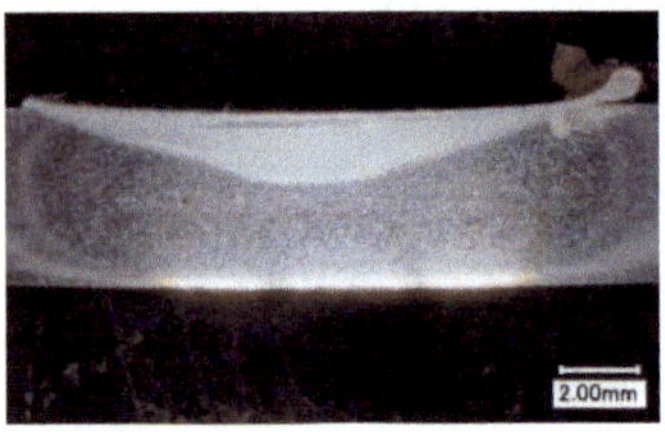

Figure 6. Cross-section of a FSW on AISI304 (2mm probe length).

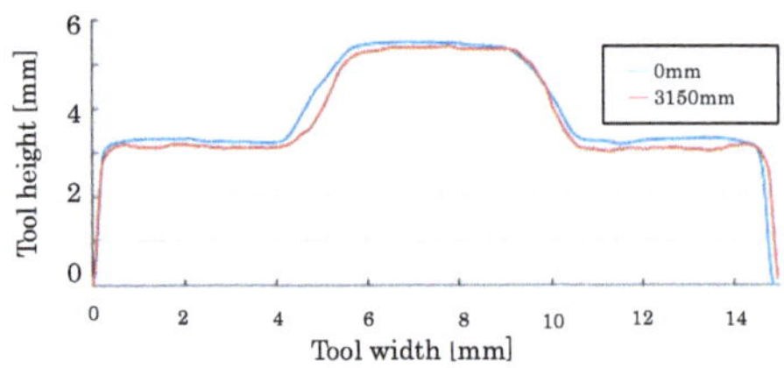

Figure 7. Tool shape profile for the FSW of AISI304.

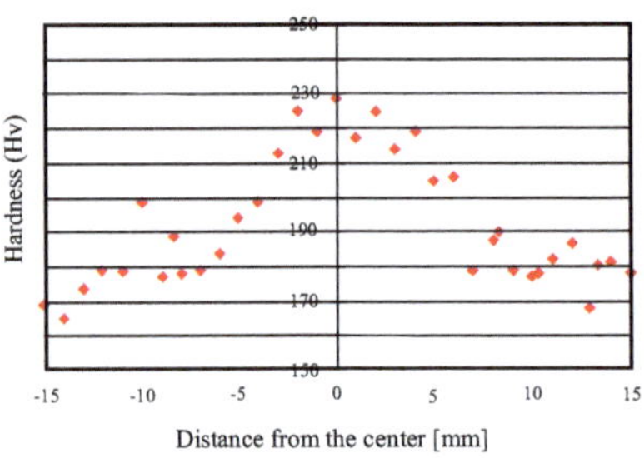

Figure 8. Vickers hardness of a FSW on stainless steel (2mm probe length).

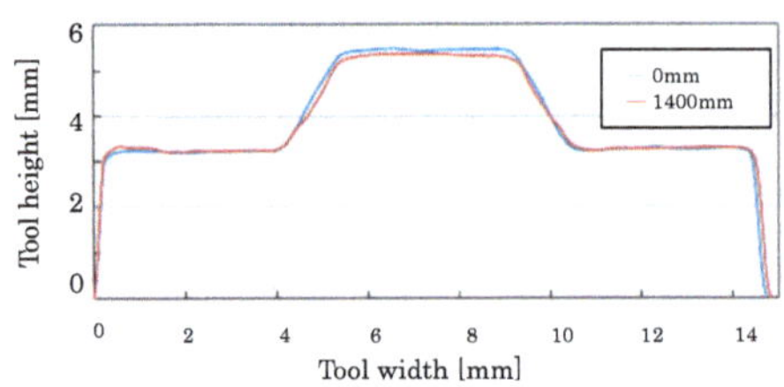

Figure 10. Tool shape profile for the FSW of ultrahigh carbon steel.

Figure 9. Weld appearance of a FSW on ultrahigh carbon steel (2mm probe length).

FSW trial with a 5 mm long probe

The weld appearance produced on AISI304 (Figure 11) is well-looking with 5 mm long probe as well as trial with 2 mm long probe. Additionally, the cross section of weld (Figure 12) reveals that finer grain structure is produced in the stir zone than in the heat-affected zone or base material. The 5 mm long probe tool experienced virtually no wear over the course of, a 500 mm long FSW (Figure 13), much like the 2 mm long probe tool, implying that this tool is capable of much longer FSW lengths. The Vickers hardness profile of the weld produced by the 5 mm long probe tool (Figure 14) also exhibits greater hardness in the stir zone, with transverse tensile testing causing fracture in the base material region rather than in the stir zone. These results suggest that an Ir-containing Ni base superalloy tool with a 5 mm long probe suitable for the FSW of AISI304 without deformation or severe tool wear.

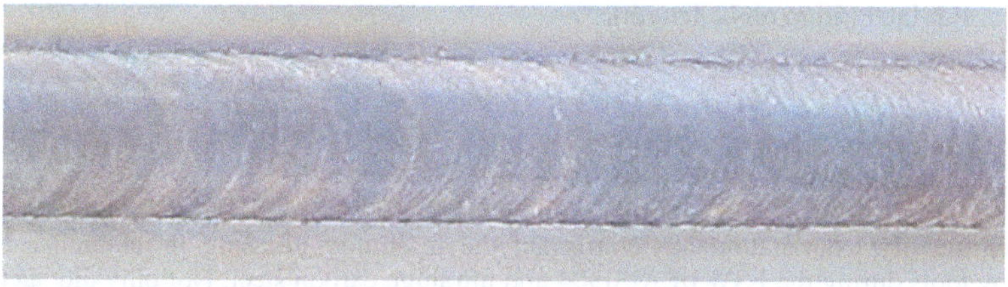

Figure 11. Weld appearance of a FSW created on AISI304 (5 mm length probe).

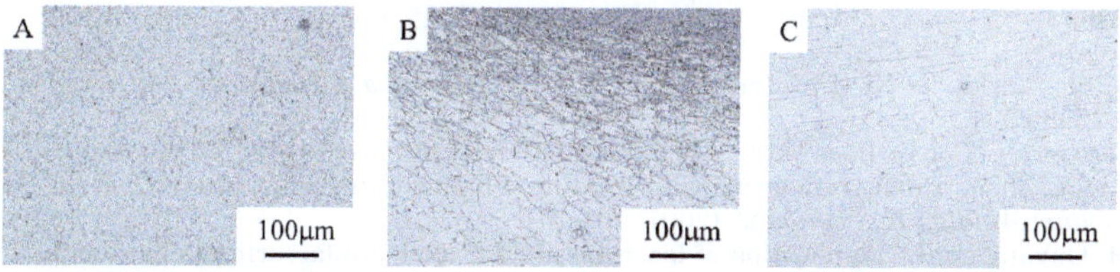

Figure 12. Microstructure in the (A) stir zone and (B) heat affected zone (C) base material.

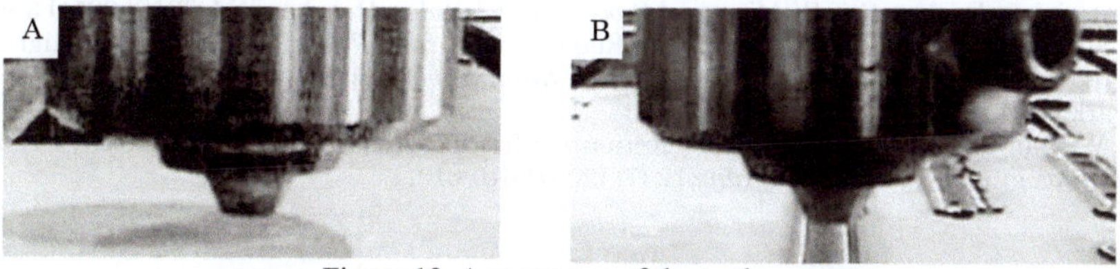

Figure 13. Appearance of the tool
(A) before FSW and (B) after a 500 mm FSW.

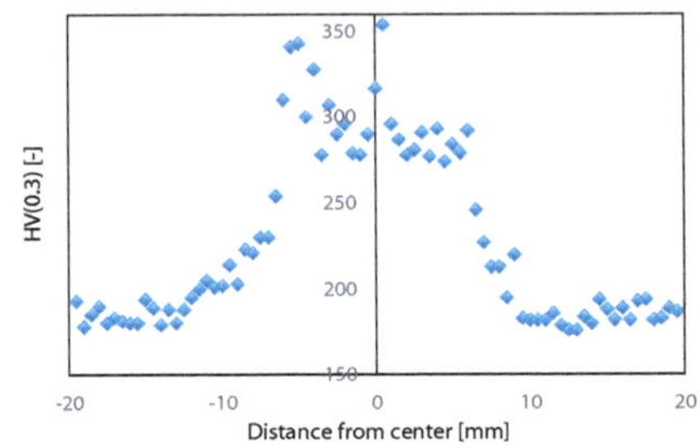

Figure 14. Vickers hardness of a FSW created on AISI304 (5mm probe length).

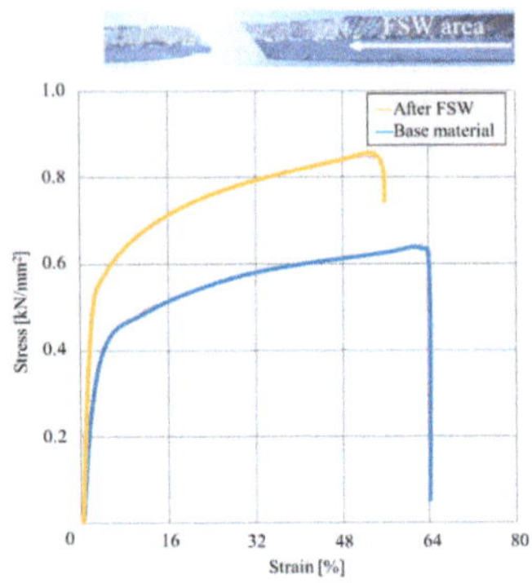

Figure 15. Tensile strength of AISI304 base material and after FSW.

Conclusion

This preliminarily examination has confirmed the feasibility of using Ir-containing Ni base superalloy tools during the FSW of AISI304 and ultrahigh carbon steel. Not only did such welds have a greater strength than the base alloy, but tools with 2 and 5 mm probes lengths were found not to undergo any significant deformation or wear during FSW.

References

1. T.B. Massalski, "Ir-Ni (Iridium-Nickel)", *Binary Alloy Phase Diagrams 2nd Edition*, vol. 3 (1990) 2333-2334.

2. J. Sato et al., "Cobalt-Base High-Temperature Alloys," *Science*, 7 (312) (2006), 90-91.

3. H. Liu, L. Zhou, "Progress in friction stir welding of high melting point materials," *Transactions of the China Welding Institution* 28 (2007) 101-104.

4. J. H. Ouyang et al., "Application of laser-based additive manufacturing to production of tools for friction stir welding", *Proceedings of Thirteenth Solid Freeform Fabrication Symposium* (Austin, TX. University of Texas: 2002) 65-72.

5. Y. S. Sato et al., "Development of a cobalt-based alloy FSW tool for high-softening-temperature materials", *Friction Stir Welding and Processing VI* (Wiley, Hoboken, NJ, TMS: 2011) 3-9.

6. S. Packer et al., "Tool and Equipment Requirements for FSW Ferrous and Other High Melting Temperature Alloys", *Proceedings of Fourth International Symposium on Friction Stir Welding* (Park City: 2003) 14-16.

7. W. Gan et al., "Tool materials selection for friction stir welding of L80 steel", *Science and Technology of Welding and Joining* 12 (7) (2007) 610-613.

Friction Stir Welding and Processing VIII
Edited by: Rajiv S. Mishra, Murray W. Mahoney, Yutaka Sato, and Yuri Hovanski
TMS (The Minerals, Metals & Materials Society), 2015

HEAT INPUT AND POST WELD HEAT TREATMENT EFFECTS ON REDUCED-ACTIVATION FERRITIC/MARTENSITIC STEEL FRICTION STIR WELDS

Wei Tang, Jian Chen, Xinghua Yu, David A. Frederick, Zhili Feng
Oak Ridge National Laboratory, Oak Ridge, TN

Keywords: Eurofer'97, friction stir welding, microstructure, microhardness

Abstract

Reduced-activation ferritic/martensitic (RAFM) steels are an important class of structural materials for fusion reactor internals developed in recent years because of their improved irradiation resistance. However, they can suffer from welding induced property degradations. In this paper, a solid phase joining technology friction stir welding (FSW) was adopted to join a RAFM steel Eurofer'97 and different FSW parameters/heat input were chosen to produce welds. FSW response parameters, joint microstructures and microhardness were investigated to reveal relationships among welding heat input, weld structure characterization and mechanical properties. In general, FSW heat input results in high hardness inside the stir zone mostly due to a martensitic transformation. It is possible to produce friction stir welds similar to but not with exactly the same base metal hardness when using low power input because of other hardening mechanisms. Further, post weld heat treatment (PWHT) is a very effective way to reduce FSW stir zone hardness values.

Introduction

An extremely hostile environment exists inside fusion reactors, such as intense heat flux, significant cyclic thermo-mechanical stresses, intense flux of high-energy neutrons, electromagnetic radiation, and reactive chemicals. Reduced-activation ferritic/martensitic (RAFM) steels were developed in recent years and are considered as the most promising structural material for use in a fusion reactor to resist damages because of the aggressive environment [1-3]. However, when joining RAFM steels with traditional fusion welding technologies, such as tungsten inert gas (TIG) welding, electron beam welding and laser welding, significant strength deterioration occurs due to the development of δ-ferrite and precipitate dissolution in the welded joint [4]. To solve this weld properties degradation, a solid phase joining technology friction stir welding was adopted to join RAFM materials of Eurofer'97 plates [3] and F82H thin sheets [2]. Hardness, similar to the base metal, was obtained by post weld heat treatment and by using a low tool rotation rate. In this paper, 7.9 mm thick Eurofer'97 plates were friction stir welded with different tool rotation rate/heat inputs. Defect free welds were obtained with multiple FSW parameter sets. FSW heat input, stir zone microstructure, weld zone microhardness and the influence of heat input in material structures and properties are discussed.

Experimental procedures

The workpiece material is a RAFM steel Eurofer'97. General chemical composition (wt. %) of this material is: 0.11C, 8.7Cr, 1W, 0.10Ta, 0.19V, 0.44Mn, 0.004S, balance Fe [1]. Plates to be welded are about 7.9 mm in thickness. All FSW processes were carried out on the ORNL FSW process development system (PDS) using a PCBN tool with a 25.4 mm diameter convex shoulder and a 6.5 mm long tapered pin. Except for one butt joint weld, all other trials were bead-on-plate. Weld parameters included three sets of tool rotation rates in coordination with welding speeds of: 400 rpm and 50.8 mm/min, 200 rpm and 50.8 mm/min, and 100 rpm and 25.4 mm/min. All FSW was carried out with position control. FSW response parameters such as resistance force and torque were recorded by the control computer.

After FSW, two metallographic specimens were cut from each weld and one specimen of each condition was post weld heat treated (PWHT) at 760°C X 1 hour + air cool. This heat treatment is similar to the Eurofer'97 base metal temper treatment (760°C X 1.5 hours [1]). The only difference is the high temperature time duration.

All of the as-welded and PWHT specimens were ground, polished, and etched for microstructural observation under an optical microscope. Vickers microhardness was performed on metallographic specimens along weld midplanes. Microhardness testing was completed using a 500 g load for 15 seconds of loading time and a 0.25 mm spacing between indentations.

Results and Discussion

Welds of all conditions are shown in Figure 1. The first two conditions, 400 rpm and 200 rpm tool rotation rates, performed quite well, and no FSW defect was observed from the top surfaces or exit holes. The third condition, 100 rpm, required higher torque than the machine could provide without the gear reduction box installed. Therefore, the tool rotation stopped and the FSW tool pin failed after less than 10 mm of travel. The tool may not fail if the machine can supply sufficient driving torque after the gear reduction box is installed.

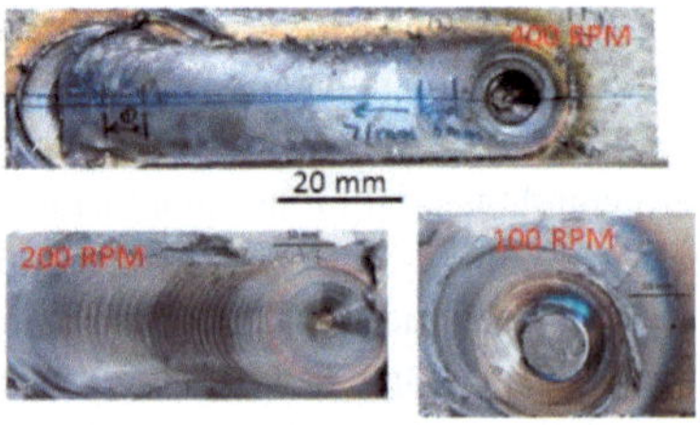

Figure 1. Eurofer'97 friction stir welds made with different parameters.

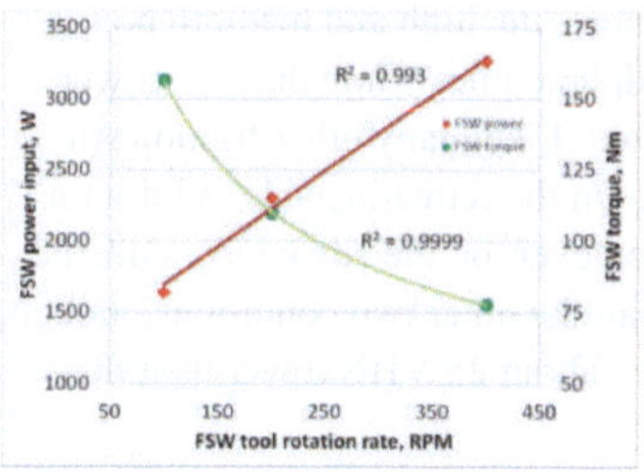

Figure 2. FSW torque and power in various conditions.

In the successful FSW trials, torques and power inputs presented the same trends with previous FSW research results, i.e., torque decreases and power increases with increasing tool rotation rate, Figure 2. In addition, very good power and linear curve fittings can be applied to these experimental results. Previous studies showed a power plateau when the tool rotation rates reach certain values at the high end. The trends may be the same here, but tool rotational speeds used here were not high enough to demonstrate this power plateau.

Base metal and stir zone center microstructures of different welds are shown in Figure 3. Similar to the as-received condition in other publications [1, 2], the base metal has a fully martensitic structure with lath-shaped martensite subgrains. However, with the FSW heat input, grain structure may change dramatically. For the high power input case (400 rpm tool rotation rate), grain size was larger, conversely, for the low power input case (100 rpm tool rotation rate), grain size was similar to the base metal, though the metallographic sample was very close to the FSW tool plunge location and definitely was affected by the plunge heat input.

Figure 3. Microstructure of Eurofer'97 base metal and stir zone of FSW made with different heat input.

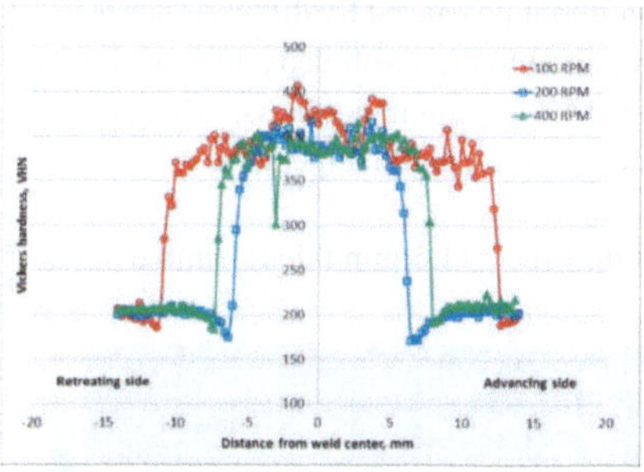

Figure 4. Microhardness of welds made with various tool rotation rates.

Vickers hardness distributions along the stir zone midplane of three power input FSW are shown in Figure 4. The FSW heat input raised the stir zone material into the austenite region during welding with subsequent quenching to martensite during cooling resulting in the stir zone hardness for all conditions increasing from 200 VHN to 400 VHN. In the HAZ, the high power input FSW (400 RPM) resulted in a wider area than the medium power input FSW (200 RPM) because of more heat generated in the stir zone. However, the low power input FSW (100 RPM) demonstrated the widest HAZ, not because of the local heat input, but from the heat input generated during the tool plunge process. Likely, that is the cause for the higher stir zone hardness compared to the other two welds. Furthermore, the high and medium power input FSW stir zone hardness are quite uniform along the midplane lines while the low power input FSW stir zone hardness values are asymmetric distributed. Like many other friction stir welds, the minimum hardness values are located in the HAZ. On the retreating side, minimum hardness values are about the same for all parameter sets. However, on the advancing side, the medium power input FSW has a lower minimum hardness than the other two conditions, though the absolute value is close to those on the retreating side. i.e., about 25 VHN lower than the base metal hardness.

Microhardness distributions of high and medium power input FSW after PWHT are shown in Figure 5. From Figure 5, stir zone hardness decreased considerably but not quite back to the base metal level, still about 50 VHN higher than that of the base metal. In Eurofer'97 base metal, carbide precipitation, presented as Cr rich ($M_{23}C_6$) and Ta/V rich ((Ta, V)C), as well as several kinds of inclusions, such as MnS and Ta-rich inclusions, were identified by X-ray diffraction and TEM studies[1]. Therefore, the PWHT applied in this paper might reduce the martensite hardening effect caused by FSW inside the stir zone but might not ease other effects such as precipitation shapes and distribution. Moreover, the FSW heat input level affected the stir zone hardness distribution after PWHT. The high heat input FSW had higher hardness than the medium heat input FSW on the retreating side but they are almost the same on advancing side. This is a different pattern compared with the hardness distribution in the as-welded condition shown in Figure 4. This phenomenon might be caused by different thermal-mechanical processes in FSW but further studies are required to draw solid conclusions. Finally, the further temper treatment of PWHT didn't affect the HAZ minimum hardness. That is not an unexpected result considering RAFM steels are designed to resist high temperature creep.

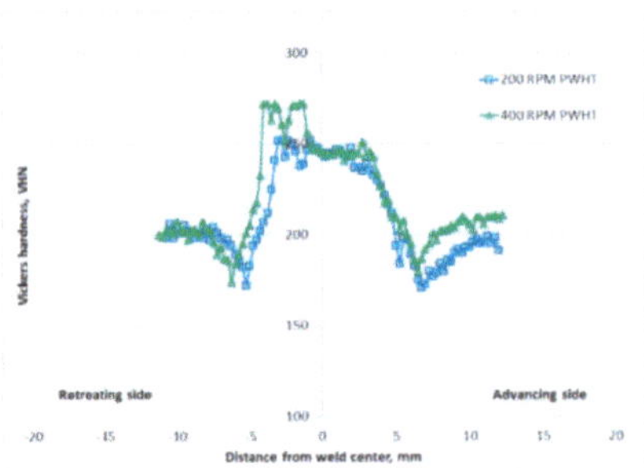

Figure 5. Post weld heat treated welds microhardness distribution

In a study carried out with another RAFM steel, F82H thin sheet (1.5 mm thick), similar hardness to base metal was realized, 240 VHN vs. 220 VHN, by using a low tool rotation rate FSW (100 RPM) without PWHT. In that investigation, uniformly distributed precipitates and very fine ferritic grains were observed in the stir zone [2]. Obviously, there are other hardening mechanisms than

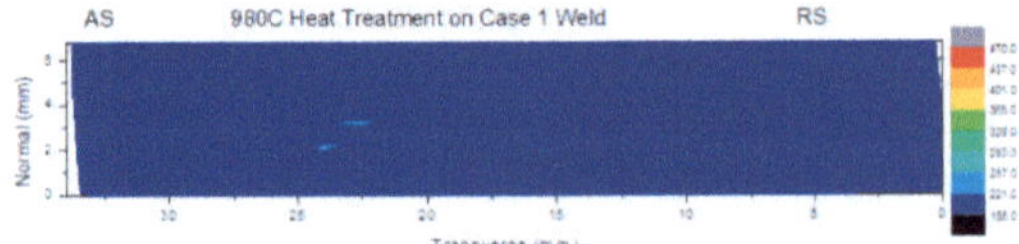

Figure 6 Microhardness of FSW and full solution and temper PWHT [3].

martensitic hardening mechanism in RAFM materials. In a previous study, a total solution + temper heat treatment was carried out with a Eurofer'97 friction stir weld, 980° X 0.5 hour + quench + 760°C X 1 hour + air cool, and a uniform hardness distribution across the welded joint was obtained [3], Figure 6. Therefore, PWHT is a very effective way to restore FSW joint mechanical properties.

Conclusions

Several hardening mechanisms exist in Eurofer'97 material, including martensitic, carbide precipitation and inclusion, and they may be all affected by FSW processes. Generally, the temperature in FSW is higher than RAFM materials austenized temperature and stir zone hardness values doubled when compared with the base metal hardness value. Conversely, very low power input FSW may weld RAFM materials with temperature lower than the Ac_1 line with medium to thick material thickness. However, even when the welding temperature is lower than the Ac_1 line, mechanical properties of the stir zone may not be exactly the same as the base metal due to other hardening mechanisms. There is a soft zone in the HAZ, and the minimum hardness value is not affected by the post weld tempering heat treatment. Finally, temper PWHT is a very effective way to decrease the stir zone hardness to a reasonable level.

Acknowledgement

This research was sponsored by the U.S. Department of Energy, Office of Fusion Energy Sciences. The author is thankful to Stan A. David for valuable discussions.

References

1. P. Fernandez et al., "Metallurgical Characterization of the Reduced Activation Ferritic/Martensitic Steel Eurofer'97 on As-received Condition," *Fusion Engineering and Design*, 58-59(2001) 787 – 792.
2. Sanghoon Noh et al., "Microstructural Evolutions of Friction Stir Welded F82H Steel for Fusion Applications," (Paper presented at the Transactions of the Korean Nuclear Society Autumn Meeting, Gyeongju, Korea, October 25 – 26, 2012), 2.
3. Zhenzhen Yu et al., "Similar and Dissimilar Friction Stir Welding of ODS and RAFM Steels," (Paper presented at)
4. Tanigawa et al., "Technical Issues of Reduced Activation Ferritic/Martensitic Steels for Fabrication of ITER Test Blanket Modules," *Fusion Engineering and Design*, 83 (2008) 1471 – 1476.

FRICTION STIR WELDING AND PROCESSING VIII

Aluminum and Magnesium Alloys

Friction Stir Welding and Processing VIII
Edited by: Rajiv S. Mishra, Murray W. Mahoney, Yutaka Sato, and Yuri Hovanski
TMS (The Minerals, Metals & Materials Society), 2015

FSW OF HIGH STRENGTH 7XXX ALUMINUM USING FOUR PROCESS VARIANTS

X. Huang[1], J. Scheuring[2], and A. P. Reynolds[1]

[1]Department of Mechanical Engineering, University of South Carolina, 300 Main Street, Columbia, SC29208, USA;
[2]Kaiser Aluminum, 15000 E Euclid Avenue, Spokane Valley, WA, 99215, USA

Keywords: 7XXX Aluminum Alloy, FSW, Conventional Shoulder, Stationary Shoulder, Single Pass, Dual Pass, Hardness, Residual Stress.

Abstract

A high strength aerospace aluminum alloy, 24~25 mm thick plate, has been friction stir welded using four different process variants. The process variants used are: stationary shoulder single pass (SSSP), conventional shoulder single pass (CSSP), stationary shoulder dual pass (SSDP), and conventional shoulder dual pass (CSDP). The goal of the reported work was to optimize properties while maintaining nugget temperature below the incipient melting temperature. Defect free welds were produced by all process variants. Properties and characteristics including residual stress, hardness distributions and process control and response parameter relationships are reported.

1. Background and Introduction

Friction stir welding (FSW) has enabled the joining of high strength aerospace aluminum alloys which were formerly considered unweldable by most fusion welding techniques. However, in order to reap the benefits of these high strength alloys in welded structure, it is important to produce welds which do not excessively degrade the strength. A typical measure of weld strength is the transverse tensile test. Joint strength of 7XXX alloy welds in transverse tension is strongly correlated with heat affected zone (HAZ) minimum hardness [1]. Minimum hardness has been shown to be increased by welding at high speeds and may also be affected by application of various thermal boundary conditions designed to increase quench rates in the HAZ [1, 2~5]. Welding speed is limited in practice primarily by two factors: (1) in-plane force on the weld tool increases with increasing welding speed and ultimately at sufficient speed the tool will fail and (2) increasing welding speed normally requires increased weld power, weld power is closely correlated with peak weld temperature, and peak weld temperature must be maintained below the incipient melting point of the alloy being welded or else the weld nugget ductility and strength will be compromised.

So, there is a tension between welding at the highest possible speed to minimize HAZ overaging and maintaining weld peak T below the incipient melting temperature. Process modifications which would enable welding at reasonably high speeds without exceeding incipient melting temperatures are desirable. In this paper we explore four process variants in the welding of a 25 mm thick high strength aerospace aluminum alloy with the goal of maximizing transverse properties while preventing overheating. The process variants used include: Conventional, full penetration FSW; Stationary shoulder, full penetration FSW; Dual pass conventional FSW; and Dual pass stationary shoulder FSW.

In both of the dual pass processes, welds are made from both sides of the plate with each pass having a weld penetration of slightly greater than half of the plate thickness. Properties resulting from each of the four process variants are reported in the foregoing and an attempt is made to correlate the properties with the corresponding process control and response variables.

2. Materials and Experimental Procedures

2.1 Base Metal

For this study, 25.4mm material was used due to its perceived potential future use in tailored blank products for many applications. The tensile strength in the longitudinal direction is 607 MPa while the yield strength is 581 MPa with a 4D elongation of 13.8%. For the longitudinal-transverse orientation, the values are 598 MPa, 569 MPa, and 13.0% for the tensile strength, yield strength, and 4D elongation, respectively. Additionally, this results in a base metal Vickers hardness of 192.

2.2 Friction Stir Welding

All the friction stir welds were produced using an MTS FSW process development system (PDS) in z-axis force control mode. The welding direction was parallel to plate rolling direction. Tool tilt was varied between 0° and 1°. Welds were made in lab air with an approximate ambient temperature of 23°C. In the dual pass processes, both passes were performed in the same direction with the same tool rotation sense, producing joints in which the advancing side of the first pass was on the same side of the joint as the retreating side of the second pass. A tool-steel backing plate was used in all cases.

Single pass welds were performed on 25.4 mm thick plates while the dual pass welds were made on 24.1 mm thick plates machined from the 25.4 mm thick base metal. The thinner plates for dual pass welding were utilized to facilitate the use of a tool originally designed for welding 12.5 mm thick plate while producing some overlap between the first and second pass weld regions. In each case, final weldment dimensions were 203 mm wide by 610 mm long. FSW control parameters are listed in Table 1.

The tools used for performing all conventional shoulder welds were of a two piece design with a scrolled shoulder and a pin with threads and 3 flats (T+3F). The tools used for performing all stationary shoulder welds were of a two piece design with a stationary shoulder and a rotating pin having threads and 3 counter-flow flutes (T+3CT). Counter flow indicates that with chosen tool rotation direction, the threads push material down while the flutes pull material up. Flats are expected to be essentially neutral with regard to vertical flow. The single scroll tool shoulder was fabricated out of H13 steel, and the tool pin in the shape of a truncated cone was fabricated out of MP-159. More detailed tool information is listed in Table 2.

Table 1. Summary of FSW Process Control and Response Parameters

Weld No.	FSW Type	Rotating Speed RPM	Welding Speed mm/s	Fz KN	Fx KN	Fy KN	Fxy KN	Torque N·m	Power *, KW	Temperature ℃
1	CSSP	160	0.85	46.7	6.5	11.1	12.9	317	5.3	484
2	CSSP	160	1.69	66.7	13.6	11.8	18	408	6.8	493
3	SSSP	160	0.85	62.3	16.3	19.7	25.6	313	5.2	477
4.1	CSDP	160	1.69	42.3	8.6	7.8	11.6	251	4.2	433
4.2	CSDP	160	1.69	40	6.1	8.1	10.2	252	4.2	444
5.1	CSDP	200	3.39	55.6	17.5	12.6	21.5	259	5.4	451
5.2	CSDP	200	3.39	53.4	15.7	9.6	18.4	256	5.4	463
6.1	SSDP	160	1.69	35.8	19.9	9.9	22.3	258	4.3	424
6.2	SSDP	160	1.69	35.8	21.6	10.1	23.8	244	4.1	430
7.1	SSDP	200	3.39	43.5	29.3	13.9	32.5	251	5.2	442
7.2	SSDP	200	3.39	43.5	29.5	12.3	32	244	5.1	458

*Power for dual pass welds is the sum of the power for both passes.

Table 2. Tool Information of FSW

FSW Type	Thread Pitch, mm/thread	Flat/Flute Depth, mm	Pin Length, mm	Pin Top Diameter, mm	Pin Tip Diameter, mm	Pin Taper, °	Shoulder Diameter, mm
SPCS	1.75	0.89	24.4	19.1	10.9	9.5	35.6
SPSS	1.75	0.89	24.9	19.1	10.9	9.5	33.0
ADCS	2.12	1.35	12.7	15.9	12.3	8	25.4
ADSS	2.12	1.35	12.7	15.9	12.3	8	31.8

Process response variables including x and y-axis in-plane forces and the spindle torque are acquired from the FSW Process Development System. Resultant in-plane forces were calculated from average values of the X axis force and Y axis force. Weld power was calculated from torque and tool rotation rate. Probe temperature during FSW was monitored and recorded by a k-type thermocouple connected to a HOBO data logger. The thermocouple was spot welded into the probe at the probe mid-plane height on the rotation axis. It is expected that the tool temperature measured in this location is an accurate relative measure of the process zone temperature.

2.3 Metallographic Examination and Mechanical Testing

Metallographic examination samples were post-weld heat treated (PWHT) at 121°C for 4 hours, then ground, polished and etched using Keller's reagent. Macro and microstructures of the weld nuggets were evaluated. Those samples were later used for hardness testing.

A Vickers hardness indenter, with a 200g load and a 10s load time was applied to measure hardness distribution along transverse mid-planes of the single pass FSW, dual pass FSW, and the mid-plane of the second pass in dual pass FSW.

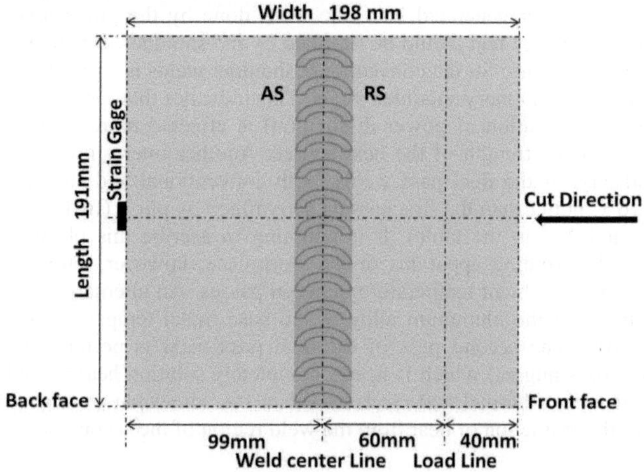

Figure 1. Schematic Diagram of Residual Stress Test

Through thickness average, longitudinal, residual stress was measured before and after post-weld heat treatment for all weld conditions. The stress was measured using the cut-compliance method developed by Cheng [6], Schindler [7] and Prime [8] as described in Canaday et al. [9].

Sample geometry and strain gage placements are shown in Figure 1. An initial slot length of 40mm was used for all tests. The weld centerline was located in the transverse mid-plane and the slot was advanced from the retreating to advancing side.

3. Results and Discussion

3.1 Process Responses

Process response parameters are listed in Table 1 along with the control parameters used to make each weld. It should be noted that for each welding method, at least one weld has been made using the same RPM and welding speed with a different welding method. For example, 160 RPM and 0.85 mm/s welds were made using both CSSP and SSSP welding methods. This was done in an attempt to facilitate comparison between methods. However, it is necessary to note the following:

(a) It is not possible to compare values of the in-plane forces between stationary and conventional shoulder welds. In conventional shoulder welding, essentially all of the in-plane force results from interactions between the pin and the workpiece (this has been shown by observation of the dramatic reduction in the forces when a tool fails). For stationary shoulder welding, there may be a very large component of x-force which results from the coulomb friction between the stationary shoulder and workpiece crown surface.

(b) Only a small effort was expended to find the minimum necessary z-force for the stationary shoulder welding. Therefore, the z-force for these welds should not be considered to be "optimum".

The most interesting comparisons among the various welding conditions are related to the torques, power levels, and measured pin temperatures. Quite surprisingly, the power required for conventional shoulder welds vs. stationary shoulder welds performed with the same RPM and welding speeds never differed by more than 5% and ranged to as low as 0%. This indicates that nearly all of the heat generated by a rotating shoulder reduces the torque required to turn the pin. In the absence of shoulder generated heat, the work done by the pin increases by nearly the amount of "missing" energy that would be supplied by the shoulder. It is interesting to note that the measured pin temperature for the conventional shoulder welds is typically nearly 10°C higher than the corresponding stationary shoulder welds. This indicates that the shape of the heat source (that is the spatial distribution of power dissipation) is affected more by the lack of shoulder heating than is the total strength of the heat source. Another interesting point is that in every case, the second pass of the dual pass welds, both conventional and stationary shoulder, was between 6 and 16°C hotter than the first pass as determined by pin T (first and second passes are denoted as X.1 and X.2 in the table). It is tempting to ascribe this phenomenon to residual heating of either the welding apparatus or the workpiece; however, care was taken to allow everything to return to ambient temperature between passes. An alternative theory relates to the thermal conductivity of the aluminum alloy in the base metal temper (T7651) and as welded (near W) conditions. The second pass of the dual pass weld is performed over a region of material (the first pass nugget) which is nearly completely solution heat treated. This will result in a lower conductivity (thermal and electrical) than the somewhat overaged T7651 condition and could reduce the extraction of heat from the weld region of the second pass.

3.2 Weld Macrostructure

Typical cross sections of single pass and dual pass FSWs are shown in Figure 2a-d. The most obvious differences between the conventional and stationary should welds can be seen at the weld crowns. The conventional shoulder welds exhibit a flaring of the weld nugget region away from the weld centerline near the weld crown which is related to the deformation induced by the shoulder. The weld nuggets of the stationary shoulder welds conform almost exactly to the outline of the pin. Also visible in the CSSP weld (figure 2a) is a wider area of heavily etched material outside the nugget in the thermomechanically affected zone as compared to the SSSP weld (figure 2b). No defects were visible in any of the welds based on macroscopic inspection however, welds #2 (CSSPFSW, T+3F, 160 RPM, 2 IPM) and #7.2 (the 2nd pass of SSDP, T+3CT, 200 RPM, 8 IPM) did contain very minor defects which were observed during examination at high magnification. In weld #2, two small wormhole defects (100μm long and 66μm wide) were observed at the AS between crown and mid-plane; in weld #7.2, a small wormhole defect (68μm long and 21μm wide) appeared at the center between mid-plane and root.

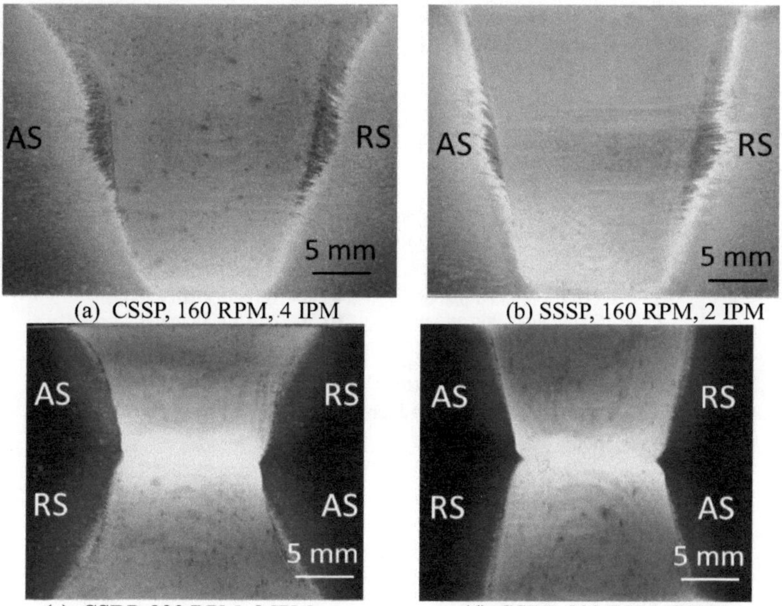

(a) CSSP, 160 RPM, 4 IPM (b) SSSP, 160 RPM, 2 IPM

(c) CSDP, 200 RPM, 8 IPM (d) SSDP, 200 RPM, 8 IPM

Figure 2. Typical Cross Sections of Single Pass and Dual Pass FSW

3.3 Hardness Distributions

Hardness results of PWHT samples are shown in Figures 3 and 4. Figure 3 shows the HAZ minimum hardness measured at the plate midplane for all welds. In addition, the open symbols are minimum hardness values for dual pass welds measured at the midplane of the second pass i.e. at approximately 3/4 of the plate thickness. In general, all of the data follows the normal trend of increasing minimum hardness with increasing welding speed. No significant difference in hardness values is observed between conventional and stationary shoulder welds made with the same welding speed. It is noteworthy that the minimum hardness in the 3/4 thickness HAZ of the dual pass welds is higher than the minimum hardness at the dual pass weld midplane (at ½ thickness). This is likely due to the fact that the region near the plate midplane of the dual pass welds is subjected to two weld thermal cycles.

Figure 4 shows nugget average hardness for the same positions as Figure 3. Again, the open symbols are for the mid planes of the 2nd passes of the dual pass welds. Here again, increased welding speed correlates strongly with hardness. In the weld nugget, high hardness is normally achieved by reaching a weld temperature near the solution treatment temperature for the alloy being welded and quenching rapidly from that temperature in order to retain solute and vacancies for subsequent precipitation hardening. Interestingly, all of the single pass weld data falls on a line with the 2nd pass midplane data for the dual pass welds. Here, as for the HAZ minimum hardness, the midplane of the dual pass weld exhibits relatively lower hardness. Overall, as for the HAZ hardness, there is little difference between the stationary and conventional shoulder welds.

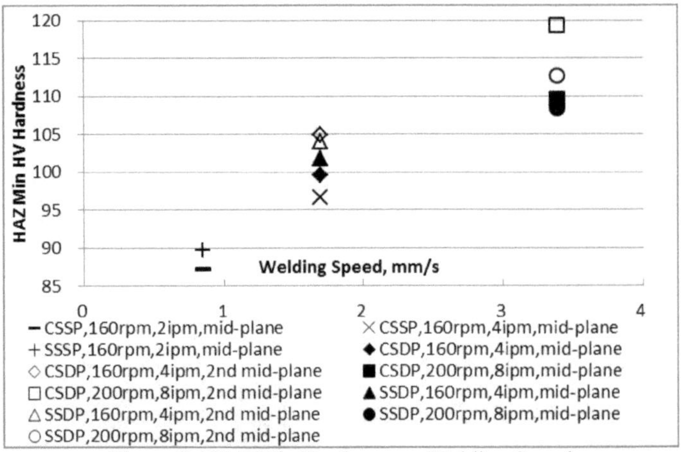

Figure 3. HAZ Min Hardness vs. Welding Speed

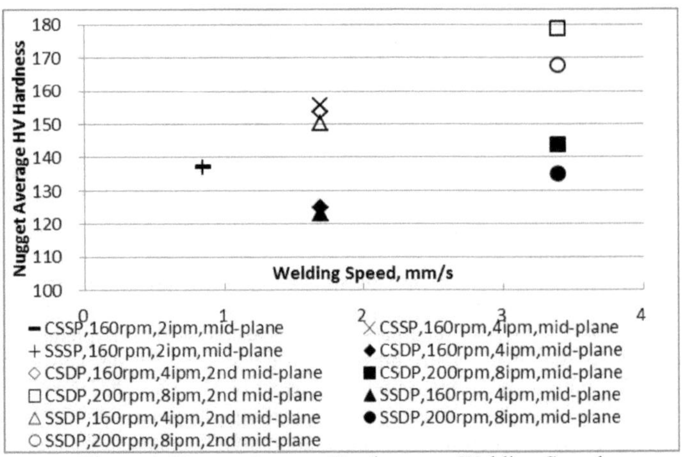

Figure 4. Nugget Average Hardness vs. Welding Speed

3.4 Residual Stress

Post weld heat treatment had essentially no effect on measured residual stress so here only results for PWHT specimens will be presented. Figure 5 shows the peak residual stress vs. welding speed for all of the welds. Peak residual stress increases nearly linearly with increasing welding speed presumably due to compression of the isotherms. As for many other properties reported in this work, there is essentially no difference in the peak stress between otherwise similar stationary and conventional shoulder welds. Figure 6 shows the full width at half maximum (FWHM) for the measured residual stress distributions plotted against welding speed. In every case, the FWHM is greater for conventional shoulder welds when compared to similar stationary shoulder welds. For dual pass and single pass welds made at the same welding speed, the single pass weld has a greater FWHM; however, this is likely due mainly to the larger tool dimensions.

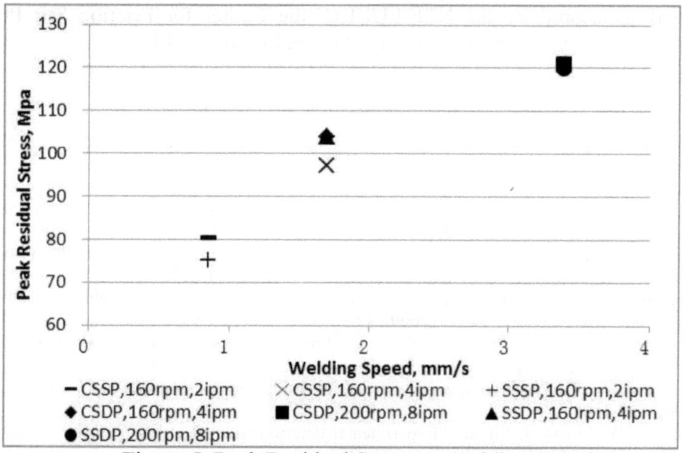

Figure 5. Peak Residual Stress vs. Welding Speed

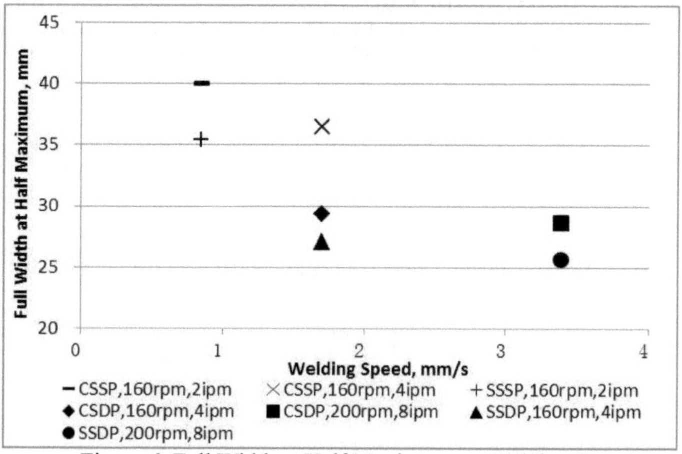

Figure 6. Full Width at Half Maximum vs. Welding Speed

4. Summary and Conclusions

It has been demonstrated that 25 mm thick high strength aluminum alloys can be welded by a variety of different FSW techniques. In agreement with a great deal of published work, the minimum hardness is observed to increase with increasing welding speed. Residual stress increases with increasing welding speed. Overall, very little difference between otherwise similar welds is observed for stationary vs. conventional friction stir welds.

Acknowledgements

This work was sponsored by the NSF-I/UCRC, the Center for Friction Stir Processing. All material was provided by Kaiser Aluminum Fabricated Products, LLC.

References

1. P. Upadhyay and A. P. Reynolds, "Effects of Thermal Boundary Conditions in Friction Stir Welded AA7050-T7 Sheets," *Materials Science and Engineering A*, 527 (2010), 1537-1543.
2. A. P. Reynolds, W. Tang, Z. Khandkar, J. A. Khan and K. Lindner, "Relationships between weld parameters, hardness distribution and temperature history in alloy 7050 friction stir welds," *Science and Technology of Welding and Joining*, 10 (2005), 190-199.
3. T Long, W Tang, AP Reynolds, "Process response parameter relationships in aluminium alloy friction stir welds," *Science and Technology of Welding & Joining*, 12 (2007), 311-317.
4. P. L. Threadgill, A. J. Leonard, H. R. Shercliff, P. J. Withers, "Friction stir welding of aluminium alloys," *International Materials Reviews*, 54 (2) (2009), 49-93.
5. H. J. Zhang, H. J. Liu, L. Yu, "Effect of Water Cooling on the Performances of Friction Stir Welding Heat-Affected Zone," *Materials Engineering and Performance*, 21 (7) (2012), 1182-1187.
6. W. Cheng, I. Finnie, "The crack compliance method for residual stress measurement," *Welding in the World*, 28 (5-6) (1990), 103-110
7. H. J. Schindler, W. Cheng, I. Finnie, "Experimental determination of stress intensity factors due to residual stresses," *Experimental Mechanics*, 37 (3) (1997), 272-277.
8. M. B. Prime, "Residual Stress Measurement by Successive Extension of a Slot: The Crack Compliance Method," *Appl. Mech. Rev.*, 52 (2) (1999), 75-96.
9. C.T. Canaday, M.A. Moore, W. Tang, and A.P. Reynolds, "Through Thickness Property Variations in a Thick Plate AA7050 Friction Stir Welded Joint," *Materials Science & Engineering A,* 559 (2013), 678-682.

Friction Stir Welding and Processing VIII
Edited by: Rajiv S. Mishra, Murray W. Mahoney, Yutaka Sato, and Yuri Hovanski
TMS (The Minerals, Metals & Materials Society), 2015

FSW OF ALUMINUM TAILOR WELDED BLANKS ACROSS MACHINE PLATFORMS

Yuri Hovanski[1], Piyush Upadhyay[1], Blair Carlson[2],
Robert Szymanski[2], Tom Luzanski[3], Dustin Marshall[3]

[1]Applied Materials Processing Group, Pacific Northwest National Laboratory
902 Battelle Blvd. Richland, WA 99352, USA
[2]General Motors Research and Development Laboratory
30500 Mound Rd, m/c 480-1069-RL2 Warren, MI 48090.
[3]TWB Company, LLC. 1600 Nadeau Rd. Monroe, MI 48162

Keywords: Friction stir welding; TWB; aluminum, machine.

Abstract

Development and characterization of friction stir welded aluminum tailor welded blanks was successfully carried out on three separate machine platforms. Each was a commercially available, gantry style, multi-axis machine designed specifically for friction stir welding. Weld parameters were developed to support high volume production of dissimilar thickness aluminum tailor welded blanks at speeds of 3 m/min and greater. Parameters originally developed on an ultra-high stiffness servo driven machine where first transferred to a high stiffness servo-hydraulic friction stir welding machine, and subsequently transferred to a purpose built machine designed to accommodate thin sheet aluminum welding. The inherent beam stiffness, bearing compliance, and control system for each machine were distinctly unique, which posed specific challenges in transferring welding parameters across machine platforms. This work documents the challenges imposed by successfully transferring weld parameters from machine to machine, produced from different manufacturers and with unique control systems and interfaces.

Introduction

Strategic research and development investments over the last few decades have moved tailor welded products from concept to reality [1]. Several companies now produce tailor-welded blanks (TWB), tailor welded coils and tailor welded tube sections in a variety of steel alloys for use in applications such as door inners, shock towers, floor pans, and a variety of other applications [2-5]. Utilization of tailor-welded technology is increasing with a greater demand for lightweight structures that reduce part count and lower cost. While automotive manufacturers would like to expand the usage of tailor-welded technologies into aluminum alloys, current joining technologies prevent such utilization by significantly reducing the formability of the as-welded materials associated with the complexities of molten aluminum [6-9]. Aluminum alloys have great potential for usage in tailor-welded structures, and preliminary assessments show that the weight of numerous high volume automotive components could be reduced by more than 40% [10].

Several fusion welding methods have been employed to join aluminum alloys for TWB applications ranging from gas tungsten arc welding to electron beam and beyond. [9, 11-13]. While each method has produced TWBs with their own unique characteristic properties, all employ local melting in some degree. Since aluminum has high reflectivity, low molten viscosity and inherent affinity to form surface oxide, fusion welding methods increase susceptibility to porosity, hot cracking and elemental loss in the weld seam. Thus volumetric defects and strength reductions are difficult to avoid [9, 11, 14].

Friction Stir Welding (FSW), a solid state welding process, does not require melting to join aluminum alloys. FSW utilizes frictional heating and plastic deformation enabled by a non-consumable rotating tool to produce a welded joint between abutting faces. Extensive research and development in industry and academia have demonstrated effective weld process parameters and tool designs that produce aluminum joints defect free and with significant improvement in strength, ductility, corrosion resistance while simultaneously reducing distortion [15-18].

While several groups have shown the preliminary potential for FSW aluminum blanks of similar and dissimilar thicknesses, no weld schedules exist that can be modified or adapted for very dissimilar thicknesses (2:1 thickness ratios and greater). Additionally, only a few report welding speeds at velocities that can support high volume production, and those only in thickness ratios of 1.5:1 or less [19, 20]. In order to achieve weight savings in excess of 40%, weld parameters must be demonstrated that enable multiple alloys to be joined in very dissimilar thickness sheet combinations while maximizing their post-weld formability.

Most of the reported data available in literature have studied formability properties of FSW joints in similar thickness configurations and without much emphasis on weld parameter optimization [21-24]. Thus there is a dearth of information on the effects various welding control parameter have on weld formability and other mechanical properties for TWB applications. Our previous work reports extensively on parameter and tool development supporting high speed FSW [25]; however, the objective of the present paper is to report on how to effectively transfer parameters across machine platforms once tool and weld parameters have been optimized for a fixed product. This approach contrasts with previously reported work suggesting that FSW should be optimized individually on different machines to optimize weld schedules and tooling to meet specifications [26].

Equipment and Experimental Details

Having previously optimized the weld schedule and tooling to join dissimilar thickness AA5182-O in thicknesses of 2-mm to 1.1-mm on an ultra-high stiffness precision FSW system at the Pacific Northwest National Laboratory (PNNL) [25], this work focuses on transferring these welding parameters, tooling and fixtures to two separate machines located at the General Motors (GM) Research and Development Technical Center in Warren, Michigan and the production facilities of TWB LLC. in Monroe, Michigan. These three unique FSW machines, shown in Figure 1, were built by different manufacturers and designed for very separate operating conditions. Table 1 enumerates several of the unique capabilities including maximum linear and rotational velocities of each FSW machine utilized in this effort.

Table 1. Operational comparison of FSW equipment used herein.

	Spindle Characteristics	Linear Velocity (m/min)	Rotational Velocity (RPM)
PNNL	Ultra-high Stiffness/Precision	3	2000
GM	High Stiffness/Multi-axis	6	2500
TWB LLC.	High Speed	10	4500

Figure 1. Visual Comparison of FSW equipment utilized for this study. High Speed FSW System at TWB LLC (left), R&D Machine at GM Technical Center (center), and R&D Machine at PNNL (right).

Welding parameters were optimized at 3m/min and 1650 RPM utilizing a 12.7-mm diameter tool fabricated from H-13 tool steel and hardened to Rockwell C-45. Development of the tool geometry was determined through use of statistical Design of Experiment, and has been documented in detail previously [25]. Figure 2 shows a representative image of the tool tip and shoulder geometry that was utilized for all welding herein.

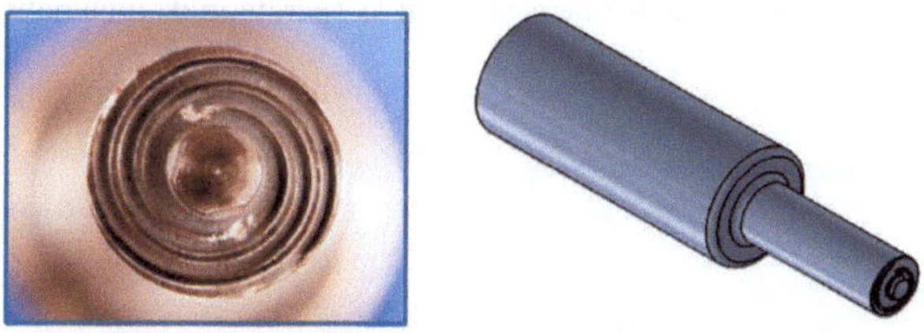

Figure 2. Representative Tool Geometry for FSW Al TWBs.

A representative weld approximately 1250-mm in length was produced to fabricate a full-size aluminum welded door inner panel. The door panel was selected per GM specifications to allow for prototypical stamping of the panels produced at both GM and TWB. Weld parameters were first developed at PNNL, and subsequently transferred to the facilities in Warren and finally to the facility in Monroe, Michigan.

Results & Discussion

While previously reported and unreported attempts at applying machine transferability have concentrated on either reporting how a fixed set of welding parameters and tooling work when applied to another machine or simply trying to independently develop machine parameters that produced a similar weld, the focus of this work was to demonstrate how we can transfer the same welding environments on different machines with unique control systems and varied stiffness. To do this, we determined that welding speed, rotational velocity and tool geometry should be fixed. As such all welding was done at 3m/min, 1650 RPM, and utilized the tool shown in Figure 2. The development on each machine then focused on achieving similar tool penetration across machines with very different beam and bearing compliance. All data reported herein was from welds that were mechanically characterized and stamped into weld panels, demonstrating defect free welds capable of withstanding rigorous post-weld formability requirements.

As weld parameters were initially developed at PNNL on a high stiffness gantry machine with servo-actuated ball screws (shown on the right of Figure 1). The preferred control methodology for this weld was position control. As such a commanded position was determined that accommodated the deflection of both beam and spindle bearings to achieve the desired tool penetration in the surface of the 2-mm to1.1-mm AA5182 butt weld. The control philosophy on the PNNL machine allowed the programmer to command a plunge depth, while the machine estimated what the actual plunge depth is based on known machine deflections at a given load. The machine inherently using the control model applied to this weld did not compensate for the difference between commanded and actual plunge depths. The actual plunge depth achieved for this weld and the respective tool forces are shown in Figure 3. While full size weld panels produced at both TWB LLC. and GM were approximately 1250mm in length, all welds produced at PNNL were on development panels with a weld length of 550mm.

In the process of transferring welding parameters to the GM machine, it was determined that for this weld scenario the servo-hydraulically controlled machine would more accurately control the position of the weld using a force control mode rather than position control used with PNNL machine. As such, a force of 14kN, the maximum force measured during the position controlled weld at PNNL was applied to the weld on the GM machine. The plot of tool position vs. weld length in Figure 3 demonstrates how the actual position of both the GM welds and the PNNL welds are very similar when each machine is allowed to operate in its own native control mode.

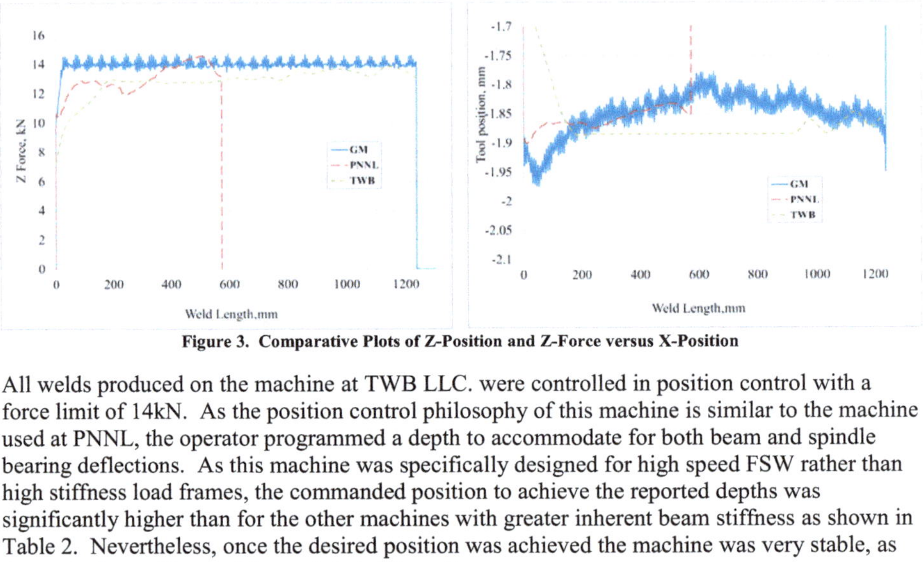

Figure 3. Comparative Plots of Z-Position and Z-Force versus X-Position

All welds produced on the machine at TWB LLC. were controlled in position control with a force limit of 14kN. As the position control philosophy of this machine is similar to the machine used at PNNL, the operator programmed a depth to accommodate for both beam and spindle bearing deflections. As this machine was specifically designed for high speed FSW rather than high stiffness load frames, the commanded position to achieve the reported depths was significantly higher than for the other machines with greater inherent beam stiffness as shown in Table 2. Nevertheless, once the desired position was achieved the machine was very stable, as shown in the position and force plots in Figure 3.

Table 2. Comparison of Z-Position Responses of Various FSW Machines.

	Commanded Position/Force	Actual Position (mm)	Actual During Ramp (mm)
PNNL	2.05-mm	1.84-1.87	1.86-1.93
GM	14 kN	1.8-1.87	1.86-1.97
TWB LLC.	2.5-mm	1.85-1.88	1.7-1.88

The comparative data presented in Table 2 demonstrate the effect that control modes (position or force control) and machine stiffness have on transferability of FSW parameters. While weld properties from each machine were shown to be essentially identical, transferring weld parameters from machine to machine required fundamentally understanding the control philosophies and inherent compliance variations between the machine platforms. While each machine evaluated in this study was produced from different manufacturers, it is clear that compliance variations between platforms of a given manufacturer would still need to be accommodated during weld parameter transfer. However, accommodating variations in control philosophies from unique machine builders added to the complexity of this study.

Conclusions

Weld parameters and tooling were successfully transferred across three machine platforms, each built from different manufacturers, for FSW of dissimilar thickness aluminum TWBs. While each machine had a unique control interface and inherent control philosophy, the data presented herein demonstrates the following:

1. FSW parameters can successfully be transferred across machine platforms with significant differences in machine compliance and control algorithms.
2. When holding linear and rotational velocity constant for a fixed tool, some informed adjustments in commanded Z- postion or Z force based on machine characteristics must be made to achieve a similar Z-position. This allows for the weld to experience the same process parameters achieved on other machine platforms thus ensuring weld repeatability.
3. Both position and force control philosophies can effectively be used to join dissimilar thickness aluminum TWBs; however, the most effective control mode may vary with different manufacturers and inherent machine construction.

Acknowledgments

The authors gratefully acknowledge funding from the Department of Energy-EERE-Vehicle Technology Office's Lightweight Materials Program under the direction of Mr. William Joost. This data shared herein was developed as part of a collaborative effort between the Pacific Northwest National Laboratory, General Motors, TWB Company, and Alcoa Inc.

References

1. Merklein, M., et al., *A review on tailored blanks-Production, applications and evaluation.* Journal of Materials Processing Technology, 2014. **214**(2): p. 151-164.
2. Pallett, R.J. and R.J. Lark, *The use of tailored blanks in the manufacture of construction components.* Journal of Materials Processing Technology, 2001. **117**(1-2): p. 249-254.
3. *Tailor welded blanks for advanced manufacturing.* Reference and Research Book News, 2011. **26**(6).

4. Baron, J.S., *A cost comparison of weld technologies for tailor welded blanks.* Welding Journal, 1997. **76**(10): p. 39-45.
5. Rooks, B., *Tailor-welded blanks bring multiple benefits to car design.* Assembly Automation, 2001. **21**(4): p. 323-328.
6. Miller, W.S., et al., *Recent development in aluminium alloys for the automotive industry.* Materials Science and Engineering a-Structural Materials Properties Microstructure and Processing, 2000. **280**(1): p. 37-49.
7. Bhagwan, A.V., G.T. Kridli, and P.A. Friedman, *Formability Improvement in Aluminum Tailor-Welded Blanks via Material Combinations.* Journal of Manufacturing Processes, 2004. **6**(2): p. 134-140.
8. Davies, R.W., et al., *Weld metal ductility in aluminum tailor welded blanks.* Metallurgical and Materials Transactions a-Physical Metallurgy and Materials Science, 2000. **31**(11): p. 2755-2763.
9. Shakeri, H.R., et al., *Study of damage initiation and fracture in aluminum tailor welded blanks made via different welding techniques.* Journal of Light Metals, 2002. **2**(2): p. 95-110.
10. Kinsey, B., V. Viswanathan, and J. Cao. *Forming of aluminum tailor welded blanks.* in *SAE 2001 World Congress, March 5, 2001 - March 8, 2001.* 2001. Detroit, MI, United states: SAE International.
11. Davies, R.W., et al., *Characterizing Al tailor-welded blanks for automotive applications.* Jom-Journal of the Minerals Metals & Materials Society, 1999. **51**(11): p. 46-50.
12. Barnes, T.A. and I.R. Pashby, *Joining techniques for aluminium spaceframes used in automobiles: Part I — solid and liquid phase welding.* Journal of Materials Processing Technology, 2000. **99**(1–3): p. 62-71.
13. Kuo, T.Y. and H.C. Lin, *Effects of pulse level of Nd-YAG laser on tensile properties and formability of laser weldments in automotive aluminum alloys.* Materials Science and Engineering a-Structural Materials Properties Microstructure and Processing, 2006. **416**(1-2): p. 281-289.
14. Davies, R.W., et al., *Mechanical properties of aluminum tailor welded blanks at superplastic temperatures.* Journal of Materials Processing Technology, 2002. **128**(1-3): p. 38-47.
15. Sato, Y.S., et al., *Effect of Microstructure on Postweld Formability in Friction Stir Welded Al Alloy 5052.* Trends in Welding Research, Proceedings, 2006: p. 387-391.
16. Steuwer, A., M.J. Peel, and P.J. Withers, *Dissimilar friction stir welds in AA5083-AA6082: The effect of process parameters on residual stress.* Materials Science and Engineering a-Structural Materials Properties Microstructure and Processing, 2006. **441**(1-2): p. 187-196.
17. Tavares, S.M.O., J.F. dos Santos, and P.M.S.T. de Castro, *Friction stir welded joints of Al-Li Alloys for aeronautical applications: butt-joints and tailor welded blanks.* Theoretical and Applied Fracture Mechanics, 2013. **65**: p. 8-13.
18. Threadgill, P.L., et al., *Friction stir welding of aluminium alloys.* International Materials Reviews, 2009. **54**(2): p. 49-93.
19. Sato, Y.S., et al., *Post-weld formability of friction stir welded Al alloy 5052.* Materials Science and Engineering a-Structural Materials Properties Microstructure and Processing, 2004. **369**(1-2): p. 138-143.
20. Grant, G., et al. *The formability of friction stir welds in automotive stamping environments.* in *2005 SAE World Congress, April 11, 2005 - April 14, 2005.* 2005. Detroit, MI, United states: SAE International.

21. Miles, M.P., B.J. Decker, and T.W. Nelson, *Formability and Strength of Friction-Stir-Welded Aluminum Sheets.* Metallurgical and Materials Transactions, 2004. **35A**(11): p. 3461-3468.
22. Kim, D., et al., *Macro-performance evaluation of friction stir welded automotive tailor-welded blank sheets: Part II - Formability.* International Journal of Solids and Structures, 2010. **47**(7-8): p. 1063-1081.
23. Lee, W., et al., *Experimental and numerical study on formability of friction stir welded TWB sheets based on hemispherical dome stretch tests.* International Journal of Plasticity, 2009. **25**(9): p. 1626-1654.
24. Leitao, C., et al., *Formability of similar and dissimilar friction stir welded AA 5182-H111 and AA 6016-T4 tailored blanks.* Materials & Design, 2009. **30**(8): p. 3235-3242.
25. Hovanski, Y., et al., *Aluminum Tailor-Welded Blanks for High Volume Automotive Applications*, in *Light Metals 2014.* 2014, John Wiley & Sons, Inc. p. 265-270.
26. Jurak, S.F.B., Dwight; McCoy, Michael. *Analysis of Mechanical and Metallurgical Properties of Friction Stir Butt Welded AA2024.* in *TMS 2013.* 2013. San Antonio: Wiley.

NATURAL AGING IN FRICTION STIR WELDED 7136-T76 ALUMINUM ALLOY

I. Kalemba[1], C. Hamilton[2], S. Dymek[1]

[1] Faculty of Metals Engineering and Industrial Computer Science
AGH University of Science and Technology, Kraków, Poland

[2]Department of Mechanical and Manufacturing Engineering
Miami University, Oxford, OH

Keywords: friction stir welding, natural aging, aluminum alloys, thermal model

Abstract

The long term natural aging behavior of friction stir welded aluminum 7136-T76 extrusions was investigated. The microstructure and mechanical properties in the as-welded, three years naturally aged, and six years naturally aged conditions were studied. Hardness profiles of the as-welded condition displayed the characteristic W-shape, but the advancing side hardness was lower than that on the retreating side. With natural aging, hardness recovery occurred, but the position of the hardness minima, particularly on the advancing side, shifted away from the weld centerline. Numerical simulations demonstrated that the processing temperatures were greater on the advancing side such that equilibrium phase dissolution was greater in this location and occurred to a greater distance from the centerline than on the retreating side. The hardness behavior upon natural aging, therefore, correlated to the temperature profile developed during welding and the degree to which phase dissolution occurs in the regions adjacent to the stir zone.

Introduction

Friction stir welding involves complex interactions between a variety of simultaneous thermo-dynamic processes including heating and cooling with different rates and plastic deformation as well as the physical flow of the processed material around the tool. As such, phenomena like dissolution and coarsening of precipitates or recovery and recrystallization occur with different intensities depending on the distance from the tool. Though much notable FSW research regarding aluminum alloys has focused on welding parameters, microstructure, mechanical properties and process modeling, only a few research studies have focused on the influence of post weld heat treatment on the microstructure and mechanical properties of friction stir weldments [1,2,3,4]. Moreover, post weld natural aging, as a special type of post weld heat treatment, has only been investigated sporadically [5,6,7,8]. Even though these limited efforts have capably sought to explain the phenomena occurring during natural aging of FSW welds, the mechanisms driving the microstructural evolution and the resultant hardness profiles remain unclear. If the properties of FSW joints are to be fully understood and practically modeled, these issues need to be more thoroughly considered and resolved. To that end, the research presented here concentrated on the long-term effect of natural aging (up to six years) in friction stir welds of 7136 aluminum alloy. The primary objective was to investigate the changes induced by natural aging in the mechanical properties, especially hardness, within various microstructural zones, and to correlate these properties to the thermal cycles predicted by a coupled thermal/flow numerical model. The present research sheds new light on the phenomena that occur during natural aging of aluminum 7136 friction stir welds.

Experimental Procedure

The investigated material, aluminum 7136-T76, was produced in the form of extrusions with a thickness of 6.35 mm and a width of 101.6 mm and friction stir welded in a butt-weld configuration along the longitudinal direction. The 7136 alloy contains higher levels of zinc than typical 7xxx alloys and primarily utilizes chromium and zirconium to control grain growth and recrystallization, akin to numerous alloys from the 7xxx family. The diameter of the FSW tool shoulder was 17.8 mm, the pin diameter tapered linearly from 10.3 mm at the tool shoulder to 7.7 mm at the tip and the pin depth was 6.1 mm. The welds were produced with a weld velocity of 2.1 mm/s, tool rotation speed of 250 rotations per minute (RPM) and a constant applied force of 26.7 kN. Following welding, the joints were subjected to natural aging for 3 and 6 years. Small samples were extracted from the naturally aged material for thermal analysis. The samples were sealed in aluminum pans and analyzed in a DuPont 910 differential scanning calorimeter, using an inert atmosphere. Samples were heated from room temperature to 500°C at a constant heating rate of 20°C/min.

Hardness measurements were made on metallographically polished surfaces across the mid-thickness of welds using a Vickers microhardness tester with a load of 9.81 N and duration time of 10 seconds. Additional microstructural characterization of friction stir welds was performed by light and transmission electron microscopy (TEM). Metallographic samples were prepared from the weld cross-section by initially grinding on sand papers, then electropolishing in a solution of $HClO_4$ and C_2H_5OH and finally chemically anodizing in an electrolyte containing 1.8 ml HBF_4 and 100 ml water. For TEM observation, thin foils were prepared from the center of the weld and the stir zone by excising 3 mm discs from the plane perpendicular to the weld direction. The thin foils were electropolished in a solution of 1 part HNO_3 and 3 parts CH_3OH at -30°C. The utilized voltage was 12 V.

Results and Discussion

Hardness profiles after welding and natural aging of the tested joints are shown in Figure 1. As observed in friction stir welds of other aluminum alloys, the shape of the hardness profile of the as-welded sample (measured as close to post-welding as practically permissible) consists of a region of higher hardness in the center (nugget), but significantly less than the hardness of the parent metal, surrounded by regions of lower microhardness values (TMAZ or HAZ). Thus, the profile exhibits the W-shape characteristic to friction stir welds of heat treatable aluminum alloys [9] with a hardness plateau in the central part of the weld (SZ). Hardness minima are present on either side of the weld centerline, the positions of which coincide with the transition between the TMAZ and HAZ. The hardness minima lay within regions of the weld that experienced process temperatures in the range of 280 – 340°C, according to the numerical model (the development and details of the model are discussed in [10] and [11]). The hardness of the recrystallized region (nugget) was less than that of the base material in the T76 temper, even after 6 years of natural aging. After natural aging for 3 and 6 years, the hardness profiles still retained the W-shape, but the hardness magnitudes within the weld zones increased. After 3 years of natural aging, hardness increased by approximately 10% in the stir zone and by 27% and 17% in TMAZ and HAZ on advancing and retreating sides, respectively. After 6 years of natural aging, the hardness increased by approximately 23% compared to the hardness values in the SZ in the as-welded samples. A substantial increase in hardness was also observed in the TMAZ and HAZ, i.e., by approximately 29% on the advancing side and approximately 20% on the retreating side.

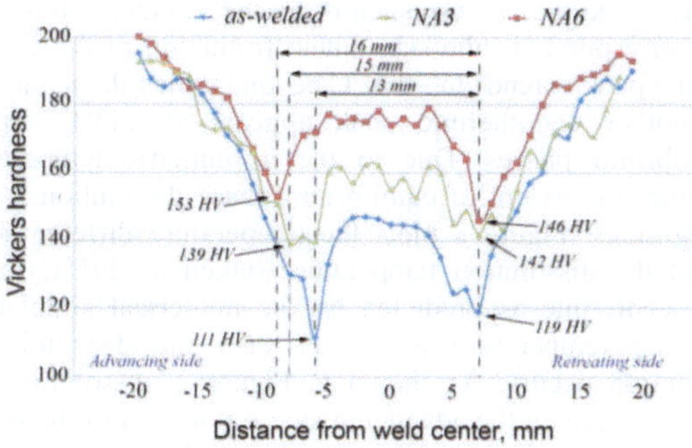

Figure 1 - As-welded, 3 years naturally aged and 6 years naturally aged hardness profiles.

The hardness profile for the as-welded condition is not symmetrical, i.e., the minimum hardness in the HAZ on the advancing side is lower than that on the retreating side. The numerical model predicts an asymmetrical temperature profile with higher welding temperatures on the advancing side. Figure 2 displays the temperature profile generated by the numerical model for the tool/workpiece interface during friction stir welding. The predicted temperatures are indicated at the leading edge of the tool (LE), the trailing edge of the tool (TE), the advancing side (AS) and the retreating side (RS). As observed in the image, the temperature profile is skewed toward the advancing side and the leading edge with the highest temperature (340°C) occurring approximately 5.25 mm from the tool center. As the tool advances during welding, cooler material in front of the tool is introduced at the leading edge and then swept to the retreating side. As the material is processed by the tool, heated material then flows toward the trailing edge and then either deposits toward the advancing side or flows into the workpiece thickness. This flow of material effectively raises the advancing side temperature above that of the retreating side and decreases the leading edge temperature below that of the trailing edge.

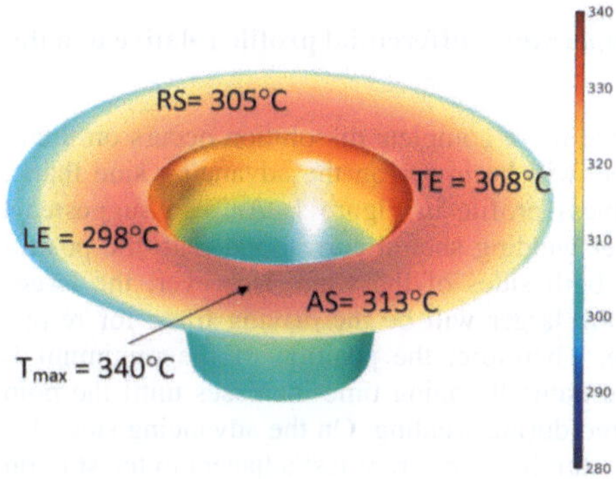

Figure 2 – Simulated temperature map of tool/workpiece interface.

A DSC trace of the friction stir welded 7136-T76 extrusions naturally aged for six years reveals two primary peaks including an endothermic peak at 157°C and an exothermic peak at

244°C. The endothermic peak likely corresponds to the dissolution of GP zones, and the exothermic peak likely corresponds to the subsequent formation of the equilibrium η and/or T phases [7]. The exothermic peak extends to ~290°C beyond which the η and/or T phases begin to dissolve. Ultimately, another endothermic peak appears at 443°C representing complete dissolution of all equilibrium phases. Due to the asymmetric temperature profile in the workpieces during welding, the extent of equilibrium phase dissolution will be unique on the retreating and advancing sides. Figure 3 plots the temperature differential (ΔT) between the welding temperature and the dissolution temperature (taken as 295°C) as a function of the distance from the weld centerline as predicted by the numerical simulation. A positive ΔT indicates that the welding temperature is greater than the dissolution temperature, and equilibrium phase dissolution occurs. As shown in Figure 3, dissolution occurs to a greater distance from the weld centerline on the advancing side (up to 9 mm) than on the retreating side (up to 7 mm). In addition, the magnitudes of the temperature differential outside the stir zone are greater on the advancing side than those on the retreating side. Therefore, equilibrium phases dissolve to a greater extent on the advancing side than on the retreating side.

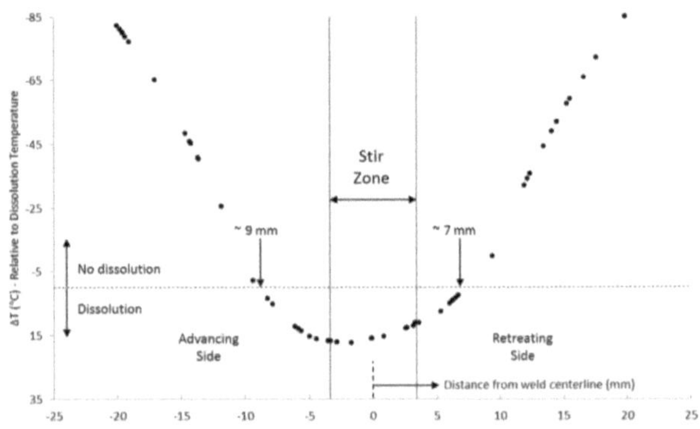

Figure 3 - Predicted temperature differential profile relative to a dissolution temperature of 295°C.

After welding, since more complete dissolution occurs on the advancing side outside of the stir zone, the hardness will be lower on the advancing side than on the retreating side in agreement with the hardness profile in Figure 1. As also suggested by Figure 3, the lowest hardness *immediately* after welding should correspond with the position of the largest ΔT just outside the stir zone on both sides of the weld. However, the larger the ΔT relative to the dissolution temperature, the larger will be the driving force for re-precipitation as the welded workpieces naturally age. Therefore, the position of the minimum hardness will follow the temperature profile as the naturally aging time increases until the point at which $\Delta T = 0$, i.e., limited dissolution occurred during welding. On the advancing side, the position of the hardness minimum will shift with naturally aging from just adjacent to the stir zone up to ~ 9 mm from the weld centerline and on the retreating side, the position of the hardness minimum will shift up to ~ 7 mm from the weld centerline. As shown in Figure 1, the position of the minimum hardness on the advancing side begins at 6 mm in the as-welded condition, but moves to 8 mm after three years natural aging and finally moves to 9 mm after six years of natural aging. On the retreating side, the position of the hardness minimum in the as-welded condition was measured at 7 mm from the weld centerline, and it remained in this position with natural aging up to six years. Similar displacements in the hardness minima with natural aging were reported by Leonard [5] in friction stir welded aluminum 2014A-T651 and 7075-T651, but the author did not elaborate on these displacements. Similar hardness behavior can also be observed in the work by Fuller et al. [7] on friction stir welded aluminum 7050 and 7075, but again, the authors did not discuss the phenomenon. Additionally, the review paper by Threadgill [9] displays hardness profiles for naturally aged friction stir welded 7075-T6 in which the positions of the minima clearly change with aging time, but the phenomenon is not addressed in the article.

Due to the asymmetrical temperature distribution, the hardness recovery during natural aging will be more substantial on the advancing side than on the retreating since the extent of dissolution is greater on the advancing side. From Figure 1, in the as-welded condition, the minimum hardness on the advancing side (111 HV) is lower than that on the retreating side (119 HV). However, after three years of natural aging, the hardness minima on each side of the weld are approximately equal (139 HV on the advancing side and 142 HV on the retreating side). Ultimately, after six years of natural aging, the minimum hardness on the retreating side (146 HV) is lower than that on the advancing side (153 HV), a reversal from the as-welded condition. Nelson et al. [6] reported that post-weld natural aging can partially restore the hardness in FSW aluminum 7075 and suggested that the recovery is attributed to the re-precipitation during natural aging. Woo et al. [12] reported the recovery of hardness and presented the natural aging kinetics of various regions in a friction stir processed aluminum 6061-T6. In another paper, Woo et al. [13] proposed a semi-empirical model to predict the hardness evolution in 6061-T6 subjected to FSW during post-weld natural aging. In Woo's work, the phenomenon of shifting hardness minima during natural aging was recognized but was not investigated microstructurally. More recently, Fuller et al. [7] reported the formation of GP(II) zones within the nugget zone and HAZ after long-term post-weld natural aging in aluminum 7050 and 7075 friction stir welds.

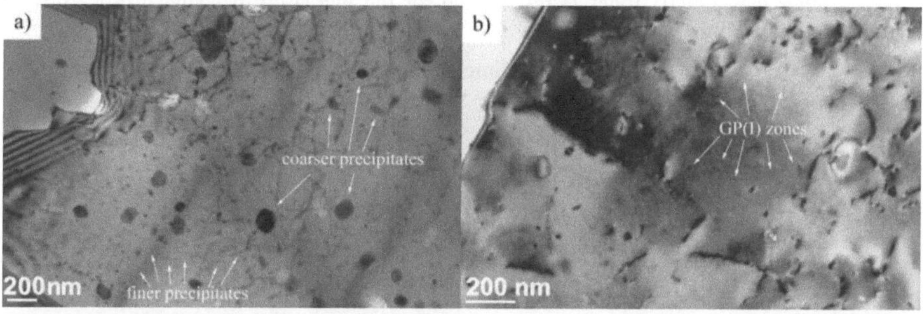

Figure 4 - Typical microstructure in investigated samples: a) the as-welded condition, b) sample naturally aged for 6 years.

Since natural aging mainly affected the stir zone, detailed examination by TEM was performed in this region. Figures 4 and 5 show the microstructure of the stir zone for the as-welded and naturally aged samples. During friction stir welding, the strengthening precipitates appear to coarsen and alternatively have gone into solution and re-precipitated according to the weld thermal cycle. This results in a bi-modal particle size distribution in the as-welded condition. The size of coarser precipitates was in the range of 50 – 150 nm, with larger particles located on grain and subgrain boundaries, while the finer precipitates exhibited a size of about 10 nm (Fig. 4a). In addition, some residual dislocations were often observed as a result of incomplete recrystallization. Selected area diffraction (SAD) patterns suggested the presence of η' and η phases, however, no indication of GP zones formation was noticed. The natural aging did not change these particles distributions, however, new tiny precipitates appeared in the microstructure (Fig. 4b and 5a). In bright-field images, the precipitates appear with bright or dark contrast depending on diffraction conditions. Some precipitates, larger in size, nucleated heterogeneously on dislocations (Fig. 5a). Selected area diffraction patterns recorded from regions containing these minute precipitates revealed diffuse maxima (spots) in addition to the aluminum matrix spots (Fig. 5b). These diffuse spots appeared at positions 1, 7/4, 0 in the Al [001] zone axis (and in symmetrical positions). According to earlier studies by Berg et al. [14] and Stiller et al. [15], these diffuse spots can be uniquely attributed to the formation of the GP zones referred to as GP(I) zones. Two types of GP-zones, i.e., solute-rich GP(I) clusters and vacancy-rich GP(II) clusters, can form during the decomposition of the supersaturated solid solution of Al-Zn-Mg alloys. The GP(II) zones also produce characteristic spots in $(001)_{Al}$ SAD patterns, very close to the 1, 1, 0 positions, however, such spots or even increased intensity was not observed in the present investigation. Thus, the formation of GP(I) zones is the only result of long-term natural aging, unlike in the study of Fuller et al., [7] who reported that the GP(I) zones, which formed after only 198 hours, were replaced after extended aging (73,300 hours) by the GP(II) zones. This discrepancy can be rationalized by the value of the highest temperature achieved during friction stir welding. In our studies, the highest temperature was only 340°C. However, according to Berg et al. [14], the GP(I) and GP(II)-zones are formed in different temperature ranges. GP(I) zones were found after aging at room temperature, and up to about 140°C, while the GP(II)-zones are formed mainly above 70°C. Moreover, the GP(II) were not found after quenching from temperatures at or below 450°C. In our study, this temperature was not approached. Thus, the analysis of microstructure in the stir zone suggests that the formation of GP(I) rather than GP(II) zones is responsible for the increased mechanical properties after long-term natural aging.

Conclusions

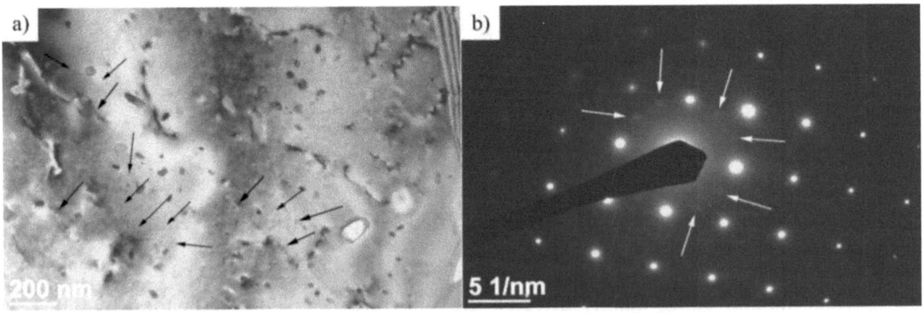

Figure 5 - TEM microstructure in sample after 6 years of natural aging: a) precipitates of GP(I) zones (indicated by arrows), b) SAD pattern with diffuse spots from GP(I) zones (indicated by arrows); [001] zone axis.

In this study, the effect of natural aging on the mechanical properties of friction stir welded 7136 aluminum alloys has been analyzed. The following important conclusions have been drawn:

(1) Natural aging notably influences the mechanical properties of friction stir welded 7136 aluminum alloy and cannot be neglected in practical applications of the alloy in the friction stir welded condition.

(2) The shape of the hardness profile can be correlated to the temperature distribution predicted by the numerical model.

(3) The hardening effect resulting from the natural aging is associated with the nucleation of the large volume fraction of GP(I) rather than GP(II) precipitates.

Acknowledgements

The authors acknowledge the AGH University of Science and Technology for the support of this research within the project no. 11.11.110.143.

References

1. Mahoney MW, Rhodes CG, Flintoff JG, Spurling RA, Bingel WH. Properties of Friction-Stir-Welded 7075 T651 Aluminum. Metall Mater Trans A 1998; 29:1955–64.
2. Sullivan A, Robson JD. Microstructural properties of friction stir welded and post-weld heat-treated 7449 aluminum alloy thick plate. Mat Sci Eng A 2008; 478:351–60.
3. Singh RKR, Sharma C, Dwivedi DK, Mehta NK, Kumar P. The microstructure and mechanical properties of friction stir welded Al-Zn-Mg alloy in as welded and heat treated conditions. Mater Design 2011; 32:682–7.
4. Sharma C, Dwivedi DK, Kumar P. Effect of post weld heat treatments on microstructure and mechanical properties of friction stir welded joints of Al-Zn-Mg alloy AA7039. Mater Design 2013; 43:134–43.
5. Leonard AJ. Microstructure and Ageing Behaviour of FSWs in Aluminium Alloys 2014A-T651 and 7075-T651. Proceedings of 2nd International Symposium on FSW; 2000 Jun; Götheborg, Sweden. paper on CD.
6. Nelson TW, Steel RJ, Arbegast WJ. In situ thermal studies and post-weld mechanical properties of friction stir welds in age hardenable aluminum alloys. Sci Technol Weld Joi 2003; 8:283–8.
7. Fuller CB, Mahoney MW, Calabrese M, Micona L. Evolution of microstructure and mechanical properties in naturally aged 7050 and 7075 Al friction stir welds. Mat Sci Eng A 2010;527: 2233-40.
8. Dong P, Sun D, Li H. Natural aging behavior of friction stir welded 6005A-T6 aluminum alloy. Mat Sci Eng A 2013; 576:29–35.
9. Threadgill PL, Leonard AJ, Sherliff HR, Withers PJ. Friction stir welding of aluminum alloys. Int Mater Rev 2009; 54:49–93.

10. Hamilton C, Kopyściański M, Senkov O, Dymek S. A coupled thermal/material flow model of friction stir welding applied to Sc-modified aluminum alloys. Metall Mater Trans A 2013; 44:1730–40.

11. Kalemba, I., Hamilton, C. Dymek, S. Natural aging in friction stir welded 7136-T76 aluminum alloy. Mater. Des. 2014; 60: 295 – 301.

12. Woo W, Choo H, Brown DW, Feng Z. Influence of the Tool Pin and Shoulder on Microstructure and Natural Aging Kinetics in a Friction-Stir-Processed 6061–T6 Aluminum Alloy. Metall Mater Trans A 2007; 38:69–76.

13. Woo W, Choo H, Withers PJ, Feng Z. Prediction of hardness minimum locations during natural aging in an aluminum alloy 6061-T6 friction stir weld. J Mater Sci 2009; 44:6302–9.

14. Berg LK, Gjønnes J, Hansen V, Li XZ, Knutson-Wedel M, Waterloo G, Schryvers D, Wallenberg LR. GP-zones in Al–Zn–Mg Alloys and Their Role in Artificial Aging. Acta Mater 2001;49: 3443–51.

15. Stiller K, Warren PJ, Hansen V, Angenete J, Gjønnes J. Investigation of precipitation in an Al–Zn–Mg alloy after two-step ageing treatment at 100° and 150°C. Mat Sci Eng A 1999; 270:55–63.

THE EFFECT OF HEAT TREATMENT ON THE PROPERTIES OF FRICTION STIR PROCESSED AA7075-O WITH AND WITHOUT NANO ALUMINA ADDITIONS

M. Refat[1], A.M.M. Abdelmotagaly[2], M.M.Z. Ahmed[2], I. El-Mahallawi[3]

[1]Materials Science Laboratory, Faculty of Engineering, The British University in Egypt, 11837 Cairo, Egypt
[2]Department of Metallurgical and Materials Engineering, Faculty of Petroleum and Mining Engineering, Suez University, Suez 43721, Egypt
[3]Department of Metallurgy and Materials Engineering, Faculty of Engineering, Cairo University, 12316 Giza, Egypt

Keywords: Friction Stir Processing, Nano-alumina, AA7075, Aluminum-nano composites, Heat treatment, Hardness, Microstructure.

Abstract

In this study, four passes friction stir processing (FSP) were applied on AA7075-O with and without the incorporation of alumina nano-particles (Al_2O_3) of average size 40nm. FSP parameters were constant at 500rpm and 40mm/min speed with tilt angle of 3^0. FSP rotation direction applied on clockwise and then counters clockwise direction every two passes. The friction stir processed (FSPed) materials were section and solution treated at 515°C for 1.5hrs that followed by age hardening at 120°C for 12, 24 and 36hrs. The effect of heat treatment regimes on the microstructure and hardness were examined. The microstructure and hardness after HT were compared with that of as FSPed material in each case. It has been found that in case of as FSPed material without any HT the effect of nano-alumina particles is significant in hardness increase relative to the one without powder. However, applying heat treatment regime has resulted in a slighter difference in the hardness between the two cases with and without nano particles additions.

Introduction

High performance aluminum alloys 7xxx are usually ideal materials used for structural application in when lightweight accompanied by high strength is required [1]. These alloys are known to exhibit a number of properties including acceptable resistance to liquation cracking, solidification and appropriateness for manual arc welding [1]. However, the low copper content in these alloys makes them susceptible to stress corrosion (SC) particularly in the heat affected zones (HAZ), as the exposure to heating conditions resulting from the fusion welding causes deviation in their structure from that obtained after heat treatment [1-4]. Indeed, synchronized developments in aluminum alloy design have indicated higher yield strength than conventional armor alloy, for example AA7055-T7751 has almost 40% yield strength higher than AA7017-T651 [1, 5].

A number of studies have focused on the ballistic failure modes including various absorbing processes of kinetic energy such as ductile hole growth, spalling, plugging and front face petaling that is based on projectile type and materials properties [1,6-9]. The interactions between various types of ballistic damage and strength and toughness levels and microstructure are still not studied enough [1,6-10]. Actually only few standard tests involving ballistic performance with relevance to materials properties and microstructure are stated [1,6]. Such studies are important since the majority of welding and heat treatment conditions result changes in the microstructure. For example, a study has found that at under-aged temper condition, the microstructure is more sensitive to slip concentration, while at over-aged temper condition, it is more susceptible to grain boundary fracture due to the occurrence of high density of precipitation of grain boundary [1, 11-12]. Also, friction stir welding results changes in the microstructure in terms of the grain structure and precipitate distribution size [1,13-16], which will have an influence on the material response to the projectile impact [1]. It is also worth mentioning that the properties of the projectile are very important such as shape, size, hardness and mass relative to the plate thickness and material [1].

Friction stir processing (FSP) is a novel effective solid state processing technology, which was developed by Mishra et al. [17, 18-19], used for (i) surface composite development on aluminum substrate [17, 18], (ii) microstructural modification of metal matrix composite (MMC) [17, 19], (iii) enhancement in properties of cast aluminum alloys [17, 20], and (iv) homogenization of powder metallurgy aluminum alloy [17, 21]. The idea of FSP is very simple and is based on the main concept of friction stir welding (FSW) where a rotating tool is integrated in a monolithic plate to obtain a microstructure modification which is used for particular feature enhancement; such as high strain superplasticity which was achieved via FSP in AA7075 [17, 22-24].

In a previous study by the authors [25] in which AA7075-O /nano-alumina composite was developed using FSP, it has been observed that, heat treatment (HT) of the nano-surface composite resulted in lower hardness in the processed region relative to the base material at the same HT condition. At that stage, this was attributed to the existence of nano particles that delays the hardening effect of T6 heat treatment in the processed region. Significant increase in surface hardness was only found to occur after four passes in the un-heat treated condition. Thus the aim of the current work was to study the effect of various heat treatment regimes on the hardening behavior of the friction stir processed (FSPed) AA7075-O with and without the addition of nano particles.

Experimental

Materials

AA7075 sheet plates and Al_2O_3 nanoparticles (40 nm average particle size) were used in this work. AA7075 was received in T6 temper condition of dimensions of $200 \times 150\ mm$ and $150\ mm$ depth with a measured Vickers hardness number of 177 Hv. AA7075-T6 was annealed to have it in the O temper condition of measured Vickers hardness number 64.3 Hv. AA7075-O was used as a metal matrix and Al_2O_3 nanoparticles as the reinforcements.

Sample Preparation For FSP

Two AA7075-O plates of 7cm width, 30cm length and 15mm thickness were cut. One was used for FSP without nano particles addition. The other was used for FSP with nano particles addition. In this one a groove of 4.5 mm depth and 3mm width was machined through the surface of the AA7075 plate. Al_2O_3 nanoparticles were added into the grooves after closing the sides of each groove to avoid escaping of nanoparticles during FSP. Then a probeless tool of 19 mm shoulder diameter was used for top closing of the grooves after packing of the nanoparticles at the same FSP parameters used that will be mentioned in the next section.

Friction Stir Processing and Heat Treatment

The plates were processed using conventional FSP using a tool made of H13 tool steel. The FSP parameters for the stirring passes were 500 rpm tool rotation rate, 40 mm/min – traverse speed and 3° tool tilt angle. The FSP tool dimensions were of 6.2 mm probe diameter, 5.3 mm probe length and 19 mm diameter shoulder. The plate was subjected to 4 passes based on results found by previous work by the authors [26]. The advancing side of the previous pass converted into retreating side in the next pass and vice-versa. After FSP, the friction stir processed (FSPed) materials were cut perpendicular to the processing direction. Then the FSPed material was age hardened to T6 temper condition by solution treatment at 515 $°C$ and water quenching followed by age hardening at 120 $°C$ for several times: 12, 24, and 36 hrs. It should be noted here that the samples were left for only 9 days before the start of heat treatment process. The samples were prepared according to the standard preparation technique starting with grinding of different grades up to 1200 size paper and then mechanically polished using 6 μm, 1 μm diamond suspension and using 0.05 μm alumina suspension. A number of material characterization tests were carried out; namely: optical microstructure and hardness investigation. The samples for metallographic examination and after final polishing were etched using diluted keller's reagent of chemical composition (100 ml distilled water, 10 ml HNO_3, 10 ml HCL and 2ml HF) for 15 sec. The microstructure investigation was carried out for all samples using optical microscopy. The Vickers hardness was measured at 1 mm intervals across the transverse cross section of FSPed materials near the top surface of the NG, using load of 1Kg for 15s dwell time.

Results and Discussion

Optical Macrographs

Figures 1 (a, b) shows the optical macrographs of the as FSPed AA7075-O without nano alumina additions without (a) and with (b) T6 heat treatment. Also Figures 1(c, d) shows the optical macrographs of the as FSPed AA7075-O with nano alumina additions without (c) and with (d) T6 heat treatment. FSP was carried out for 4 passes using 500rpm rotation rate and 40mm/min traverse speed two passes in clockwise direction and two passes in counterclockwise direction. The typical nugget (NG) shape of such a FSPed material is observed in **a,** but after HT abnormal grain growth (AGG) is observed at the top and bottom areas of the NG. Also some white spots at the interface between the AGG region and the fine grains region in the NG are observed also. After addition of the nano alumina particles it can be observed from the macrograph in **c** that some agglomerates of nano particles exist near the top surface of the NG in addition to some small voids near the base of the NG. However, after T6 HT it can be observed that the AGG is limited to very small areas in the NG as it can be seen in macrograph d in Figure 1.

Optical Microstructure

The optical microscopy was used to further clarify the observed features in the macrographs above. Figure2 shows the optical microstructure of the as FSPed sample without nano alumina additions. The optical microstructure in the base material (BM) in **a,** clearly shows the coarse and stable second phase θ as the material is fully annealed. Inside the NG the newly recrystallized fine grain structure can be seen in **b** of the figure. Figure 3 shows the optical microstructure of the FSPed sample without nano alumina additions after T6 heat treatment and the AGG at the base of the NG is shown in a and the white spots observed in the macrograph at the interface between fine and AGG grain near the top surface is shown in b of the figure.

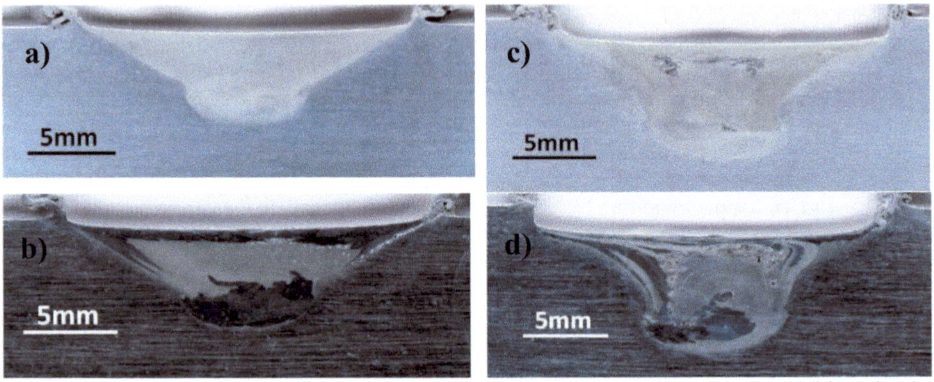

Fig. 1: Optical macrographs of the as FSPed AA7075-O without nano alumina additions in a) and after heat treatment in b), optical macrographs of the as FSPed AA7075-O with nano alumina additions in c) and after heat treatment in d).

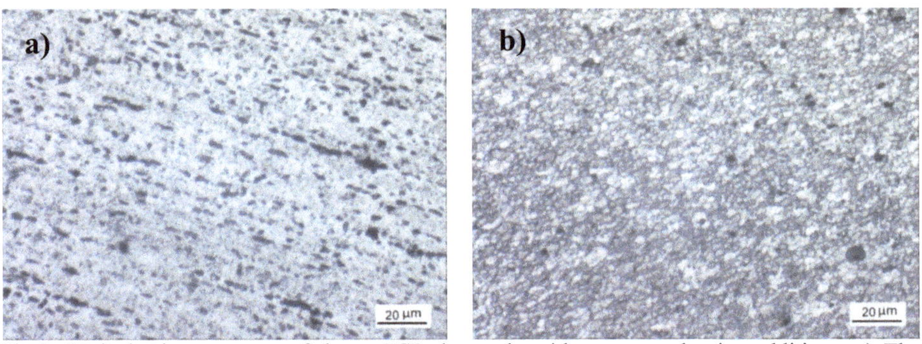

Fig.2: Optical microstructure of the as FSPed sample without nano alumina additions. a) The base material (BM), and b) inside the NG region.

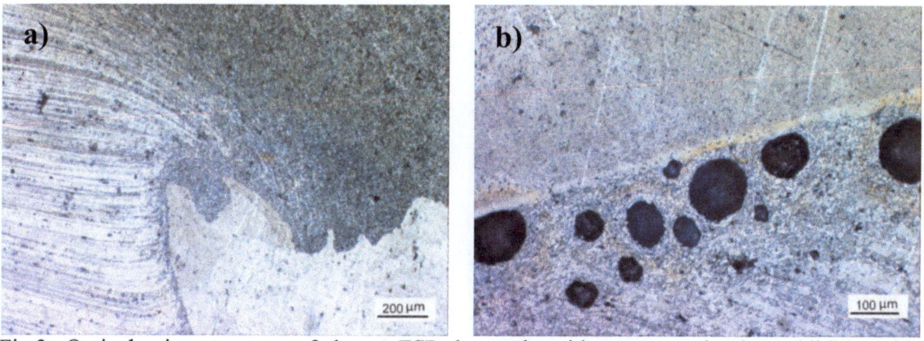

Fig.3: Optical microstructure of the as FSPed sample without nano alumina additions after solution treatment at 515 C for 1.5hrs and aging at 120C for 12hrs. a) Near the base of the NG, and b)near the top surface of the NG region.

Figures 4(a,b) show the optical microstructure of the as FSPed sample with nano alumina additions. This figure clearly shows that the nano particles after 4passes are still not uniformly distributed across the NG as some areas have agglomerates as it can be seen in a and others almost free of nano-particles such as the micrograph in b. Figure 5(a,b) show the optical microstructure of the as FSPed sample with nano alumina additions after solution treatment at 515 C for 1.5hrs and aging at 120C for 12hrs. This shows the selective AGG at the areas contains nano-particles such as the micrograph in (a) and fine grain structure without any AGG indication as it can be seen in b.

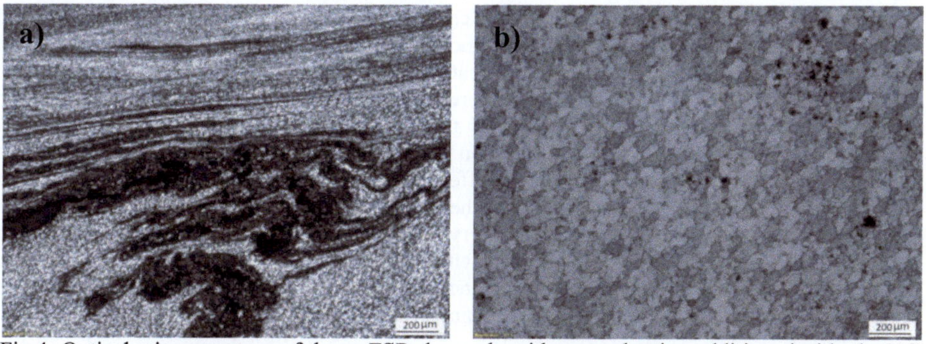

Fig.4: Optical microstructure of the as FSPed sample with nano alumina additions inside the NG region. a) The agglomerates of nano particles, and b) Fine grain structure

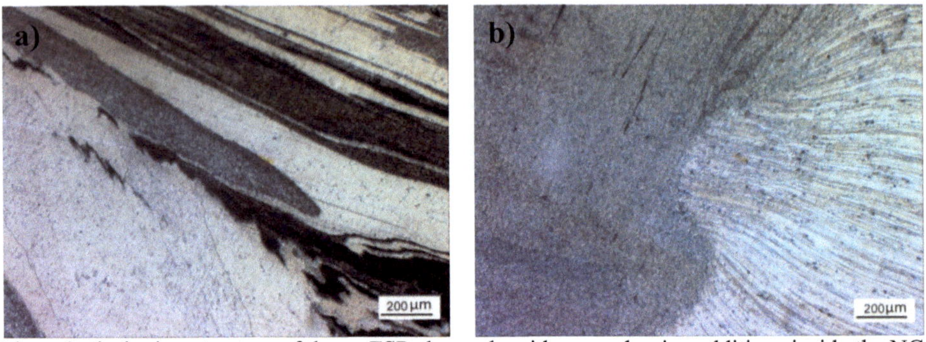

Fig.5: Optical microstructure of the as FSPed sample with nano alumina additions inside the NG region after HT. a) The selective AGG, and b) The interface between NG with Fine grains and BM structure.

Vickers Hardness

For the sake of understanding the effect of nanodispersion on the surface hardness after T6 heat treatment, the hardness was measured at the same position for all samples and after all aging times. Figure 6 shows the Vickers hardness curves measured near the top surface of the FSPed materials at an interval of 1mm across the transverse cross section. a) As FSPed, b) 12hrs, c)24hrs, d)36hrs, f) collective curves for all condition and g) comparison of BM hardness with the FSPed before HT .. The samples from b to d were subjected to solution treatment at 515C for 1.5hrs and then aging at the indicated times. It can be observed that in case of as FSPed material without any HT the effect of nano-alumina particles is significant in hardness increase relative to the one without powder. However, applying heat treatment regime of solution treatment at 515C for 1.5hrs and age hardening at 120C for different times has resulted in a slighter difference in the hardness between the two cases with and without nano particles additions, though the scatter in the hardness results across the tested region has decreased. This suggests that adding nano particles during FSP will tend to conceal the effect suppress the formation of softened regions in the treated zones. The figure also shows that the highest hardness point in the hardness profiles was also reached at the nugget zone. The observed increase in hardness after FSP without nano particles addition and before heat treatment is attributed to a number of causes: finer grain sizes and work hardening effect [27]. Whereas, the increase in hardness associated with nano particles addition can be attributed to Orawan strengthening effect [27].

Conclusions

Friction stir processing can be utilized to develop surface-nanocomposites with Al_2O_3 nanodispersions embedded in the matrix. The new surface nanocomposites showed enhancement in the hardness of the surface of A7075-O to almost double of the starting material after 4 passes FSP with incorporating nano-alumina particles before heat treatment. After T6 heat treatment up to 36 aging hours the hardness increased to the standard values in all cases with slight increase in case of nanodispersions.

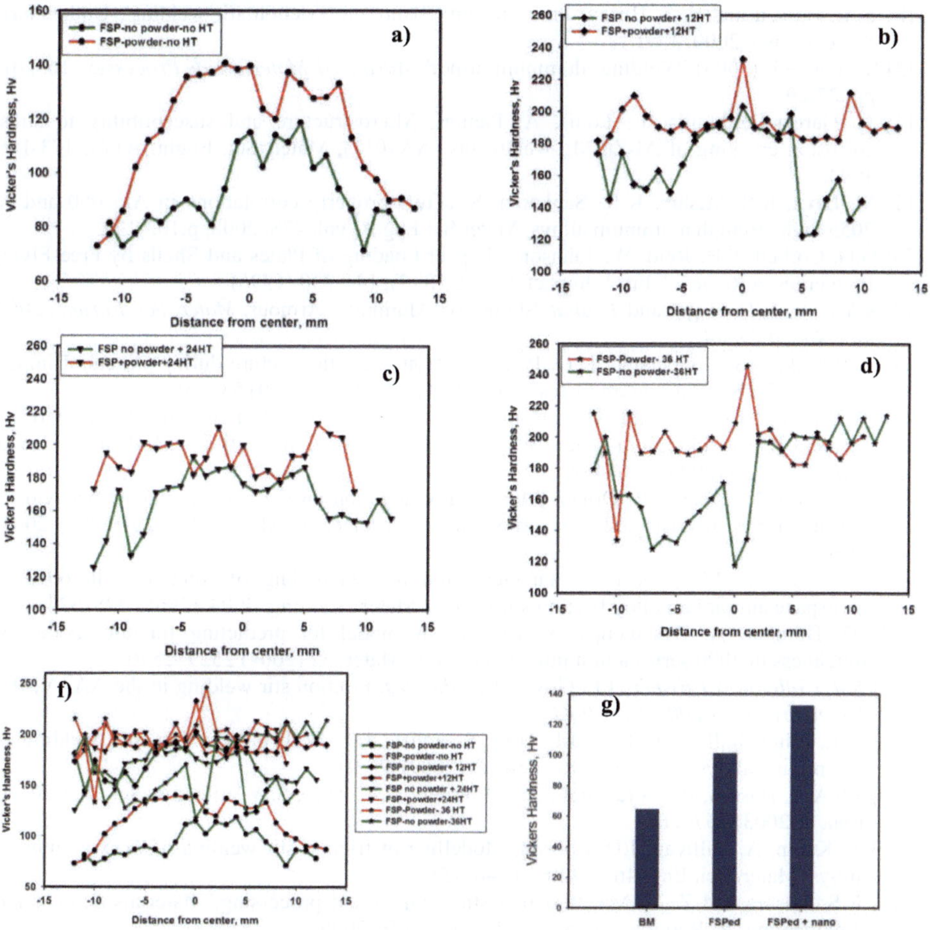

Fig. 6: average Vickers hardness measured near the top surface of the FSPed materials at an interval of 1mm across the transverse cross section. a)As FSPed, b)12hrs, c)24hrs, d)36hrs and f) collective curves for all condition and g)comparison of BM hardness with the FSPed before HT .

Acknowledgment

The science and technology development fund (STDF), Ministry of scientific research (Egypt) is deeply acknowledged for funding this work with grant number of ID: 3926.

References

[1] A. Sullivan, C. Derry, J.D. Robson, I. Horsfall, P.B. Prangnell. Microstructure simulation and ballistic behaviour of weld zones in friction stir welds in high strength aluminium 7xxx plate. Materials Science and Engineering: A. 2011 April; 528(9): 13.

[2] J. C. Bassett and S. S. Birley: Proc 2nd Int. Symp. on 'Friction stir welding', Gothenburg, Sweden, June 2000, TWI.
[3] K. Sampath (2005) 'Welding aluminum armor', *Advanced Materials & Processes,* 163(10), pp. 27-29.
[4] C. Garcia-Cordovilla, E. Louis, A. Pamies, Microstructure and susceptibility to stress corrosion cracking of Al-Zn-Mg weldments (AA7017), Mater. Sci. Engin. A174, 173-186, 1994.
[5] M. Dixit, R.S. Mishra, K.K. Sankaran. Structure-property correlations in Al 7050 and Al 7055 high-strength aluminum alloys, Mater Sci Eng A, vol. 478, 2008, p.163-172.
[6] G.G. Corbett, S.R. Reid, W. Johnson, "Impact Loading of Plates and Shells by Free-Flying Projectiles: a Review," Int. J. Impact. Engng 18 (2), 141-230 (1996).
[7] *K.S. Kumar*, *D. Singh*, and *T. Bhat*, Studies on Aluminum Armour. *Mater. Sci. Forum., 2004, 465–466*, p *79–84.*
[8] T. Børvik, O.S. Hopperstad, K.O. Pedersen, "Quasi-brittle fracture during structural impact of AA7075-T651 aluminium plates" Int. J. Impact Eng. 37 (2010) 537–551.
[9] T. Børvik, M.J. Forrestal, O.S. Hopperstad, T.L. Warren, M. Langseth, "Perforation of AA5083-H116 aluminium plates with conical-nose steel projectiles–calculations" Int. J. Impact Eng. 36 (2009) 426–437.
[10] T. Demir, M. Ubeyli, R. Orhan-Yildirim. Investigation on the ballistic impact behavior of various alloys against 7.62 mm armor piercing projectile" Mater. Des. 29 (2008) 2009–2016.
[11] C.G. Derry, J.D. Robson. Characterisation and modelling of toughness in 6013-T6 aerospace aluminium alloy friction stir welds." Mater. Sci. Eng. 490A (2008) 328–334.
[12] D. Dumont, A. Deschamps, Y. Brechet. A model for predicting fracture mode and toughness in 7000 series aluminium alloys, Acta Mater. 52 (2004) 2529–2540.
[13] *K.J. Colligan*, J.J *Fisher*, J.E. Gover, J.R. *Pickens*. Friction stir welding in the AAAV. *Adv Mater Processes 2002*:*160*:*39-41*
[14] P.L. Threadgill, A.J. Leonard, H.R. Shercliff, J.P. Withers. Friction stir welding of aluminium alloys. Int. Mater. Rev. 54 (2) (2009) 49–93.
[15] Kh.A.A. Hassan, P.B. Prangnell, A.F. Norman, D.A. Price, S.W.Williams, Sci. Tech.Weld. Join. 8 (2003) 257–268.
[16] N. Kamp, A. Sullivan, J.D. Robson. Modelling of friction stir welding of 7xxx aluminium alloys. Mater. Sci. Eng. 467A (2007) 246–255.
[17] R.S. Mishra and Z. Y. Ma, "Friction stir welding and processing,"Materials Science and Engineering: R: Reports, vol. 50, no. 1–2, pp. 1-78, 2005.
[18] R.S. Mishra, Z.Y. Ma, I. Charit, "Friction stir processing: a novel technique for fabrication of surface composite", Mater Sci Eng A, 341 (1-2) (2003), 307-310.
[19] J.E. Spowart, Z.Y. Ma, R.S. Mishra, in: K.V. Jata, M.W. Mahoney, R.S. Mishra, S.L. Semiatin, T. Lienert (Eds.), Friction Stir Welding and Processing II, TMS, 2003, pp. 243–252.
[20] R.S. Mishra, Z.Y. Ma, S.R. Sharma, M.W. Manohey, Microstructural modification of cast Aluminum alloys via friction stir processing, Materials Science Forum 426-432 (2003) 2891-2896.
[21] P.B. Berbon, W.H. Bingel, R.S. Mishra, C.C. Bampton, M.W. Mahoney, "Friction stir processing: a tool to homogenize nanocomposite aluminum alloys", Scripta Mater, 44 (1) (2001), 61-66.

[22] R.S Mishra, M.W Mahoney, S.X McFadden, N.A Mara, A.K Mukherjee, High strain rate superplasticity in a friction stir processed 7075 Al alloy, Scripta Materialia, Volume 42, Issue 2, 31 December 1999, Pages 163-168, ISSN 1359-6462.
[23] Mishra, R.S.; Mahoney, M.W. Friction stir processing: A new grain refinement technique to achieve high strain rate superplasticity in commercial alloys. Mater. Sci. Forum 2001, 357–359, 507–512.
[24] Z.Y. Ma, R.S. Mishra, M.W. Mahoney. Acta Mater. 50 (2002) 4419.
[25] M.M.Z. Ahmed, M. Refat, I. El-Mahallawi "Manufacturing of Nano-Surface AA7075 Composites By Friction Stir Processing " Characterization of Minerals, Metals, and Materials 2014, TMS2014, Jon Wiley& Sons.
[26] S.F. Kashani-Bozorg and K. Jazayeri: Formation of Al/B4C Surface Nanocomposite Layers on 7075 Al Alloy employing Friction Stir Processing, AIP Conf. Proc. 1136,(2009), pp 715-719.
[27] S. V. Emani, J. Benedyk, P. Nash, D. Chen; "Double Aging and Thermomechanical Heat Treatment of AA7075 Aluminum Alloy Extrusions," J. Mater. Sci. (2009) 44:6384-6391.

FRICTION STIR WELDING AND PROCESSING VIII

Dissimilar Materials

Friction Stir Welding and Processing VIII
Edited by: Rajiv S. Mishra, Murray W. Mahoney, Yutaka Sato, and Yuri Hovanski
TMS (The Minerals, Metals & Materials Society), 2015

FRICTION STIR WELDING OF DISSIMILAR LIGHTWEIGHT METALS WITH ADDITION OF ADHESIVE

W. Yuan[1], K. Shah[1], B. Ghaffari[2], and H. Badarinarayan[1]

[1]Automotive Products Research Laboratory, Hitachi America Ltd., R&D Division, Farmington Hills, MI
[2] Research and Advanced Engineering, Ford Motor Company, Dearborn, MI

Keywords: Friction stir welding, dissimilar metals, aluminum alloy, magnesium alloy, intermetallic compounds, materials flow, strength, load

Abstract

Dissimilar welding of aluminum to magnesium alloy sheets is desired for automotive body structures, but challenging due to the formation of brittle intermetallics at the interface. In a previous study, friction stir welding was utilized to evaluate the joining potential of a cast magnesium alloy to aluminum alloy sheets, and specifically to produce robust and reliable friction stir welds between sheets of wrought AA6022-T4 aluminum alloy and cast AM60B magnesium alloy. The welds are generally stronger when welding is conducted from the aluminum-sheet side, given the thickness of aluminum is half that of magnesium. In the present study, friction stir welding was further developed to produce structural welds between these sheets from the aluminum side, with and without adhesive at the faying surface. The effect of adhesive on the interface was examined for resultant material mixing and microstructure. Mechanical performance of welds was also evaluated.

Introduction

Joining dissimilar lightweight metals has become a critical challenge for achieving optimum part design in transportation industries. Friction stir welding (FSW), a solid-state joining technique, has demonstrated promising capability and high efficiency in joining similar lightweight alloys, such as Al alloys or Mg alloys separately [1,2]. Recently, many attempts have been made to FSW dissimilar alloys, such as Al to Mg alloys [3-7], Al alloys to steels [8, 9] and Mg alloys to steels [10,11]. Formation of brittle intermetallic compounds (IMCs) is still the major barrier to obtaining sound dissimilar welds [4,5,10]. Sato et al. [5] concluded that the distribution of IMCs, other than the thickness, played an important role on lap-shear strength of friction stir spot welded AA5083 and AZ31.

To reduce the formation of IMCs, submerged friction stir welding under water was conducted [12]. It is reported that the formation of IMCs was significantly suppressed due to a decrease of peak temperature from 403°C to 378 °C. Nickel foil used as filler material between Al and Mg has also been adopted during FSW and laser welding, and successfully reduced the formation of IMCs [13,14]. Also, adhesive shows promising impact on strength of FSW and laser welding of dissimilar Al and Mg alloy [13,15,16].

Previous research indicated that sound welds between Al and Mg alloys using FSW can be achieved via interrupting and dispersing the brittle IMCs [17,18]. The aim of this research is to investigate the influence of adhesive on dissimilar friction stir linear welding (FSLW) between

Al and Mg alloys. The joining process was evaluated on the basis of material mixing, microstructure, and strength by varying the adhesive application.

Experimental Procedure

The United States Automotive Materials Partnership (USAMP) Magnesium Front End R&D project is tasked with design and fabrication of magnesium-intensive "demonstration" automotive front end sub-structures consisting of a cast Mg shock tower and Al rails (wrought and extruded). The Mg shock tower and wrought Al rail will be joined using FSLW. An extensive coupon study from previous work showed robust and reliable welds for this stackup [17-19]. In the current study, the same wrought Al alloy AA6022-T4 and cast Mg alloy AM60B sheets, having a thickness of 1.5 mm and 3.1 mm respectively, were selected. Both sheets were sheared to a dimension of 152.4 mm long and 101.6 mm wide and overlapped by 152.4 mm × 30 mm during welding, as schematically shown in Figure 1. The Al alloy sheet was placed on top of the Mg sheet in a lap joint configuration. Unlike the previous study, an evenly distributed thin layer of Terokal 5089 adhesive (epoxy based, highlighted in purple in Figure 1) was applied in the overlapping area between the Al and Mg after both sheet surfaces were degreased by acetone. FSLW was then performed at the center of the overlapped area with a tool rotation rate of 1500 rpm and tool traverse speed of 50 and 75 mm/min. For each setup, two stitch welds with a length of 55 mm were produced. A 2.4-mm long triangular pin tool with a 12-mm diameter concave shoulder was employed. All welds were produced by securing the overlapped sheets with a specially-designed clamp, as shown in Figure 1b. The above welding parameters and setup had been shown to produce the strongest FSLW lap-shear joints without adhesive. After FSLW, some welded pieces were tested in the as-weld condition, and the remaining were oven cured at 180 °C for 30 min. For comparison purposes, welds made without adhesive were also evaluated.

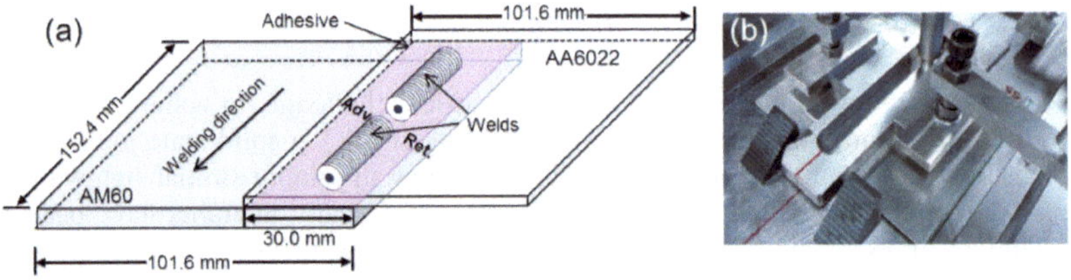

Figure 1. FSLW of Al to Mg sheets (a) schematic of welding configuration (b) clamping fixture used to secure the two sheets during welding.

Investigations were also made to monitor and compare the process temperature and thermal cycle during welding, especially between welds with and without adhesive. For a single workpiece, the temperature profiles for the first and the second stitch welds were compared. K-type thermocouples were embedded in the workpiece right underneath the weld bead in the lower sheet, as well as the top surface of the upper sheet on the advancing side of the welds. The locations of the thermocouples with respect to the workpiece are detailed in Figure 2 in plan and sectional views. A computer-based data acquisition system was employed to record temperatures during welding at a frequency of 10 Hz.

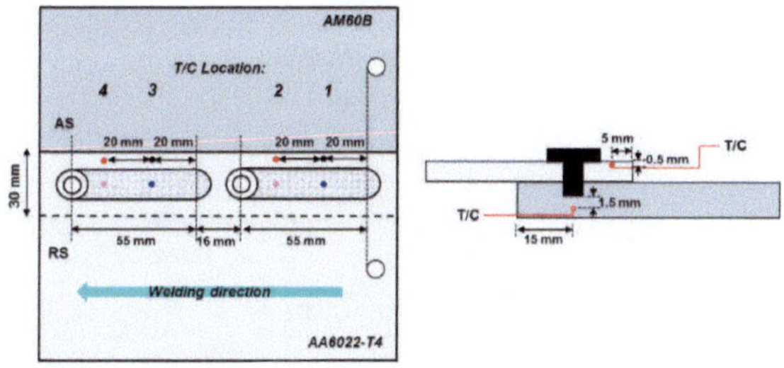

Figure 2. Schematic illustration of thermocouple positions.

Microstructural characterization of weld cross sections and comparison of mechanical properties of the welded parts were carried out in order to investigate the influence of adhesive addition and subsequent post-weld curing. Specimens for metallographic study were cross sectioned perpendicular to the welds. All samples were mounted and mechanically polished using diamond suspensions down to 1 μm and further fine polished with 0.05 μm alumina. Macro and microstructures were examined using optical microscopy and scanning electron microscopy (SEM) with energy-dispersive X-ray spectroscopy (EDX). An Instron screw-driven testing machine was used to evaluate the lap-shear strength of welds. Lap-shear specimens having a width of 30 mm were extracted from weld beads via water-jet cutting. Shims were used to center the specimen in the Instron, and tested at a crosshead speed of 2 mm/min. The fracture surface of the tested specimens was investigated under SEM.

Results and Discussion

3.1 Temperature Profile

The temperature profiles for two welding conditions, with and without adhesive, are shown in Figure 3. The temperatures were monitored for both stitch welds in each condition. Figures 3a and 3b illustrate the temperature history of welds without adhesive and Figures 3c and 3d show that of welds with adhesive. Several remarks can be made: First, peak temperature underneath the weld bead is between 300 °C and 350 °C, and between 200 °C and 300 °C for the upper sheet on the advancing side. Second, peak temperatures underneath the welds with adhesive are slightly lower (by ~25 °C in general) than for the welds without adhesive. Last, the surface temperature for the second weld (location 4) is higher than that of the first weld (location 2), regardless of the existence of adhesive.

The peak temperatures in the current study are relatively lower compared to those reported by others [4,12]. This is directly related to locations selected for embedding the thermocouples. Higher temperature is generally noticed when thermocouples are placed closer to the heat source (i.e., rotating pin and tool shoulder). This difference can also be seen in the observed delay for the advancing side to achieve peak temperatures. As the rotating pin passed by the thermocouple embedded underneath the weld beads, the temperature at this point in the sample rose to a peak ~340 °C, however, the temperature peak was delayed ~3 s for thermocouples on the advancing side. The relatively lower peak temperature in the lower sheet for welds with adhesive, compared to those without adhesive, is likely due to the thermal insulation impact of the adhesive, which separated the upper and lower sheet alloys.

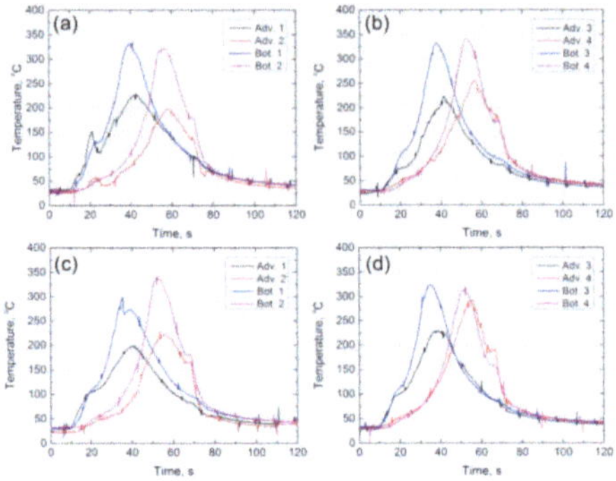

Figure 3. Temperature profiles for: (a) (b) welds without adhesive and (c) (d) with adhesive. The graph legends correspond to the thermocouple locations marked in Figure 2.

3.2 Microstructure

Figure 4 presents a typical cross section and microstructures of welds without adhesive. Detailed microstructure evaluation during optimization of the FSLW process has been discussed elsewhere [17,18,19]. The key findings are summarized: (1) the weld zone can be divided into various regions with the upper region occupied by the Al alloy, the lower region consisting of the Mg alloy, and a transition region in between. (2) Microvoids on the order of 50-200 μm are occasionally visible at the end of the cold lap feature on the retreating side and at the bottom of the stir zone on the advancing side where material deposition occurs. (3) In the transition region, a layered structure with two intermingled IMCs were observed using EDX analysis. The lighter-etching layers consisting of Al3Mg2 appear mostly near the top, and the darker ones consisting of Al12Mg17 are near the bottom. (4) The faying surface adjacent to the stir zone, but under the tool shoulder, also consists of IMCs.

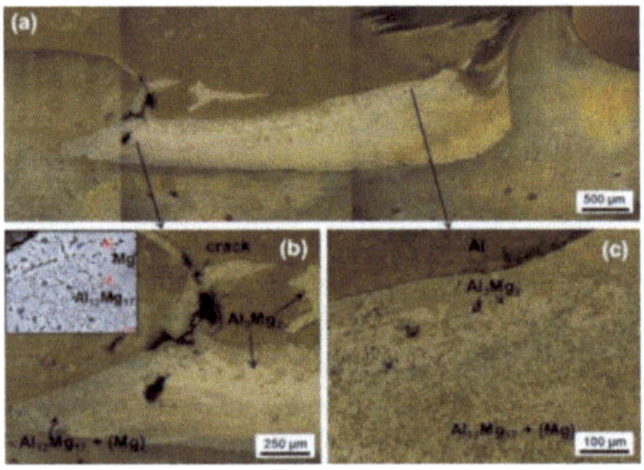

Figure 4. Macro and micro structures of a typical weld showing microvoids and IMCs [17].

Figure 5 shows typical SEM images of faying surfaces adjacent to the stir zone on the advancing side of welds with and without adhesive. Distinct features can be observed. For the weld without adhesive (Figure 5a), an interfacial layer, including two kinds of IMCs, was identified based on

EDX analysis. The dark layer close to the bottom Mg sheet is composed mainly of $Al_{12}Mg_{17}$, while the light layer close to the top Al sheet consists mainly of Al_3Mg_2. A similar observation has been reported during friction stir spot welding of Al and Mg alloys [5]. The formation of these IMCs at the interface away from the stir zone was attributed to the diffusion process occurring during welding. As measured in the temperature profiles in Figure 3, the peak temperature underneath the weld bead is approaching 350°C and the actual temperatures in and adjacent to the stir zone could be even higher. Also, the upper and lower sheets were tightly clamped during welding to prevent any sliding. Both of the above factors contributed to the acceleration of diffusion and the formation of IMCs. For the weld with adhesive (Figure 5b), the interface away from the weld stir zone was instead filled with a layer of adhesive (~10-13 μm) which was verified by EDX elemental analysis as shown in Figure 5d. A high amount of elements such as Si, Ca, P and O were detected in the layer. However, the quantitative analysis indicates the weight percentage for Al or Mg is less than 25%. Unlike the weld without adhesive, for which thin brittle IMCs form away from the stir zone, this area is adhesively bonded in the weld with adhesive.

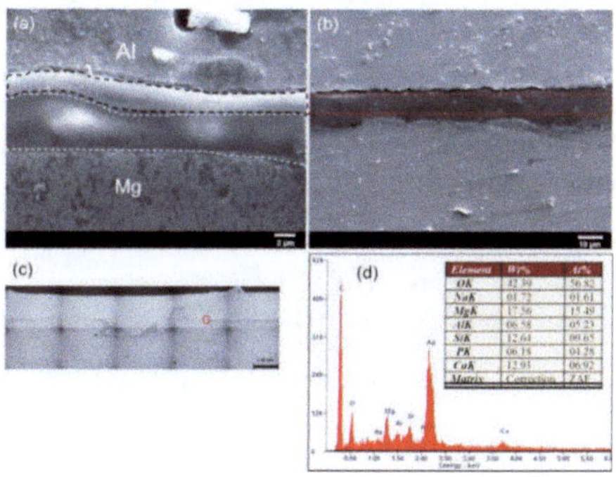

Figure 5. SEM images of the interface outside the stir zone, but under the tool shoulder, for welds made (a) without adhesive and (b) with adhesive. The examined location (b) is highlighted in weld cross section (c) and the corresponding EDX analysis result is plotted (d).

Figure 6 presents the SEM images and EDX analysis of the stir zone of a weld with adhesive. Figure 6a represents the stir zone on the retreating side. The red box in Figure 6a highlights the end of the cold lap feature, which extends into the stir zone. EDX analysis does not show any of those elements previously observed in the adhesive except for oxygen. This might suggest possible adhesive decomposition during welding or sufficient mixing between the upper and lower sheets. Epoxy adhesive decomposition has been reported during laser welding of Al and Mg alloys [16]. The high temperature in the weld zone and the vigorous stirring likely lead to adhesive decomposition in the current study. Figure 6b shows the very bottom of the stir zone on the advancing side. The noticeable amounts of Si, Ca, P and O from EDX elemental analysis indicate the presence of adhesive. During FSLW, the material in front of the tool is driven upward. The upward-moving material on the leading side then sweeps around the pin in the rotation direction. However, on the trailing side, the layer rotating with the tool is decelerated and gets deposited in the void in the wake of the pin. The adhesive material identified at the bottom of the stir zone on the advancing side might experience direct deposition and less mixing. The mutual presence of Al and Mg in both EDX analyses might be associated with the IMCs near the bi-material interface.

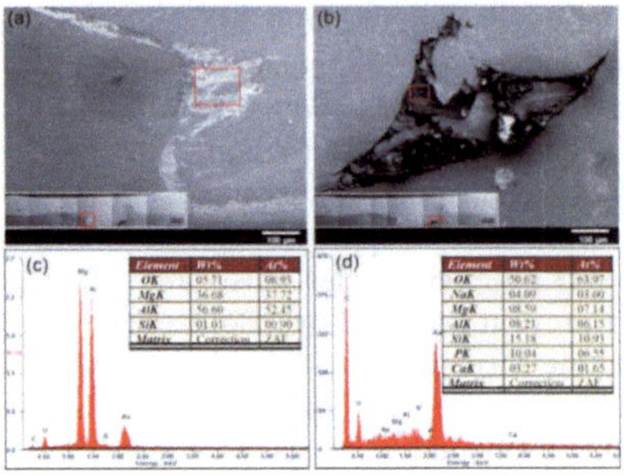

Figure 6. SEM images and EDX analyses of the stir zone on the (a)(c) retreating side and (b)(d) advancing side of a weld with adhesive.

The average adhesive thickness at the weld faying surface was observed to be ~18 μm for locations at least 6 mm away from the weld centerline, and remained constant for up to 12 mm away (Figure 7). The adhesive thickness drops significantly to about ~12 μm at locations about 4 mm away and becomes negligible when the distance is smaller than 3 mm. The reduced thickness and undetected adhesive could be a synergy of the forging pressure applied by the tool shoulder and adhesive decomposition discussed previously.

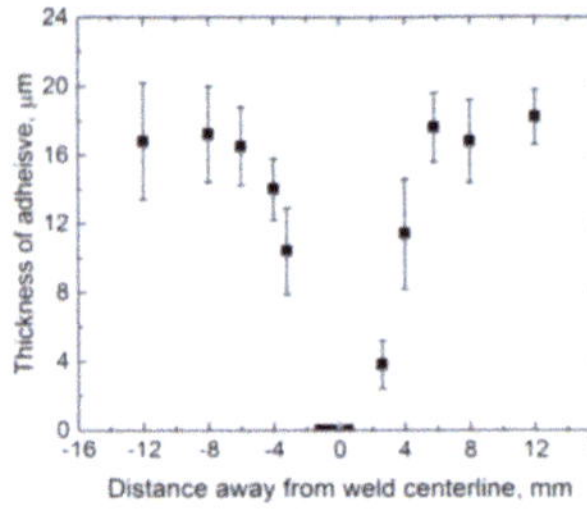

Figure 7. Adhesive thickness at the faying surfaces as a function of distance away from the weld centerline.

3.3 Lap-shear Strength

Lap-shear tests were performed in order to compare the strength of welds with three sets of welding conditions, i.e., without adhesive, with adhesive but without curing, and with adhesive as well as post-weld curing. At least four samples were tested for each welding condition. The typical load-displacement curves of welds tested are shown in Figure 8a. The differences are quite distinct. Compared to welds without adhesive, applying adhesive improved the lap-shear failure load as well as the total displacement before weld separation. Curing of welds made with adhesive further enhanced the failure load and displacement. As statistically summarized in Figure 8b, the average lap-shear failure load increased from 3.1 kN to 4.4 kN due to the application of adhesive and further to 6.3 kN after curing the adhesive. The load increment for welds with adhesive compared to those without could be due to curing of the adhesive adjacent to the weld by the frictional heat, as seen from the temperature profiles in Figure 4. An important conclusion from this higher load is that the presence of adhesive at the interface did not greatly reduce the integrity of FSLW, if at all. The post-weld curing further hardened the remaining adhesive not impacted by welding and led to a further load increment. Higher lap-shear failure loads for an addition of adhesive to friction stir spot welding and laser welding of AZ31 to AA5754 or AA6061 alloys have also been reported [15,20,21].

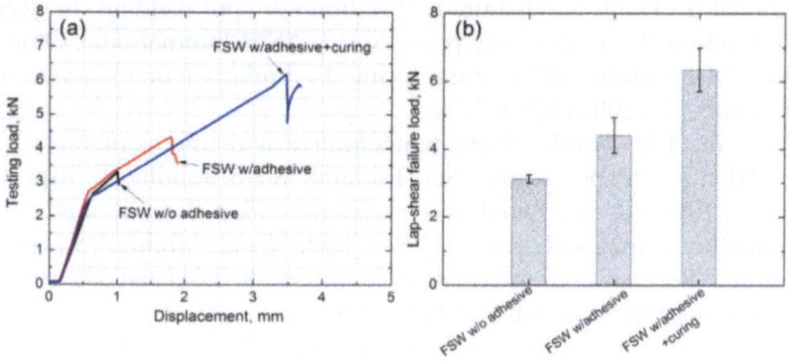

Figure 8. Lap-shear test data plot for FSLW of Al to Mg with/without introduction of adhesive and curing condition.

Conclusions

Dissimilar Al and Mg alloy sheets with adhesive at the interface were successfully friction stir linear welded. The introduction of adhesive improved the lap-shear failure load from 3.1 kN, for welds without adhesive, to 4.4 kN. The failure load was further enhanced to 6.3 kN after curing the welds with adhesive. The load increment for welds with adhesive was most likely due to the minimal curing associated with frictional heat of welding. Post-weld curing further hardened the remaining adhesive not impacted by heat from welding and led to a further load increment. Although the introduction of adhesive does not appear to influence the weld stir zone in terms of material mixing between the upper and lower sheet materials and intermetallic compound formation, the adhesive prevented the formation of intermetallic compounds at the faying surface adjacent to the stir zone and provided added bonding, which also contributed to the overall strength improvement.

Acknowledgments

The authors acknowledge the United States Automotive Materials Partnership Magnesium Front End R&D project for support. This material is based upon work supported by the Department of Energy, Office of Energy Efficiency and Renewable Energy Vehicle Technologies Program under Award Number No. DE-EE0005660. This report was prepared as an account of work sponsored by an agency of the United States Government. Neither the United States Government nor any agency thereof, nor any of their employees, makes any warranty, express or implied, or assumes any legal liability or responsibility for the accuracy, completeness, or usefulness of any information, apparatus, product, or process disclosed, or represents that its use would not infringe privately owned rights. Reference herein to any specific commercial product, process, or service by trade name, trademark, manufacturer, or otherwise does not necessarily constitute or imply its endorsement, recommendation, or favoring by the United States Government or any agency thereof. The views and opinions of authors expressed herein do not necessarily state or reflect those of the United States Government or any agency thereof.

References

[1] R.S. Mishra, Z.Y. Ma, "Friction stir welding and processing," Mater. Sci. Eng. R, 50 (2005) 1-78.

[2] H. Badarinarayan, F. Hunt, K. Okamoto, "Friction stir spot welding, In: R.S. Mishra, M.W. Mahoney (Eds.), Friction stir welding and processing", ASM International, Ohio, 2007, 235-272.

[3] Y.S. Sato et al., "Constitutional liquation during dissimilar friction stir welding of Al and Mg alloys," Scripta Mater., 50 (2004) 1233–1236.

[4] A. Gerlich, P. Su, T.H. North, "Peak temperatures and microstructures in aluminum and magnesium alloy friction stir spot welds," Sci. Technol. Weld. Joining, 10 (6) (2005) 647-652.

[5] Y.S. Sato et al., "Effect of interfacial microstructure on lap shear strength of friction stir spot weld of aluminum alloy to magnesium alloy," Sci. Tech. Weld. Joining, 15 (4) (2010) 319-324.

[6] D.-H. Choi et al., "Formation of intermetallic compounds in Al and Mg alloy interface during friction stir spot welding," Intermetallics, 19 (2) (2011) 125-130.

[7] S.H. Chowdhury et al., "Lap shear strength and fatigue life of friction stir spot welded AZ31 magnesium and 5754 aluminum alloys," Mater. Sci. Eng. A, 556 (2012) 500-509.

[8] C.-Y. Lee et al., "Dissimilar friction stir spot welding of low carbon steel and Al-Mg alloy by formation of IMCs," Sci. Technol. Weld. Joining, 14 (3) (2009) 216-220.

[9] S. Bozzi et al., "Intermetallic compounds in Al 6016/IF-steel friction stir spot welds," Mater. Sci. Eng. A, 527 (16-17) (2010) 4505-4509.

[10] Y.C. Chen, K. Nakata, "Effect of tool geometry on microstructure and mechanical properties of friction stir lap welded magnesium alloy and steel," Mater. Design, 30 (9) (2009) 3913-3919.

[11] T. Liyanage et al., "Joint formation in dissimilar Al alloy/steel and Mg alloy/steel friction stir spot welds," Sci. Technol. Weld. Joining 14 (6) (2009) 500-508.

[12] M.A. Mofid, A. Abdollah-zadeh, F. Malek Ghaini, "The effect of water cooling during dissimilar friction stir welding of Al alloy to Mg alloy," Mater. Design, 36 (2012) 161-167.

[13] H.Y. Wang, L.M. Liu, F. Liu, "The characterization investigation of laser-arc-adhesive hybrid welding of Mg to Al joint using Ni interlayer," Mater. Design, 50 (2013) 463-466.

[14] W.S. Chang et al., "Microstructure and mechanical properties of hybrid laser-friction stir welding between AA6061-T6 Al alloy and AZ31 Mg alloy", J. Mater. Sci. Technol. 2011, 27, 199–204.

[15] S.H. Chowdhury et al., "Lap shear strength and fatigue behavior of friction stir spot welded dissimilar magnesium-to-aluminum joints with adhesive," Mater. Sci. Eng. A, 562 (2013) 53-60.

[16] H.Y. Wang, L.M. Liu, "Analysis of the influence of adhesive in laser weld bonded joints," Inter. J. of Adhesion & Adhesive, 52 (2014) 77-81.

[17] W. Yuan, H. Badarinarayan, H. Rao, "Friction stir welding of dissimilar aluminum and magnesium alloys for automotive sub-structure," The 10th Friction Stir Welding Symposium, Beijing China, May 20-22, 2014.

[18] H.M. Rao, W. Yuan, H. Badarinarayan, "Effect of process parameters on mechanical properties of friction stir spot welded magnesium to aluminum alloys," Mater. Design, DOI: 10.1016/j.matdes.2014.10.065.

[19] H.M. Rao et al., "Effect of process parameters on lap-shear strength of friction stir linear welded aluminum 6022-T4 to magnesium AM60B," to be submitted.

[20] H.Y. Wang et al., "Laser weld bonding of A6061Al alloy to AZ31B Mg alloy," Sci. Technol. Weld. Joining 12 (3) (2007) 261–265.

[22] L. Liu et al., "Microstructure characteristics and mechanical properties of laser weld bonding of magnesium alloy to aluminum alloy," J. Mater. Sci. 42 (2007) 565–572.

mm/min. A cylindrical threaded probe with a flat tip of 3 mm in diameter was used to perform all welding tests with a probe length of 3,1 mm. This probe was made of WC-Co material in order to minimize the wear at the tip of the tool. The shoulder was made in M42 tool steel and had a diameter of 12 mm. 3 different welded coupons were prepared by using 1, 2 or 3 welding passes, as shown in Table 1, so that the bonded area at the aluminum-steel interface was varied. From the welded coupons 4 specimens being 25 mm wide were extracted, 3 for shear tensile testing and 1 for microstructural characterization and microhardness analysis.

Specimens for microstructural characterization were prepared following standard metallographic procedures and etched using a chemical reagent based on a NaOH solution with the aim of revealing the microstructural features present at the aluminum-steel interfaces. An Olympus GX51 light optical microscope and a Zeiss Ultraplus FEG-SEM equipped with a Oxford Instruments EDS system were used to investigate the microstructural features of the dissimilar joints. A Zwick Roell Z100 tensile testing machine was employed for shear tensile testing of FSW lap joint specimens. Clamps of the machine were displaced in order to avoid initial realignments and induce pure shear stresses. A Emcotest Durascan machine was used for microhardness testing. Continuous scans of microhardness indentations separated by 0,5 mm were performed in the aluminum sheet at a distance of 0,5 mm from the aluminum-steel interface. The scan measurements were performed in parallel to that interface as shown in the white lines in Figure 1 and Figure 2 (HV scan).

Table 1: Details of FSW coupons used in this investigation.

Weld Nº	Nº of FSW passes	Aprox. sample width (mm)	Aprox. bonded interface section (mm^2)
FSW 1pass	1	25	75
FSW 2passes	2	25	150
FSW 3passes	3	25	225

Results and discussion

Microstructural characterization

A microstructural characterization of the welded samples was carried out as shown in Figure 1 and Figure 2. In Figure 1 macro and microstructural features of the sample welded with 1 FSW pass are shown. No volumetric defects were found suggesting that sound welds were obtained with the welding parameters used for this test. A detailed analysis of the aluminum-steel interface showed the generation of relatively small sized waves of steel material generated as a consequence of the frictional contact between the tool and the steel surface. In addition to that, the forging effect produced by the trailing edge of the shoulder was found to be effective eliminating any potential wormhole or tunneling defect that could be expected in the advancing side of the weld [9]. A very clean aluminum-steel interface was found where evidences of intermetallic compound formation were not observed by optical microscopy.
The cross-section and main microstructural features of the sample welded with 3 FSW passes are shown in Figure 2. Overlapped nuggets can be observed where the central nugget corresponds to the 1st FSW pass, the nugget in the right to the 2nd FSW pass and the one in the left corresponds

Friction Stir Welding and Processing VIII
Edited by: Rajiv S. Mishra, Murray W. Mahoney, Yutaka Sato, and Yuri Hovanski
TMS (The Minerals, Metals & Materials Society), 2015

DISSIMILAR ALUMINUM-STEEL FSW LAP JOINTS

Egoitz Aldanondo[1], Ekaitz Arruti[1], Jorge Garagorri[1], Alberto Echeverria[1]

[1]IK4 LORTEK Centro de Investigación en Tecnologías de Unión;
Arranomendia kalea 4A, Ordizia, 20240, Spain

Keywords: Dissimilar, aluminum, steel, intermetallic compounds

Abstract

The properties of dissimilar aluminum-steel lap joints performed by FSW were studied in this investigation, using AA6082-T6 aluminum alloy and S355JR steel. The effect of the bonded area on the mechanical and microstructural properties of the joints was analyzed combining different numbers of overlap passes. Microstructural features were analyzed by optical and scanning electron microscopy; while mechanical performance was investigated in terms of microhardness and shear tensile testing. Different fracture modes of shear tensile specimens such as interfacial failure and aluminum HAZ failure were observed, being related to the bonded interface section. No volumetric defects were found in the metallographic examination. The formation of intermetallic compounds was analyzed by SEM. Encouraging results were obtained regarding the wear and durability of FSW tools using WC–Co as probe material.

Introduction

The increasing demands of fuel efficiency for energy saving and emission reduction purposes is prompting the use of complex material combinations, especially in the transport vehicle construction. One of those material combinations are dissimilar aluminum-steel joints which are employed in the automotive, shipbuilding or railway sectors. Therefore a number of joining technologies have been developed capable to produce this type of dissimilar joints [1-4], although the possibilities for further improvements are huge.
One of the technologies with potential to produce aluminum-steel dissimilar joints is Friction Stir Welding (FSW). Previous works have shown the potential of this and related techniques to produce dissimilar joints [4-10], based on its solid state nature and the possibilities to limit the generation of brittle intermetallic compounds that could reduce the quality of the joints.
The present study was focused to the development of dissimilar aluminum-steel joint using AA6082-T6 as aluminum alloy and S355JR as steel.

Experimental details

The materials employed in this investigation were a 3 mm thick AA6082-T6 aluminum alloy sheet and a 5 mm thick S355JR steel sheet. Lap joints were performed by placing the aluminum sheet on top of the steel sheet. 300 mm long and 100 mm wide coupons were used with an overlap distance of 70 mm for each welding condition.

FSW welding tests were carried out in the I-STIR PDS 4 FSW machine available at IK4 LORTEK research centre (Ordizia, Spain) which allows for a precise control of the main FSW process parameters in both position or force control. For the present study all tests were performed in force control using a rotational speed of 800 rpm and a travel speed of 305 to the 3rd FSW pass. No volumetric defects were found in any of the three nuggets suggesting a good selection of welding parameters and the suitability of performing several overlapped FSW passes to produce sound welded joints. No intermetallic compound formation could be appreciated by optical microscopy as the analyzed samples showed a very clean aluminum-steel interface. As well as in the microstructural analysis of the "FSW 1pass" sample, generation of steel waves were observed in each aluminum-steel interface region beneath the nuggets, suggesting that frictional contact between the FSW tool and the steel interface occurred in all FSW passes. Therefore it could be confirmed the bonded interface section in the "FSW 3passes" samples was approximately 3 times larger than in the "FSW 1pass" sample, as shown in Table 1.

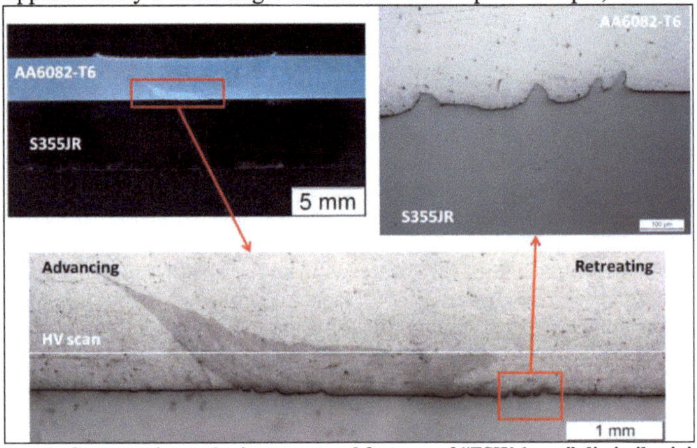

Figure 1: Cross-section and microstructural features of "FSW 1pass" dissimilar joint.

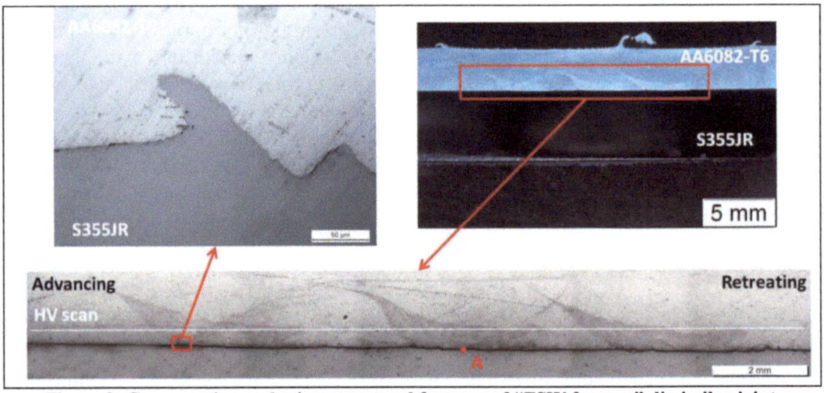

Figure 2: Cross-section and microstructural features of "FSW 3passes" dissimilar joint.

In order to study the joint formation mechanisms at the aluminum-steel interfaces in more detail, FSW samples were subjected to SEM observation and EDS analysis. Figure 3 shows representative results obtained in aluminum-steel interface regions where frictional contact between the FSW tool and the steel surface occurred (point A in Figure 2). A high magnification image taken at that point (left image in Figure 3) showed an incipient generation of an intermetallic compound layer in the aluminum-steel interface, although the presence of a continuous layer is not very clear. An element mapping performed by EDS analysis showed the presence of Al (middle image in Figure 3) and Fe (right image in Figure 3) elements in that region. Aluminum alloy in the top of the image and steel at the bottom could be clearly distinguished. In addition to that, a very thin layer (~1μm) where both Al and Fe atoms were present could be observed, suggesting the presence of a diffusion region (DR) and the generation of a thin, continuous intermetallic compound layer. This layer would be the main joint formation mechanism.

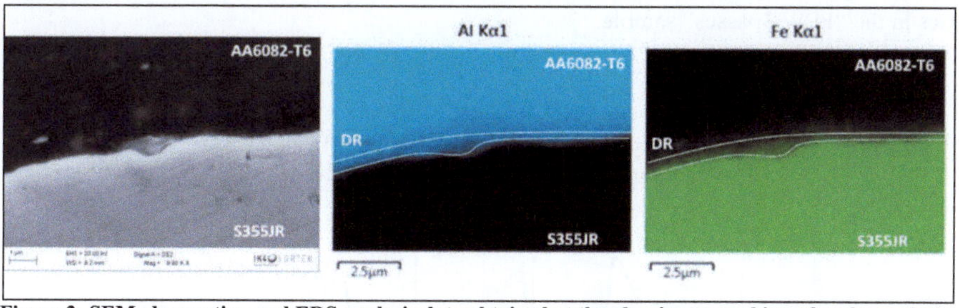

Figure 3: SEM observation and EDS analysis data obtained at the aluminum-steel interface beneath the nugget of the 1st FSW pass in the "FSW 3passes" sample (represented by point A in Figure 2).

Microhardness analysis

The results obtained in the microhardness measurements performed on the FSW samples "FSW 1pass" and "FSW 3passes" are shown in Figure 4 and Figure 5 respectively. The nugget region in close proximities to the aluminum-steel interface showed lowest hardness values of approximately 65HV0,5 in the "FSW 1pass" sample (Figure 4). A decrease of hardness in the nugget and especially in the HAZ is a well-known effect in FSW of precipitation hardenable aluminum alloys such us AA6082-T6. Such a pronounced decrease in the microhardness values in the nugget of "FSW 1pass" sample could be attributed to the high temperatures achieved in the frictional contact region of the FSW tool with the steel interface and the overaging effects produced in the aluminum material of its proximities.

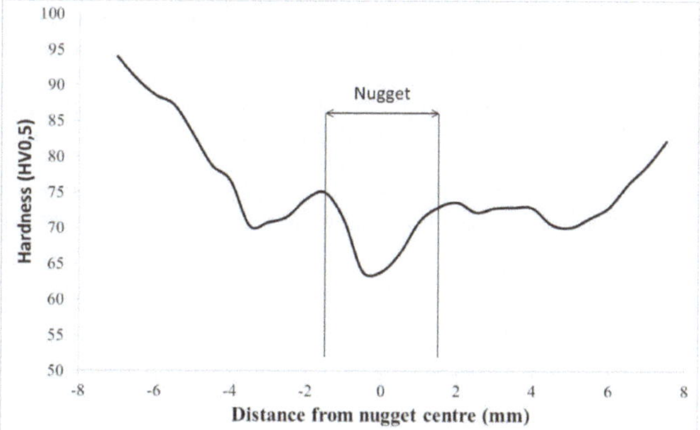

Figure 4: Microhardness scan at 0,5 mm from the Al-Fe interface in "FSW 1pass" sample.

Generally similar effects in the microhardness evolution could be observed in the "FSW 3passes" sample (Figure 5). Lowest values were measured in the nuggets at the proximities of the regions where frictional contact between the FSW tool and the steel surface took place. Due to the higher heat input suffered by the nugget produced in the 1st pass over the two nuggets in the sides, minimum microhardness values of ~55HV0,5 were measured in this region. The lower minimum microhardness value in sample "FSW 3passes" in comparison with "FSW 1pass" (65HV0,5 Vs 55HV0,5) was attributed to the larger overaging effect produced by the 3 FSW passes in the "FSW 3passes" sample.

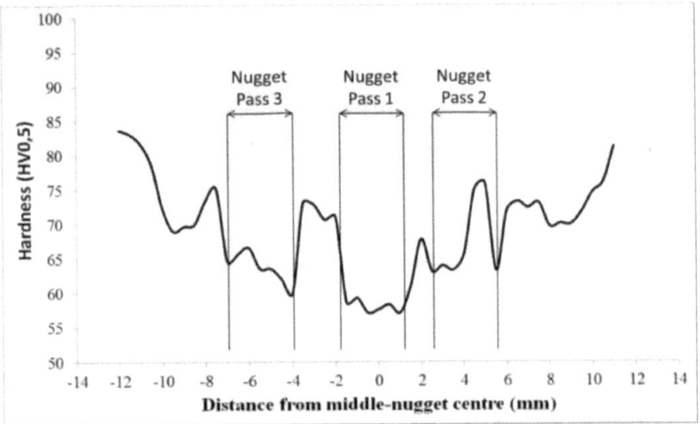

Figure 5: Microhardness scan at 0,5 mm from the Al-Fe interface in "FSW 3passes" sample.

<u>Shear tensile testing</u>

The mechanical strength of the dissimilar "FSW 1pass", "FSW 2passes" and "FSW 3passes" joints was evaluated by shear tensile tests and the results of those tests are shown in Figure 6. It could be observed that larger bonded interface sections achieved increasing the number of FSW passes result in higher strength values. Thus specimens obtained from the "FSW 1pass" sample showed shear strength values of approximately 460N/mm while "FSW 2passes" specimens showed approximately 650N/mm value levels. The fracture mode exhibited by all those specimens was a fracture mode A (Figure 6) corresponding to an interfacial failure as shown in Figure 7. Specimens extracted from sample "FSW 3passes" showed a different fracture mode (fracture mode B in Figure 6) which corresponded to a HAZ failure in AA6082-T6 material as shown in Figure 8. This type of failure was attributed to a higher strength of the Al-Fe interfacial joint than the AA6082-T6 HAZ strength so that shear strength values higher than 710N/mm were shown by the "FSW 3passes" sample. Thus, failure of the joints occurred at tensile strengths of ~235MPa.

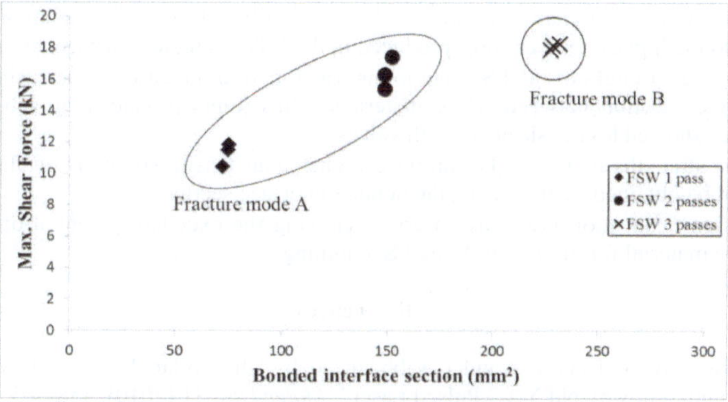

Figure 6: Shear tensile testing results for FSW dissimilar Al-Fe joints.

Figure 7: Interfacial failure in the Al-Fe interface (samples "FSW 1pass" and "FSW 2passes").

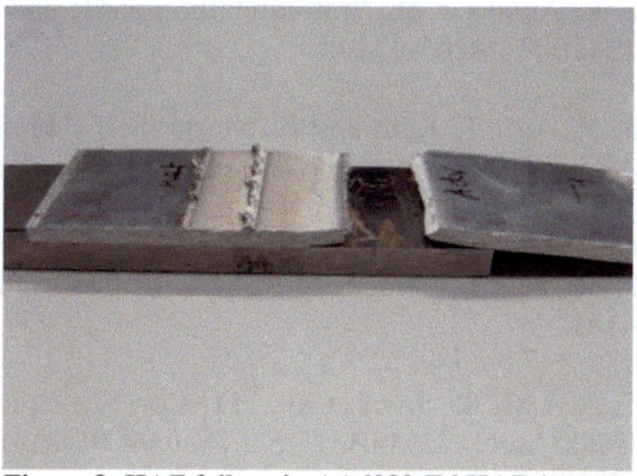

Figure 8: HAZ failure in AA6082-T6 HAZ (sample "FSW 3passes").

Tool wear behavior

Tool wear was evaluated using optical microscopy and an image treatment software. No significant wear of the employed FSW tool was observed after all FSW testing carried out in this study. Thus after approximately 5m of FSW length no wear was detected in the employed WC-Co probes showing the potential suitability of this material for continuous and repetitive FSW of dissimilar aluminum-steel joints. However further investigation is needed to evaluate the durability of these FSW tools more thoroughly.

Conclusions

Dissimilar FSW lap joints between AA6082-T6 aluminum alloy and S355JR steel were investigated in this work. The following conclusions could be obtained:

- Sound FSW lap joints were obtained by the used FSW parameters. No volumetric defects were found in the performed FSW joints using different numbers of FSW passes.
- The generation of a very thin intermetallic compound layer at the Al-Fe interface was found to be the main joint formation mechanism.
- The aluminum material of the nugget in the close proximities to the region where the frictional contact between the FSW tool and the steel surface takes place was found to

show the minimum hardness values. This was attributed to the overaging of that material due to the highest temperatures produced in the FSW tool-steel surface contact region.

- The shear strength of the FSW lap joints was found to increase by increasing the bonded interface section between Al-Fe materials. Thus joints produced by subsequent FSW passes showed higher shear strength values.
- Shear strengths of the Al-Fe interface higher than tensile strengths of aluminum HAZ could be obtained by increasing the bonded interface section.
- No significant tool wear was observed showing the excellent potential of WC-Co as a probe material for dissimilar Al-Fe FSW joining.

References

1. M.J. Torkamany, S. Tahamtan and J. Sabbaghzadeh, "Dissimilar Welding of Carbon Steel to 5754 Aluminum Alloy by Nd:YAG Pulsed Laser", *Mater Des*, 31 (2010), 458-465.

2. L. Agudo et al., "Intermetallic Fe_xAl_y-phases in a Steel/Al Alloy Fusion Weld", *J Mater Sci*, 42 (2007), 4205-4214.

3. Y. Abe, T. Kato and K. Mori, "Self-Piercing Riveting of High Tensile Strength Steel and Al Alloy Sheets using Conventional Rivet/Die", *J Mater Process Technol* 209 (2009) 3914-3922.

4. R. Qiu, C. Iwamoto and S. Satonaka. "Interfacial Microstructure and Strength of Steel/Al Alloy Joints Welded by RSW with Cover Plate", *J Mater Process Technol* 209 (2009), 4186-4193.

5. A.A.M. da Silva et al., "Friction Stir Spot Welding of AA1050 Al Alloy and Hot Stamped Boron Steel (22MnB5)", *Sci Technol Weld Joining*, DOI 10.1179/ 136217110X12785889549462 (2010).

6. E. Aldanondo et al., "Mechanical and microstructural investigation of dissimilar resistance and friction stir spot welds in AA5754-H22 and AA6082-T6 Al alloys and 22MnB5 hot-stamped boron steel", *FSW&P VI*, 140th TMS annual meeting and exhibition, San Diego, USA, March 2011.

7. A. Elrefaey et al., "Characterization of Aluminium/Steel Lap Joint by friction Stir Welding", *J. Mater. Eng. Perform.* (2005):14; 10-17.

8. Y.C. Chen et al., "Effect of the Surface State of Steel on the Microstructure and Mechanical Properties of Dissimilar Metal Lap Joints of Al and Steel by Friction Stir Welding", *Metall. Mater. Trans. A* (2008):39; 1985-1992.

9. E. Aldanondo et al., "Dissimilar joining of AA1050 aluminum alloy and hot-stamped boron steels using friction stir welding and friction stir spot welding", *Proceedings of the 8th International Friction Stir Welding Symposium*, Timmerdorfer Strand, Germany, 18-20 May 2010.

10. M. Kokabi et al. "Grouth kinetics of Al-Fe intermetallic compounds during annealing treatment of friction stir lap welds", *Materials characterization*, (2014) 90, 121-126.

11. H. Das et al., “Study on the formation and characterization of the intermetallics in Friction Stir Welding of aluminum alloy to coated steel lap joint”, *Metall. Mater. Trans. A-Phys. Metall. Mater. Sci.*, (2014) 45A (11): 5098-5106.

Friction Stir Welding and Processing VIII
Edited by: Rajiv S. Mishra, Murray W. Mahoney, Yutaka Sato, and Yuri Hovanski
TMS (The Minerals, Metals & Materials Society), 2015

FATIGUE BEHAVIOR OF FRICTION STIR LINEAR WELDED DISSIMILAR ALUMINUM-TO-MAGNESIUM ALLOYS

H.M. Rao[1], J.B. Jordon[1], W. Yuan[2], B. Ghaffari[3], X. Su[3], A.K. Khosrovaneh[4], and Y.L. Lee[5]

[1]Department of Mechanical Engineering, The University of Alabama, AL 35487, USA
[2]Hitachi America Ltd., R&D Division, Automotive Products Research Laboratory, Farmington Hills, MI 48335, USA
[3]Ford Research and Innovation Center, Ford Motor Company, Dearborn, MI 48124, USA
[4]General Motors Research and Development Center, Warren, MI 48090
[5]Chrysler Group LLC, MI 48326, USA

Keywords: Friction Stir Welding; Dissimilar, Fatigue

Abstract

In this paper, we present the results of fatigue testing and analysis of friction stir linear welded dissimilar aluminum-to-magnesium alloys in lap-shear configuration. The overlap linear welds were created by joining AA6022 aluminum alloy to AM60 magnesium alloy. In general, the test data exhibited significant scatter in the fatigue life results and the corresponding failure modes. In fact, observations from fractography analysis revealed two distinct modes of failure. In the first mode of failure observed, fracture occurred when the dominant fatigue crack propagated into either the magnesium or aluminum sheet in a kinked crack formation. Interestingly, fretting-like debris was observed at the initiation sites for this failure mode. In the second mode of failure observed, fracture occurred by interfacial weld separation. In this mode, fractography analysis suggests that the fatigue cracks initiated at weld defects and then propagated through the intermetallic phase.

Introduction

Although aluminum has been widely used, the use of magnesium in the automotive industry is hindered by the fact that common welding techniques are not favorable for mass production [1–3]. Over the years, research studies have established the friction stir welding (FSW) technique as an efficient methodology to weld magnesium alloys. Most research on FSW has been relatively limited to the joining of similar magnesium-to-magnesium alloys or aluminum-to-aluminum alloys. Previous studies have established the influence of welding process parameters on the micro and macro features of the weld and how these features affect the weld strength in a friction stir linear welded (FSLW) joint [4–14].

A key challenge in using dissimilar materials arises during the assembly and joining process. Applying heat and force during the welding process leads to the formation of brittle and structurally frail intermetallic compounds (IMCs). During the FSW of magnesium and aluminum alloys, constitutional liquation occurs in the weldment, which leads the formation of these IMCs ($Al_{12}Mg_{17}$, Al_3Mg_2) in the stir zone [15–24].

Most studies on FSW of dissimilar alloys have discussed the formation of IMCs and the corresponding effect on the static strength of the welds [16,18,21,22,25–28]. Typically, in real-world applications, these welded joints are subjected to fatigue loading and hence, it is imperative to study the fatigue behavior of these dissimilar welded joints. In one of the few existing studies on the fatigue of dissimilar FSW joints, Chowdhury et al. [29] found that the fatigue life of friction stir spot welded (FSSW) dissimilar magnesium-to-aluminum joints was much lower than the fatigue life of similar FSSW joints. The other work on the fatigue behavior of dissimilar friction stir welded joints is largely based on the joining of dissimilar aluminum-to-aluminum alloys [30–35]. Thus, an attempt has been made in this study to analyze the fatigue life properties of FSLW magnesium-to-aluminum alloys. As such, this is the first study to investigate the fatigue life properties and failure modes in dissimilar FSLW of AA6022-T4 to Mg alloys.

Materials and Experiments

Rolled AA6022-T4 aluminum alloy sheets of 1.5 mm in thickness were FSLW to cast AM60B magnesium alloy sheets of thickness 3.1 mm. The FSLW tool used to produce the lap-shear coupons used in this study is shown in Fig. 1a. The weld tool was made of standard tool steel (H13) and constituted a concave tool shoulder with a diameter of 12 mm and a smooth-surface triangular pin with an equivalent triangle diameter of 5.4 mm and a pin length of 2.4 mm. The aluminum sheet was stacked on top of the magnesium sheet in lap-shear configuration with an overlap area of 152.4 mm ×30 mm, as shown in Fig. 1b. Two stitch welds, each measuring 65 mm in length, were produced. Henceforth, these welds will be addressed as weld #1 (W1) and weld #2 (W2) throughout this paper. The stitch welds were produced in one single step, meaning, after producing W1, the tool would retract and traverse 15 mm in air and plunge back into the top aluminum sheet to produce W2. All the welds were produced at a tool rotation rate of 1500 rpm, a tool traverse speed of 75 mm/min, a tool shoulder plunge depth of 0.3 mm and the tool tilt angle was maintained at 2.5 degrees. The tool traversed from the right of the weld setup to the left, as indicated by the bold arrows in Fig. 1b; whereas the dotted arrows around W1 and W2 indicate the tool rotation direction (counterclockwise). This was to ensure the welds produced would have the advancing side (AS) at the free end of the coupon and the retreating side (RS) was on the loading side of the coupon, as shown in Fig. 1b and Fig. 1c. This particular overlap orientation with regards to the RS has been shown to give better static strength [5]. Only the lap-shear coupons without keyholes were studied for fatigue properties. The final dimensions of the lap-shear test coupons are shown in Fig. 1c. Since these welds could potentially be part of a larger component in an automobile, which would be subjected to painting and baking, the lap-shear test coupons were baked in a forced convection oven at 180° C for 30 minutes.

For the fatigue tests, the lap-shear coupons were cyclically tested on a MTS servo-hydraulic load frame in load control with a sinusoidal wave profile. The lap-shear coupons were all tested at a load ratio of 0.1 and below the average quasi-static ultimate tensile strength of the lap-shear coupons, which is approximately 3500 N. The load range tested for both weld W1 and W2 was between a maximum of 3150 N (90% of 3500 N) and a minimum of 1155 N (33% of 3500 N). The fatigue tests were conducted at a frequency of 20 Hz. Magnesium and aluminum shims were also employed on the lap-shear coupons during the testing (refer to Fig. 1(c)) to avoid additional bending moments and loads. The grip-to-grip distance was maintained at 63 mm for each lap-shear coupon.

For weld characterization, samples were sectioned through the center of the nugget and parallel to the loading direction. The samples were cold mounted in epoxy and were mechanically ground and fine polished using 0.05 μm alumina. Since two different materials were polished simultaneously, only de-ionized water was used throughout (or during) the polishing steps. In order to reveal the microstructure and IMCs, AM60B was etched using an acetic glycol solution (20 ml acetic acid, 1 ml HNO_3, 20 ml H_2O, and 60 ml ethylene glycol), while the AA6022-T4 was etched with a 20% NaOH solution (caustic etching). The microstructures of the welds were analyzed using a Keyence VHX-1000 digital optical microscope. Typical optical micrographs of the stir zone (SZ), thermomechanically-affected zone (TMAZ), heat-affected zone (HAZ) and base metal were obtained. Representative fatigued lap-shear coupons from W1 and W2 were also observed under the Keyence microscope to identify the failure modes and fracture surfaces. In addition, the fracture surfaces of fatigue-tested lap-shear coupons were examined under a Jeol 7000 scanning electron microscope (SEM) to investigate the fracture surface, crack propagation and obtain a chemical analysis using energy-dispersive X-ray spectroscopy (EDX).

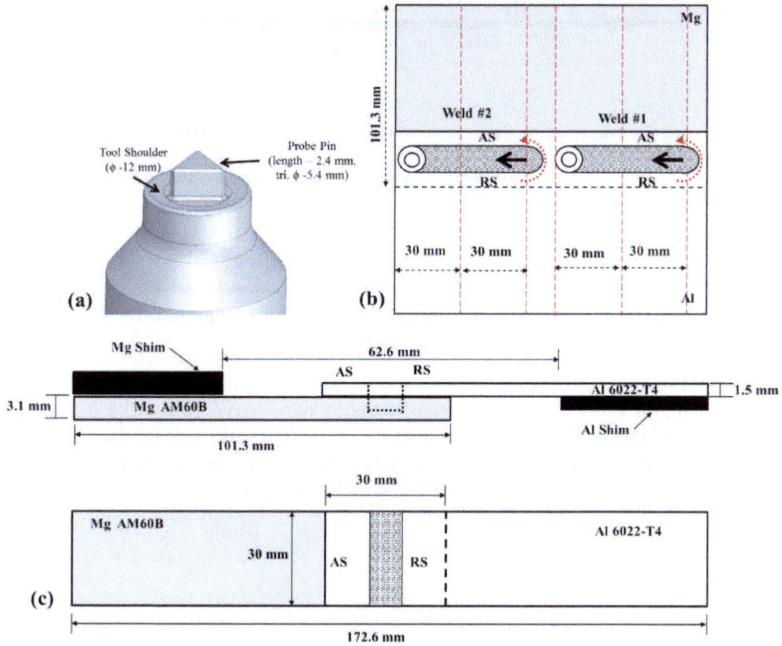

Figure 1: Representations of (a) the FSLW tool profile, (b) the FSLW methodology indicating the tool traverse direction, tool rotation direction and geometrical dimensions of the materials used, and (c) the geometrical dimensions of the FSLW test coupon after water-jet cutting.

Results and Discussion

Representative etched cross sections of the untested FSLW dissimilar aluminum-to-magnesium coupons are shown in Fig. 2, where W1 is shown in Fig. 2a and W2 is shown in Fig. 2b. The cross sections reveal a significant difference in the geometrical features of W1 and W2. The average RS effective sheet thickness (H1) in W1 and W2 were 11 mm and 10 mm, respectively. The average AS effective sheet thickness (H2) in W1 and W2 were 10.25 mm and 4.50 mm, respectively. The average weld bond width were 3.28 mm and 4.52 mm in W1 and W2, respectively. While we acknowledge that identical welding parameters produced significantly different weld features, the FSLW parameters used in this work where based on an extensive study that determined the optimal static weld strength for these dissimilar aluminum-to-magnesium coupons. In addition, we note that the observed differences in the weld features highlights the inherent difficulty in joining magnesium and aluminum alloys.

Regarding the microstructure of the FSLW coupons, the SZ, TMAZ, and HAZ were identified for both W1 and W2. Both the aluminum and magnesium sheets exhibited all three of these microstructural zones. Since there is no significant difference between the microstructures of W1 and W2, only a representative microstructure (Fig. 3) of W1 is presented in this paper. In Fig. 3a, the dotted line indicates the sharp transition of the SZ to TMAZ on the RS of the weld. A clear distinction between the fine grains of the SZ and the large recrystallized grains of the TMAZ is also clearly visible.

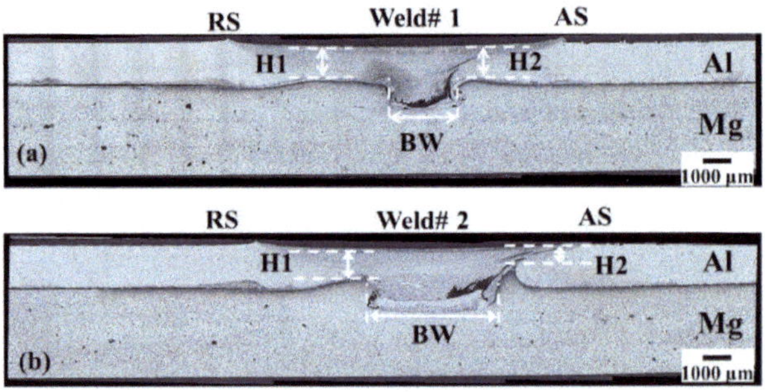

Figure 2: Cross-sectional views of the untested dissimilar FSLW aluminum-to-magnesium (a) weld 1 and (b) weld 2.

Due to the dynamically recrystallized grains in the SZ, fine grains were observed as shown in Fig. 3b. On the AS of the weld, the SZ transforms to the TMAZ across the black dotted lines shown in Fig. 3c. Regarding the microstructure in the magnesium side of the weld, only minor changes were observed due to the shallow plunge of the tool. Figure 3d shows the representative microstructural regions of SZ and TMAZ directly underneath the tool pin.

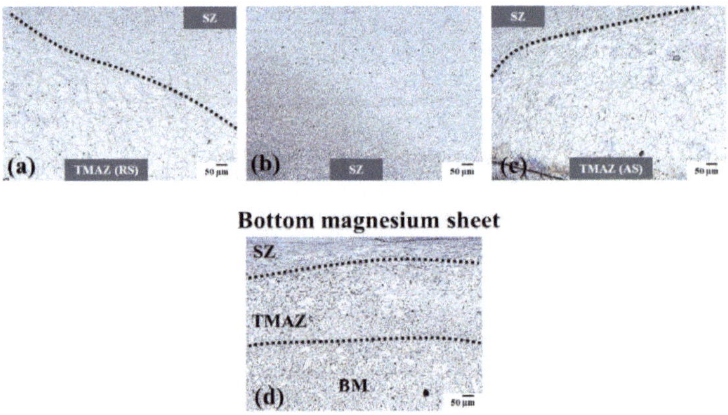

Figure 3: Microstructure of various regions in FSLW coupons (a) transition of SZ to TMAZ on RS of weld in aluminum sheet (b) SZ in the aluminum sheet (c) transition of SZ to TMAZ on AS of weld in aluminum sheet (d) TMAZ in the magnesium sheet (e) SZ, TMAZ and BM region in magnesium sheet.

The fatigue results for the lap-shear fatigue testing of W1 and W2 are shown in Fig. 4a. In general, the results indicate a wide scatter in fatigue life and no relationship between the fatigue life properties of welds W1 and W2 were observed. Overall, a comparison of the fatigue life results between W1 and W2 tested at the same load level showed a large disparity, which was more prominent at the higher fatigue loads (> 2500 N). As such, analysis of the test results suggests that any geometrical and/or microstructure differences between W1 and W2 were not significant enough to affect the number of cycles to failure.

Under fatigue testing, the dissimilar FSLW lap-shear welds were observed to fail in two main fracture modes. In the first fracture mode, the dominant crack was observed to initiate at the faying surface and then propagated either downwards through the magnesium sheet on the AS of the weld (Mode A) or propagated upward into the aluminum sheet on the RS of the weld (Mode C). In contrast, the other and more commonly observed failure mode occurred when the dominant crack propagated through the SZ of the weld, which resulted in an interfacial failure (Mode B). As shown in Fig. 4b, at low applied fatigue loads, the lap-shear welds were observed to fail in all three modes, but at higher applied fatigue loads, Mode B was the dominant failure mode.

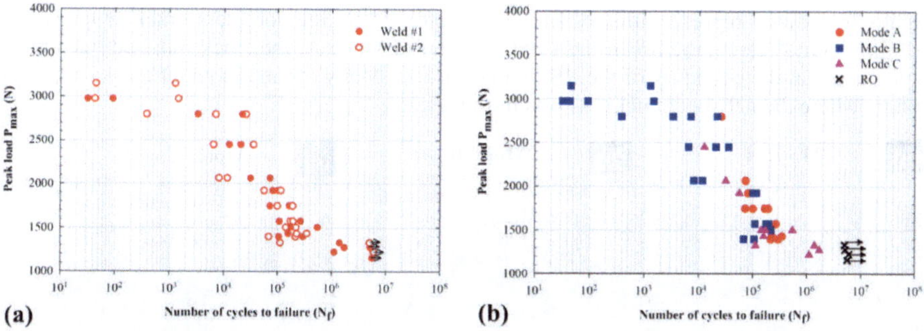

Figure 4: (a) Comparison of the experimental results of the lap-shear FSLW fatigue tests of weld #1 and weld #2. (b) Fatigue life results for both welds and the corresponding modes of failure. Mode A is the failure of the magnesium alloy sheet, Mode B is the interfacial failure, and Mode C is the aluminum alloy sheet failure.

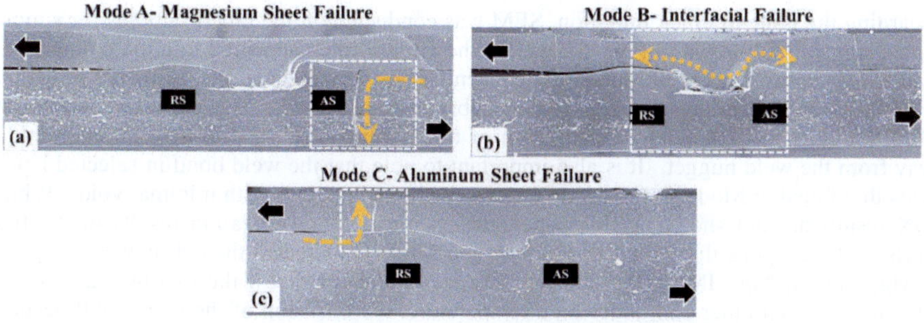

Figure 5. Cross-sectional views of representative failed coupons: (a) Mode A- magnesium sheet failure; (b) Mode B- interfacial shear failure; and (c) Mode C- aluminum sheet failure.

Figure 5 shows cross-sectional views of representative failure modes of the FSLW dissimilar lap-shear joints. The samples shown in Fig. 5 were tested to failure and then were sectioned through the center of the weld (parallel to the loading direction). Fig. 5a shows a representation of Mode A failure, where the dominant crack propagated into the magnesium sheet. It is important to note that the crack initiated at the boundary of the AS faying surface. The likely source of the fatigue crack initiation will be discussed later in this section. While the magnesium sheet contained measurable sized pores, it is unclear what influence the porosity had on the total number of cycles.

Figure 5b shows a representation of Mode B failure. In this failure mode, the crack initiated at the faying surface and propagated through the brittle intermetallic region. As shown in Fig. 4b, many of the lower bounds of the scatter are represented by Mode B. As shown in Fig. 5b, the fractured sample shows evidence of pores and/or channel defects. This suggests the presence of voids could likely lead to early crack initiation in the brittle intermetallic region and rapid propagation, resulting in a lower number of cycles compared to the other modes of failure.

Figure 5c shows a representation of Mode C failure, where the dominant crack propagated into the aluminum sheet on the RS of the weld. Similar to Mode A, the crack initiated outside of the weld nugget at the boundary of the RS faying surface. Many of the dominant cracks appeared to initiate where the net sheet thickness was the smallest.

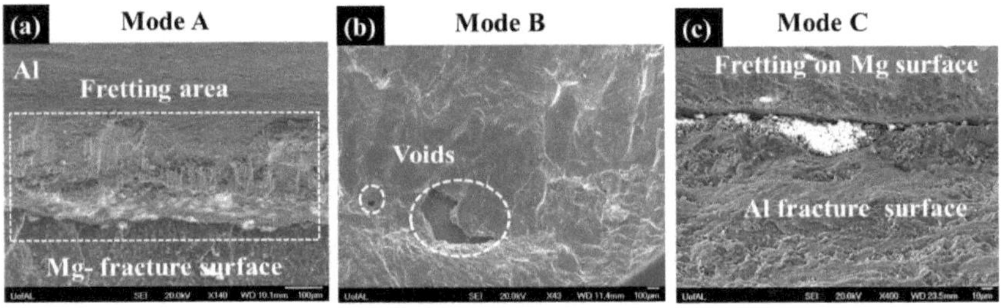

Figure 6. SEM images of (a) Fretting on Aluminum surface and magnesium fracture surface in Mode A (b) Voids in weld nugget in Mode B (c) Fretting on magnesium surface and aluminum fracture surface in Mode C.

Regarding the issue of crack initiation, SEM was conducted on the failed samples for a more in-depth analysis. As shown in Figs. 5a and 5c, the fatigue crack appeared to initiate outside of the weld and away from areas of high stress concentration. Figure 6a and 6c show the SEM images of Mode A and C failures, where fretting debris was observed. The presence of the fretting debris suggests that the relative oscillation of the matting sheets led to fatigue crack initiation away from the weld nugget. It is also important to note that the weld bond in selected lap-shear welds that failed in Mode A and Mode C possessed weld nuggets with minimal voids. While the EDX results are not shown in this paper, the chemical characterization results of the fretting debris in the coupons that failed in Mode A and Mode C indicated the debris was composed of Al_3Mg_2 and $Al_{12}Mg_{17}$ IMCs. Previous studies have indicated that if the two faying surfaces are held close to each other and under high temperature, the diffusion of the atoms will occur, thus forming the IMCs [15,24]. In this study to produce the lap-shear FSLW coupons, the top aluminum sheet was stacked on the bottom magnesium sheet and firmly clamped. The downward force of the tool shoulder and high temperature that is produced around the weld nugget leads to the formation of a thin layer of IMCs along the faying surface on either side of the weld nugget. The IMCs on these faying surfaces initiated the formation of the fretting debris. However, crack initiation for Mode B was likely due to voids in the weld due to poor material mixing. SEM of the fracture surface of welds that failed in mode B exhibited the presence of several large voids, as shown in Fig. 6b. The absence of any secondary cracks along the weldment (Fig. 5b) also strengthens the argument that the drop in load bearing area of the weld nugget due to the presence of voids resulted in premature crack initiation in the brittle IMC regions and thus ultimately lead to a shear overload along the interface.

Conclusions

The fatigue life and failure modes of dissimilar FSLW welded with the conditions and tooling described herein of aluminum-to-magnesium alloy joints were studied. The following conclusions were made for these welds:

- FSLW of the dissimilar lap-shear coupons in this study exhibited a wide scatter in the fatigue life results.
- Under fatigue loading of the coupons described herein, distinct failure modes were observed. At higher applied fatigue loads (>2500 N) the weld coupons failed due to in-plane shear along the IMCs in the stir zone. At lower fatigue loads, the weld coupons typically failed when the dominant crack propagated either into the aluminum or magnesium sheet away from the stir zone.
- The in-plane shear failure mode was also observed at fatigue loads below 2500 N. However, this mode of failure was observed in sections of the welds described herein that exhibited inferior material mixture in the stir zone, which resulted in voids and thereby reduced the number of cycles to initiate a fatigue crack.
- It is possible to achieve acceptable fatigue performance from dissimilar FSLW aluminum-to-magnesium joints, provided good material mixing is obtained in the stir zone.

Acknowledgments

This material is based upon work supported by the Department of Energy, National Energy Technology Laboratory under Award Number No. DE-EE0005660. This report was prepared as an account of work sponsored by an agency of the United States Government. Neither the United States Government nor any agency thereof, nor any of their employees, makes any warranty, express or implied, or assumes any legal liability or responsibility for the accuracy, completeness, or usefulness of any information, apparatus, product, or process disclosed, or represents that its use would not infringe privately owned rights. Reference herein to any specific commercial product, process, or service by trade name, trademark, manufacturer, or otherwise does not necessarily constitute or imply its endorsement, recommendation, or favoring by the United States Government or any agency thereof. The views and opinions of authors expressed herein do not necessarily state or reflect those of the United States Government or any agency thereof. Such support does not constitute an endorsement by the Department of Energy of the work or the views expressed herein. The authors thank the U.S. Automotive Materials Partnership (USAMP) and the U.S. Department of Energy for their support of this work. This work utilized resources owned and maintained by the Central Analytical Facility, which is supported by The University of Alabama.

References

[1] G.S. Cole, A.M. Sherman, Mater. Charact. (1995) 3.
[2] P.. Thornton, A.R. Krause, R.G. Davies, Weld. J. 75 (1996) 101s.

[3] A. Gean, S.A. Westgate, J.C. Kucza, J.C. Ehrstrom, Weld. Journal- New York 78 (1999) 80s.
[4] Q. Yang, X. Li, K. Chen, Y.J. Shi, Mater. Sci. Eng. A 528 (2011) 2463.
[5] W. Yuan, B. Carlson, R. Verma, R. Szymanski, Sci. Technol. Weld. Join. 17 (2012) 375.
[6] G. Buffa, G. Campanile, L. Fratini, a. Prisco, Mater. Sci. Eng. A 519 (2009) 19.
[7] N. Afrin, D.L.L. Chen, X. Cao, M. Jahazi, Mater. Sci. Eng. A 472 (2008) 179.
[8] W. Xunhong, W. Kuaishe, Mater. Sci. Eng. A 431 (2006) 114.
[9] M.K. Yadava, R.S. Mishra, Y.L. Chen, B. Carlson, G.J. Grant, Sci. Technol. Weld. Join. 15 (2010) 70.
[10] S.H. Chowdhury, D.L. Chen, S.D. Bhole, X. Cao, P. Wanjara, Metall. Mater. Trans. A 44 (2012) 41.
[11] S.M. Chowdhury, D.L. Chen, S.D. Bhole, X. Cao, Mater. Sci. Eng. A 527 (2010) 6064.
[12] H.. Rao, J.. Jordon, M.. Barkey, Y.. Guo, X. Su, H. Badarinarayan, Mater. Sci. Eng. A 564 (2013) 369.
[13] H.M. Rao, R.I. Rodriguez, J.B. Jordon, M.E. Barkey, Y.B. Guo, H. Badarinarayan, W. Yuan, Mater. Des. 56 (2014) 750.
[14] R.I. Rodriguez, J.B. Jordon, H.M. Rao, H. Badarinarayan, W. Yuan, H. El Kadiri, P.G. Allison, Mater. Sci. Eng. A 618 (2014) 637.
[15] D.-H. Choi, B.-W. Ahn, C.-Y. Lee, Y.-M. Yeon, K. Song, S.-B. Jung, Intermetallics 19 (2011) 125.
[16] N. Yamamoto, J. Liao, S. Watanabe, K. Nakata, Mater. Trans. 50 (2009) 2833.
[17] A.. McLean, G.L.F. Powell, I.H. Brown, V.M. Linton, Sci. Technol. Weld. Join. 8 (2003) 462.
[18] Y.C. Chen, K. Nakata, Scr. Mater. 58 (2008) 433.
[19] Y.J. Kwon, I. Shigematsu, N. Saito, Mater. Lett. 62 (2008) 3827.
[20] V. Firouzdor, S. Kou, Metall. Mater. Trans. A 41 (2010) 3238.
[21] S. Hirano, K. Okamoto, M. Doi, H. Okamura, M. Inagaki, Y. Aono, Q. J. Japan Weld. Soc. 21 (2003) 539.
[22] A. Gerlich, P. Su, T.H. North, G.J. Bendzsak, in:, Mater. Forum, 2005, pp. 290–294.
[23] S. Kou, Weld. J. 88 (2009) 213.
[24] Y.S. Sato, S.H.C. Park, M. Michiuchi, H. Kokawa, Scr. Mater. 50 (2004) 1233.
[25] A. Kostka, R.. Coelho, J. dos Santos, A.. Pyzalla, Scr. Mater. 60 (2009) 953.
[26] X.J. Cao, M. Jahazi, Mater. Sci. Forum 638-642 (2010) 3661.
[27] K. Kimapong, T. Watanabe, Mater. Trans. 46 (2005) 2211.
[28] Y.C. Chen, K. Nakata, Mater. Des. 30 (2009) 3913.
[29] S.H. Chowdhury, D.L. Chen, S.D. Bhole, X. Cao, P. Wanjara, Mater. Sci. Eng. A 556 (2012) 500.
[30] P. Cavaliere, F. Panella, J. Mater. Process. Technol. 206 (2008) 249.
[31] P. Cavaliere, R. Nobile, F.W. Panella, a. Squillace, Int. J. Mach. Tools Manuf. 46 (2006) 588.
[32] V. Tran, J. Pan, T. Pan, Int. J. Fatigue 30 (2008) 2175.
[33] a. Scialpi, M. De Giorgi, L. a. C. De Filippis, R. Nobile, F.W. Panella, Mater. Des. 29 (2008) 928.
[34] H. Lombard, D.G. Hattingh, a. Steuwer, M.N. James, Eng. Fract. Mech. 75 (2008) 341.
[35] Y. Uematsu, Y. Tozaki, K. Tokaji, N. Nakamura, 40 (2008) 138.
[36] J.B. Jordon, M.F. Horstemeyer, S.R. Daniewicz, H. Badarinarayan, J. Grantham, J. Eng. Mater. Technol. 132 (2010) 041008.
[37] D.-A. Wang, C.-H. Chen, J. Mater. Process. Technol. 209 (2009) 367.

Friction Stir Welding and Processing VIII
Edited by: Rajiv S. Mishra, Murray W. Mahoney, Yutaka Sato, and Yuri Hovanski
TMS (The Minerals, Metals & Materials Society), 2015

FRICTION STIR LAP WELDING OF ALUMINUM - POLYMER USING SCRIBE TECHNOLOGY

Piyush Upadhyay, Yuri Hovanski, Leonard S. Fifield and Kevin L. Simmons

Pacific Northwest National Laboratory,
902 Battelle Boulevard Richland, WA 99354, USA

Keywords: Friction Stir Scribe; Polymer to Metal Joining; Nylon; HDPE; CFRP

Abstract

Friction Stir Scribe (FSS) technology is a relatively new variant of Friction Stir Welding (FSW) which enables lap joining of dissimilar material with very different melting points and different high temperature flow behaviors. The cutter scribe attached at the tip of FSW tool pin effectively cuts the high melting point material such that a mechanically interlocking feature is created between the dissimilar materials. The geometric shape of this interlocking feature determines the shear strength attained by the lap joint. This work presents first use of scribe technology in joining polymers to aluminum alloy. Details of the several runs of scribe welding performed in lap joining of ~3.175mm thick polymers including high density polyethylene (HDPE), filled and unfilled polyamide 6,6 (Nylon 66) to 2mm thick AA5182 are presented. The effect of scribe geometry and length on weld interlocking features is presented along with lap shear strength evaluations.

1. Background and Introduction

Polymer / metal structures are being increasingly sought-after in the automotive industry primarily due to cost and weight savings associated with the use of polymers in resulting components. There is a significant interest in incorporating composite materials such as Carbon-Fiber Reinforced Polymers (CFRP) and Glass Fiber Reinforced Polymers (GFRP) in multi-material structures with metals like aluminum for light weight designs. However, due to large differences in material properties, joining low melting point polymers to metals has been problematic and is a primary concern in development of hybrid structures.

Joining methods available for this purpose include mechanical fastening, adhesive bonding and welding. Mechanical joining consists of fasteners like bolts and rivets that increase overall mass of the assembly. Although mechanical joining offers flexibility of disassembly for repair, stress concentration points around the holes for rivets can be sites for crack initiation and propagation. Adhesives are known to have high bonding strength but suffer from drawbacks like heat sensitivity in bond lines, strict surface preparation requirements, variation in adhesion characteristics among different material combinations and long assembly line time.

Development of a robust method of joining polymers to metals may provide solutions to the issues stated above. Over the years, several successful demonstration of joining polymer to polymer using FSW in various modifications has been made. Recognizing that a standard aluminum style rotating tool shoulder will impart unnecessarily greater heat into the weld region causing seizures, Nelson et. al. proposed and patented a modified FSW tool consisting of a "hot shoe" instead of FSW shoulder [1,2]. The "hot shoe" provided both the forging force and uniform heating, while the rotating pin plasticized and fused the butted sheets together. The group demonstrated joint efficiency almost equal to that of the base material in the case of HDPE and acrylonitrile butadiene styrene (ABS) polymers. Similarly Sciapli et al. have proposed a new variant of FSW for joining polymers together "using a blade and a shoulder which vibrate with alternating motion parallel to the weld line"[3]. Promising results were obtained, however the process is limited to linear welds and short blade life and complicated tooling needs are some of its drawbacks. Aydin [4] used a conventional FSW tool with root side heating to produce void free welds in polyethylene over a large range of welding and rotational speeds. Welded surfaces however were rough probably owing to excessive heating and deformation due to the use of rotating shoulder. Similar results have been reported for Nylon 6 by Panneerselvam and Lenin recently using a large threaded pin profile [5]. Payganesh et al[6] and Bozkurt et al.[7] have studied the influence of welding parameters on HDPE and glass fiber filled polypropylene to show that higher rotational speed and lower welding speed welds made with larger tool tilt provide better weld properties.

All of these efforts provide useful information for polymer to polymer joining, however joining polymers to metals comes with its own sets of challenges. For the polymer/aluminum pair, the high temperature required to flow aluminum will lead to undesired melting of the polymer resulting in deteriorated joints. Owing to this limitation very limited work has been reported in joining metals to polymers using FSW. One such study was carried out by Ratanathavorn [8] with a variety of polymers, wherein FSW lap welding was made with aluminum as the top sheet. Since aluminum was fully plasticized during welding, the polymer underneath invariably melted and vigorously mixed with aluminum matrix forming separate clusters of polymer and aluminum in the plasticized region. This resulted in large voids, flash and deterioration of joints resulting in poor joint strength.

Friction Stir Scribe (FSS) Technology used in the present work circumvents the need to fully plastically deform aluminum to join it with polymer thus avoiding above issues. Invented and developed at the Pacific Northwest National Laboratory (PNNL), FSS has enabled joining of materials with vastly different melting points by producing in-situ, rivet like fastening between the material interface with the use of a modified FSW pin [9]. (Detailed description of the Friction Stir Scribe Process can be found in Experimental Details Section). This technique has been successfully utilized in the past in lap joining AZ31 Magnesium to HSLA steel [10,11] demonstrating a joint efficiency of up to 90%. Here we present the first work to demonstrate the use of FSS technology to join metals to polymers.

2. Experimental Details

Friction Stir Scribe Technique: Tooling and Setup Figure 1 shows a schematic of the FSS tool and weld material setup. Although the setup is shown for the specific case of lap joining of polymer to aluminum, the idea of FSS is generic and can be applied to many material combinations. The materials to be joined are assembled in a lap configuration with the low melting point material on top. While the FSW pin plastically deforms the top sheet (low melting point material) in a similar fashion as conventional FSW, a cutter scribe attached axisymmetrically to the tip of the pin cuts the bottom sheet (high melting point material). As the tool rotates a rivet-like mechanically interlocking feature is created at the interface. Note that there is a predetermined gap maintained between the tip of the FSW pin and the joint interface to avoid any contact between the FSW pin and the bottom sheet. It is only the cutting scribe that contacts the bottom sheet.

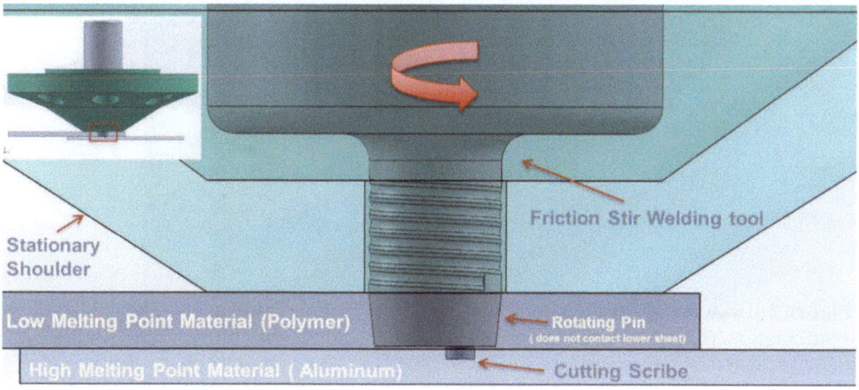

Figure 1: Schematic of Friction Stir Scribe (FSS) welding setup. The full view of the stationary shoulder is shown in the inset. Note that it is only the cutting scribe that contacts the bottom sheet. The rotating pin has predetermined clearance from the interface.

This particular FSS setup was used with a stationary shoulder in order minimize frictional heating at the surface of the polymer and thereby improve the weld quality and surface roughness of the welded plaque. Note that FSS welding can be used with either a rotating or stationary shoulder setup. The stationary shoulder was fabricated from stock steel in the shape of a frustum (See Figure 2) that consists of a recess hole for FSS pin to extend through into the work piece. The recess hole in the stationary shoulder piece was slightly larger than the FSS pin diameter to provide clearance and avoid any contact between the rotating tool and stationary shoulder.

The FSW tool pin was equipped with a shallow thread in the direction of rotation so that the plasticized polymer matrix escaping the weld region will be forced back downward. A larger flange piece above the stationary shoulder was installed for the purpose of instrumentation like in-situ temperature control and inert gas delivery for future application.

Figure 2: Stationary shoulder setup attached to the Friction Stir Welding System at PNNL (Top). Polymer to aluminum lap weld clamped down and ready for the weld pass. (Bottom)

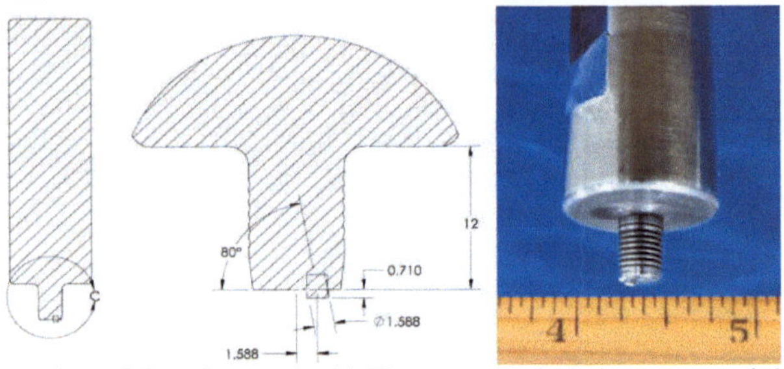

Figure 3: Drawing of the FSW tool with Tungsten Carbide insert at straight and tilted orientation. (Left). Picture of a tool used for the welding with tilted scribe (Right)

A simple cylindrical FSW tool (See **Figure 3**) was used for this preliminary study. The feature that makes FSS welding possible is the tungsten carbide (WC) insert that was installed halfway between the center and circumference of the FSW tip. Note that two types of scribe orientations were used to make joints: straight scribe (aligned with the tool axis) and a 10° degree, outward tilted scribe. Electron discharge machining was used to make an offset hole in the FSW tip. The WC tip was press fitted into the hole and cut to required length. Tool dimensions are tabulated in Table 1.

Table 1: Tool Dimensions used, mm

Stationary shoulder Diameter	Pin Diameter	Pin Length	Scribe Height
25.4	7.3	2.75-3.15	0.71-0.91

Weld Runs and Characterization

A series of FSS welds were performed to lap join aluminum alloys with a variety of polymers including unfilled polyamide (6,6) (Nylon), High Density Polypropylene (HDPE), 50wt% Long carbon fiber reinforced nylon (LCF50/PA66) and 40wt% Long glass fiber reinforced Nylon (LGF40/PA66). Polymer plaques with thickness ranging from 2.9-3.9 mm and AA5182 sheets of 2mm were used in this study. As shown in Figure 2, the plates were securely clamped to ensure intimate contact between the two surfaces and avoid any sheet lift-up during the process. The overlap between the top and the bottom sheet was 50mm. The process parameters used are shown in Table 2.

Table 2: List of Process Parameter

Tool Rotation	Travel Speed	Tool Tilt
1950 RPM	1000 mm/min	1°

Joints were produced on a high precision FSW machine at PNNL. The FSW system can measure several process responses in real time including tool forces in all three directions, tool torque and position. After welding, the joints were transversely cut into 12.5mm wide samples using water jet milling. Conventional cutting methods were not used to avoid any deformation to the samples during the cutting process. Select weld cross-sections were then ground and polished for optical macro/micrographs. A series of unguided lap shear tensile tests were performed using a universal test frame.

3. Results and Discussions

Representative lap joints made between HDPE and aluminum, and LCF 50/PA66 (CFRP) and aluminum are shown in Figure 4. After successive experimentation a tool rotation speed of 1950 RPM, travelling at the traverse rate of 1m/min yielded an acceptable weld quality and surface finish. Since a stationary shoulder was used, the extent of the deformation in the polymer is limited to the dimension of FSW pin only. The stationary shoulder imparted necessary forging pressure without adding any additional heat to the surface of the polymer; thus avoiding non-uniform through thickness heating, which can lead to overheating and ejection of plasticized material from the weld region. In the case of unfilled polymers, some change in the color of the unfilled polymer is observed which can be attributed to inclusion of aluminum chips into the polymer matrix.

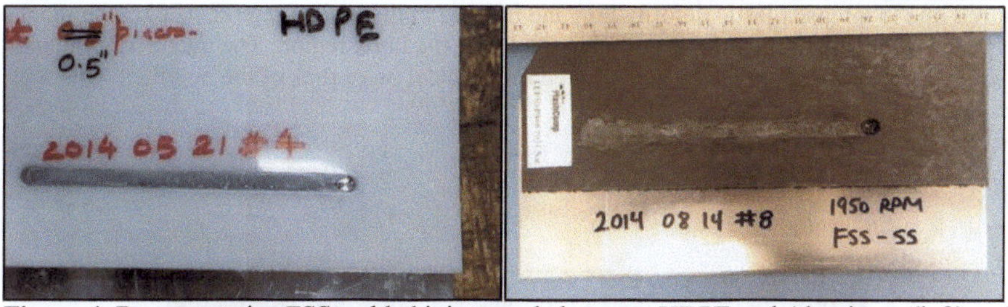

Figure 4: Representative FSS welded joints made between HDPE and Aluminum (left) and LCF40 and Aluminum (right)

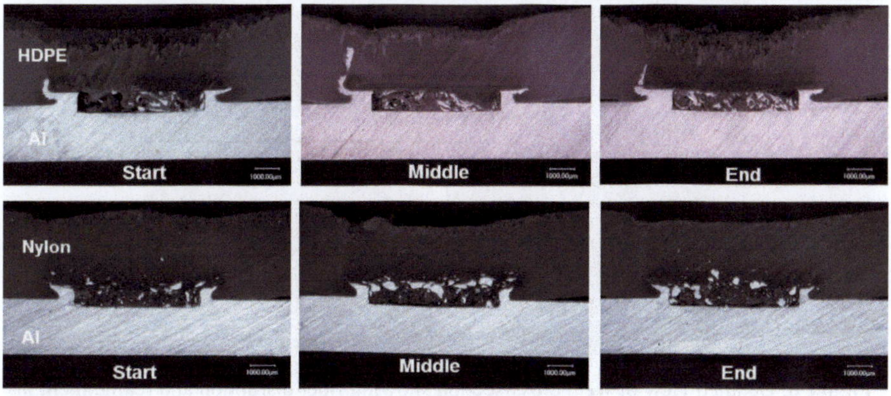

Figure 5: Transverse weld cross-section revealing the FSS joint between HDPE and aluminum (top) and Nylon 66 and aluminum (bottom). The cross-sections were obtained from near the start, middle and end of the each 250mm long joint

Cross-sections that reveal the scribe joint region between polymer and aluminum for unfilled nylon and HDPE are shown in Figure 5. The cross-sections in the figure were obtained from near the start, middle and the end of the weld. The scribe machined out a trough in the bottom aluminum sheet right underneath the FSW pin, thus creating two symmetrical hook-like features embedded to the top polymer sheet. The trough is back-filled with the polymer material and some residual aluminum. For a given joint, the hook features appears to be fairly uniform along the length. This feature is very similar to what was observed in the past for FSS joining of Mg-Zn coated Steel [10]. This hook like interlocking feature can be understood as an *in situ*, continuous and permanent rivet-like fastening. Note that the scribe has engaged in aluminum to a greater depth in the case of HDPE than in Nylon. This has led to different hook geometry in the two cases despite being welded with same set of welding parameters. This difference may be attributed to the fact that the Nylon plaque was marginally thicker than the HDPE plaque (3.18mm vs. 3.10mm). Since the FSW machine controls to the Z- position from the surface of the plaque, the tool plunged deeper into aluminum sheet for the same commanded plunge depth in HDPE. This is illustrative of the criticality of plunge depth and scribe height in the FSS process.

Figure 6 shows detailed macro/microstructure of a representative FSS joint between CFRP and aluminum made with tilted scribe. Significant tearing and reorientation of carbon fiber has occurred due to intense plastic deformation and heating in the nugget region. The realignment of fibers is predominant at the nugget boundary and near the hook region. The quantification and realignment of carbon fiber may have affected mechanical properties of the polymer. This will be explored in subsequent work.

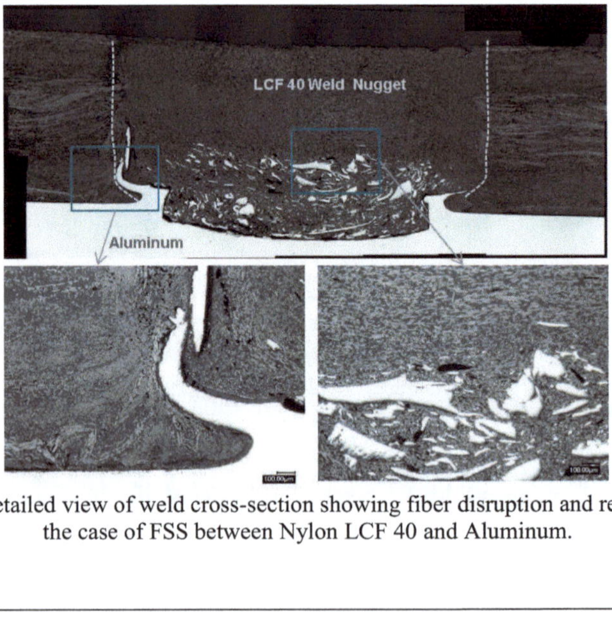

Figure 6: Detailed view of weld cross-section showing fiber disruption and realignment in the case of FSS between Nylon LCF 40 and Aluminum.

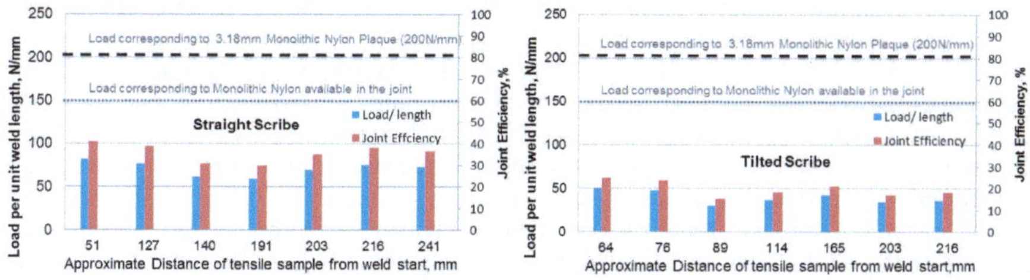

Figure 7: Load per unit joint length (right Y-axis) and Joint Efficiency (left Y-axis) plotted against the location of sample along the length of the joint for unfilled Nylon-66 to Aluminum FSS joint. The datasets for joints made with straight and tilted scribe are shown.

For each case several samples from different regions of the joints were tensile tested. Shims were added to the grips to avoid introduction of any bending stress in the samples. Since the precise area over which bonding occurs in FSS lap joint is not known, it is difficult to assign strength value (load per unit area) to the joint. For this reason load bearing capacity per unit length of the joint is obtained by dividing the tensile load at failure by the length of the weld in the test sample. Based on the calculated load per unit joint length, "Joint Efficiency" was then calculated in the basis of monolithic polymer strength. Figure 7 shows the tensile test data set for the unfilled Nylon case. Both load per unit weld length and joint efficiency are plotted against the location of sample for the joints made with straight and tilted scribe. The load carrying capacity of the joints appears to be fairly consistent along the length of the joint, which corroborates with the earlier observation of consistent hook geometry. In the case of straight scribe the joint efficiency ranges from 30-40%. Note that this joint efficiency was calculated on the basis of a full thickness monolithic Nylon plaque shown by the thick bottom line in the graph. If the reduced thickness of polymer available in the joint is considered, joint efficiency calculated (Not shown in the graph) is close to 50%. The Figure on the right shows the corresponding dataset for joints made with tilted scribe. The load carrying capacity is clearly reduced; revealing a full thickness joint efficiency of only between 15-30%.

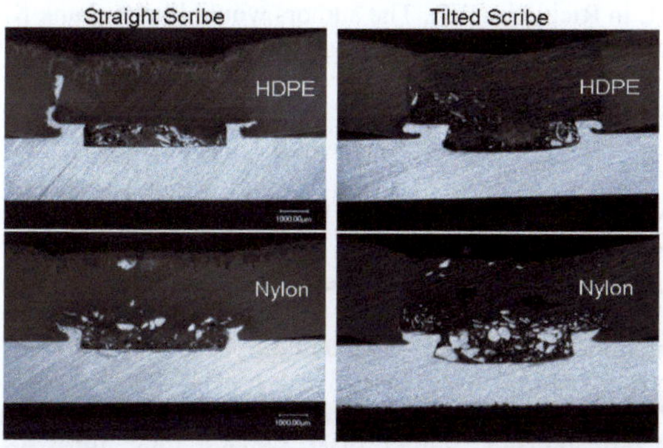

Figure 8: Comparison between FSS joint cross-section among straight and tilted scribe for HDPE and Nylon-66 showing difference in hook morphology

This difference in hook geometry between straight and tilted scribe can be seen in Figure 8. The outward tilted scribe resulted in a machined trough feature in aluminum inclined towards the polymer matrix as anticipated. It is intuitive that this would result in additional barrier in separation of the hook during tensile tests. However for both the HDPE and Nylon it appears that the upper hook region is flattened and a relatively small extent of polymer matrix is "hooked" compared to straight scribe joints (left). This might be the reason for lower load bearing capacity observed in the tilted scribe joints. The load bearing capacity of tilted scribe perhaps will increase and might surpass straight scribe if the flattening of the hooks can be avoided. This flattening of the hook might be avoided by increasing the effective length of the scribe. The effectiveness of the scribe in creating a hook morphology that results in stronger joints is dependent on several variables including plunge depth, clearance between the tip of the FSW pin and material interface and tool rotation and traverse speeds. More detailed studies are ongoing to understand the interactions and effects of these variables. The aim of the current work was to provide proof of concept and demonstrate the feasibility of FSS welding in the previously unexplored area of joining metals to polymers using a variant of Friction Stir Welding.

4. Conclusion

Friction Stir Scribe technology was successfully performed to lap join aluminum with various polymeric materials for the first time. The continuous rivet like *in situ* mechanical interlocking feature produced during the process was found to provide promising joint strength of nearly 50% of the bulk material in this preliminary effort. While fairly consistent along the length of the joint, several joining and tooling parameters can influence the effectiveness of the hook morphology. It is anticipated that this capability of readily joining metals to polymers will enable new avenues in the use of hybrid structures in conjunction with already existing joining techniques like adhesive bonding.

5. Acknowledgements

The Pacific Northwest National Laboratory is operated by Battelle Memorial Institute for the U.S. Department of Energy under contract DE-AC05-76RL01830. This work was made possible by Laboratory Directed Research and Development (LDRD) from PNNL. All research herein were performed at PNNL in Richland, WA. The authors would like to thank Karl Mattlin, Matthew P. Westman, Natalio Saenz , Shelley Carlson and Tyler Kafentzis for their assistance in material characterization and testing.

6. References

[1] Nelson, T. W., Sorenson, C. D., and Johns, C. J., 2004, "Friction stir welding of polymeric materials." US Patent 6811632 B2

[2] Strand, Seth, 2004, "Effects of Friction Stir Welding on Polymer Microstructure," Brigham Young University.

[3] Scialpi, A., Troughton, M., Andrews, S., and De Filippis, L. A. C., 2009, "Viblade^{TM}: friction stir welding for plastics," Weld. Int., **23**(11), pp. 846–855.

[4] Aydin, M., 2010, "Effects of Welding Parameters and Pre-Heating on the Friction Stir Welding of UHMW-Polyethylene," Polym.-Plast. Technol. Eng., **49**(6), pp. 595–601.

[5] Panneerselvam, K., and Lenin, K., 2014, "Joining of Nylon 6 plate by friction stir welding process using threaded pin profile," Mater. Des., **53**, pp. 302–307.

[6] Payganeh, G. H., Arab, N. M., Asl, Y. D., Ghasemi, F. A., and Boroujeni, M. S., 2011, "Effects of friction stir welding process parameters on appearance and strength of polypropylene composite welds," Int. J. Phys. Sci., **6**(19), pp. 4595–4601.
[7] Bozkurt, Y., 2012, "The optimization of friction stir welding process parameters to achieve maximum tensile strength in polyethylene sheets," Mater. Des., **35**, pp. 440–445.
[8] Ratanathavorn, W., 2012, Hybrid Joining of Aluminum to Thermoplastics with Friction Stir Welding.
[9] Hovanski, Y., Grant, G. J., Jana, S., and Mattlin, K. F., 2013, "Friction stir welding tool and process for welding dissimilar materials." US Patent US8434661 B2
[10] Jana, S., Hovanski, Y., and Grant, G. J., 2010, "Friction Stir Lap Welding of Magnesium Alloy to Steel: A Preliminary Investigation," Metall. Mater. Trans. A, **41**(12), pp. 3173–3182.
[11] Jana, S., and Hovanski, Y., 2012, "Fatigue behaviour of magnesium to steel dissimilar friction stir lap joints," Sci. Technol. Weld. Join., **17**(2), pp. 141–145.

Friction Stir Welding and Processing VIII
Edited by: Rajiv S. Mishra, Murray W. Mahoney, Yutaka Sato, and Yuri Hovanski
TMS (The Minerals, Metals & Materials Society), 2015

Friction Stir Scribe Welding of Dissimilar Aluminum to Steel Lap Joints

Todd Curtis[1,2], Christian Widener[1], Michael West[2], Bharat Jasthi[2], Yuri Hovanski[3], Blair Carlson[4], Robert Szymanski[4], and William Bane[1]

[1]Arbegast Advanced Materials Processing and Joining Laboratory
South Dakota School of Mines and Technology
501 East Saint Joseph St. Rapid City, SD 57701, USA

[2]Department of Materials and Metallurgical Engineering
South Dakota School of Mines and Technology
501 East Saint Joseph St. Rapid City, SD 57701, USA

[3]Pacific Northwest National Laboratory, Richland, WA 99352, USA

[4]General Motors R&D Center, Warren, MI 48090, USA

Keywords: Friction Stir Scribe Welding, Lap Joint, Dissimilar Joining

Abstract

The use of dissimilar material combinations such as aluminum to steel has been increasing in automobile and aerospace industries due to its potential for energy savings. Achieving an acceptable joint quality with fusion welding can be problematic due to the significant differences in physical and thermal properties between materials. One alternative to conventional fusion welding is friction stir scribe welding and because of the nature of the process, it can be a viable option for joining dissimilar metal combinations. The primary emphasis of this work is to investigate the feasibility of using FSS welding as a possible option for joining 1.0 mm thick 6022 aluminum to 0.7mm electro galvanized steel sheets in a dissimilar lap weld configuration. An H13 steel pin tool featuring a tungsten carbide scribe insert was used. An investigation on the optimum size of the scribe insert was conducted to evaluate the effects of microstructure and mechanical properties.

Introduction

As the demand for increasing fuel efficiencies in automotive technologies increases, the avenue of reducing vehicle mass poses potential energy savings and greenhouse gas emission reduction. The U.S. Department of Energy (DOE), United States Council for Automotive Research (USCAR), U.S. Department of Defense (DOD), and U.S. Department of Transportation (DOT) have shown interest in, and contributed investments to producing safer and lighter weight vehicles for the purpose of improved energy efficiencies. It is said that a 10% reduction in vehicle mass, can improve fuel efficiency by as much as 6 %[1]. To support the manufacturing of more fuel efficient vehicles, the use of stronger and lighter weight materials is necessary. However, stronger and lighter aren't always two categories met within a single material; therefore the need to use dissimilar material combinations arises. Typical materials which are combined include advanced steels, magnesium, and aluminum in a dissimilar combination such as magnesium to steel and aluminum to steel. Conventional joining techniques of dissimilar metals can sometimes be problematic and cost prohibitive. Fusion welding of aluminum to steel with an acceptable joint quality can be problematic due to the large differences in their melting points, very low solid solubility of iron in aluminum, and the formation of brittle intermetallics[2]. Because of large differences in their thermally related properties such as conductivity, thermal expansion, and heat capacity, high internal stress levels can develop in such joints. As a result fusion welding of this combination often suffers from brittle fracture in service. Because of the solid state nature of Friction Stir Welding (FSW) or more recently Friction Stir Scribe Welding (FSS) these methods lend themselves as a viable joining method for dissimilar combinations such as magnesium[3] or aluminum[4-8] to steel. Friction Stir Scribe Welding (FSS), developed at the Pacific Northwest National Laboratory[9], is a variation of the Friction Stir Welding process, in which a scribe is inserted into the pin tip. The scribe is typically fabricated from tungsten carbide, and acts as an additional tool geometry, by affecting the bottom sheet of material in a dissimilar material lap weld. The primary emphasis of this work was to investigate the use of Friction Stir Scribe Welding as a possible option for joining 1.0mm thick 6022-T4 Aluminum to 0.7mm thick low carbon electro galvanized steel sheets in a dissimilar lap weld configuration. Coupon level experiments were conducted to understand process parameters, metallurgical characterization, and mechanical properties.

Experimental

For this study 1.0mm thick 6022-T4 Aluminum was to be joined to 0.7mm thick low carbon electro galvanized steel sheets by Friction Stir Scribe Welding. 6022-T4 aluminum was placed on top of the 0.7mm galvanized steel to be joined in a lap configuration shown in Figure 1. The sheets were overlapped 25.4 mm (1 in) and the weld was made in the center of the overlapped section. Large sheets of material were cut into 100x300mm rectangular sheets for welding. Each panel was degreased with isopropyl alcohol prior to welding. Welds made for this study were performed using the MTS ISTIR 10 Friction Stir Welding system at the Arbegast Advanced Materials Processing and Joining Lab at the South Dakota School of Mines and Technology.

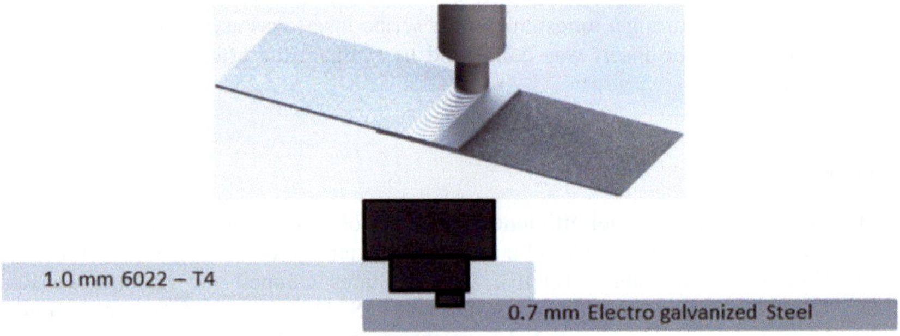

Figure 1: Lap Weld Configuration

Friction Stir Scribe (FSS) pin tools were machined from H13 tool steel featuring a flat scrolled shoulder and a tapered, tri-flat pin. The pin tool featured a 9.5 mm (0.375in) shoulder diameter and a pin length of 0.89 mm (0.035in); welding trials were conducted to vary the distance a 1 mm (0.0395 in) diameter tungsten carbide scribe protruded from the bottom of the pin. The protrusion distance was referred to in these trials as the percentage of the bottom steel sheet thickness the scribe penetrated into. Conditions consisted of either a 10 or 20 percent penetration of the scribe into the steel. The pin tool design is shown in Figure 2 and a summary of processing parameters is provided in Table 1. For placement of the tungsten carbide scribe, a hole was drilled into the end of the pin prior to heat treatment; after hardening the tool to ~48 HRC, a 1 mm (0.0395 in) diameter cylindrical tungsten carbide scribe was inserted into the end of the pin and ground to the appropriate length. A series of 3 conditioning welds was made in aluminum in order to "set" the scribe, based on the idea that small amounts of aluminum would extrude up into the interface between the tool and scribe and fill any possible loose fit up.

Figure 2: Friction Stir Scribe (FSS) Pin tool

Table 1: Pin Tool Processing Parameters

Control Mode	Position
Rotational Speed, RPM	1200
Travel Speed, IPM	40
Tool Tilt, Degree	0.5°
Scribe Length (From tip of pin)	10% - 0.15 mm (0.006 in) 20% - 0.23 mm (0.009 in)
Total Tool Penetration (Shoulder to Scribe Tip Length)	10% - 1.07 mm (0.042 in) 20% - 1.14 mm (0.045 in)

Welded panels were sectioned transverse to the weld joint using abrasive waterjet cutting for lap shear tensile evaluation and metallographic examination. Lap shear strength coupons of 38 mm (1.5 in) width, were excised from each welded panel in two different configurations in order to vary the direction of loading through either the advancing or retreating sides of the weld, as shown in Figure 3. Lap shear tests were carried out at room temperature and at a crosshead displacement speed of 2.5 mm/min (0.1in/min). The total failure load was reported as an average based on a minimum of 5 samples per weld condition. The as-polished joint interfaces were examined using optical microscopy for macro cross section evaluation as well as micro examination of detailed regions of the weld joint. Fractured cross sections as-tested lap shear coupons were also examined to determine the fracture based upon cross sectional features. Scanning electron microscope images were also taken to observe any mixing taking place between the dissimilar materials as well as any possible intermetallic formation that may have occurred.

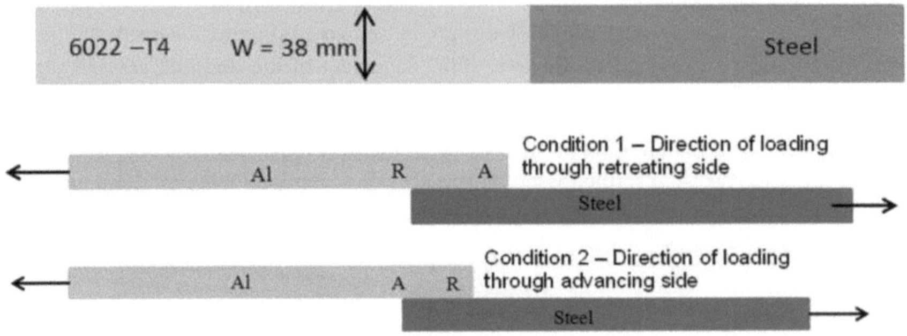

Figure 3: Lap Shear Coupon Configuration

Results and Discussion

The strategy of joining Al to steel using the FSS process is based upon the scribe just engaging the upper surface of the steel and upsetting a small amount of material which then creates a mechanical interlocking condition. For this process, the purpose of the scribe is to create an "upturn" or "hook" feature with the steel, while using the pin tool to plastically deform the aluminum and extrude it into the deformation created from the scribe. Figure 4 is a microstructure of a typical joint interface produced by the scribe pin tool. It can be observed that an interlocking hook feature is produced on the retreating side and another upturned build-up of steel material on the advancing side. The forging action produced by the general shape of the pin tool forces aluminum down into the deformation zone in the steel created by the scribe. It can also be observed that zinc from the galvanized coating on the steel mixes with the aluminum within the stir zone near the faying interface.

Figure 4: SEM representing typical cross section of dissimilar joint

The scribe insert for each tool, showed insignificant tool wear at the 10 and 20 percent conditions. An extreme case of 50 percent was attempted, but not reported as survival of the scribe was never observed with this length. Table 2 presents the average lap shear strengths corresponding to each tool condition and joint configuration. The higher lap shear strengths observed with the 20 percent scribe condition can be related to fewer defects when compared to the 10 percent scribe condition.

Table 2: Lap Shear Properties

Tool and Joint Configuration	Peak Load		Failure Location
	kN	lbf	
20% Scribe Length - Retreating Loaded	5.8 ± 0.6	1304.1 ± 124.2	Ret Side Hook
20% Scribe Length - Advancing Loaded	5.4 ± 0.7	1208.1 ± 163.7	Adv Side Upturn
10% Scribe Length - Retreating Loaded	4.2 ± 0.8	945.3 ± 174.9	Ret Side Hook
10% Scribe Length - Advancing Loaded	4.5 ± 1.1	1011.4 ± 254.0	Adv Upturn/Interface

As seen from Figure 5, for the 20 percent scribe condition, fractures were observed on either the retreating side at the interlocking hook feature or on the advancing side at the upturn build-up of steel, in respect of which side the direction of loading was placed. For the 10 percent scribe condition, fractures were observed emanating from both the advancing side up-turn as well as joint interfacial fracture. This suggests that the interface is weaker for welds made with the 10 percent scribe condition. It should be noted that the large standard deviation associated with the lap shear strengths may be attributed in the cross section along the length of a given weld. Such variation may be associated with variations in weld parameters such as tool depth, material flow around the tool, or interfacial material reactions between zinc and Al.

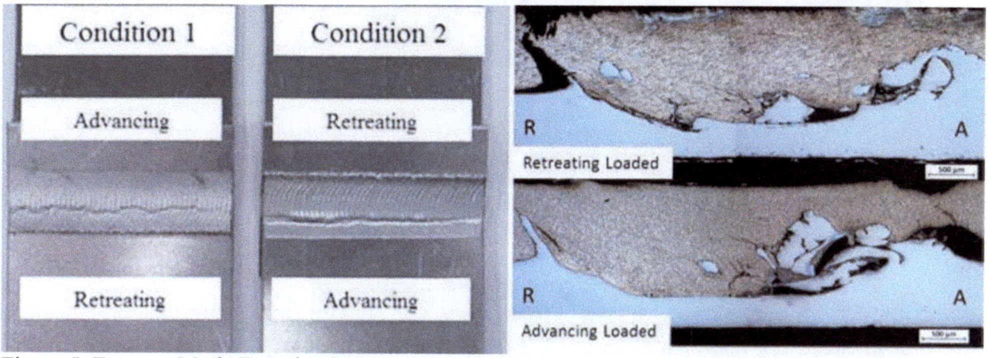

Figure 5: Fracture Mode Examination

Because the presence of the scribe created an interrupted cut condition based upon rotational positioning of the scribe, differences exist in the overall appearance of the cross section, as shown in Figure 6. This is observed between welds of similar process parameters or even within the same weld, with respect to where the specimen was sectioned. This difference can manifest itself in the size of upturn or hook created, or even the amount of "stray" steel fragments mixed within the aluminum.

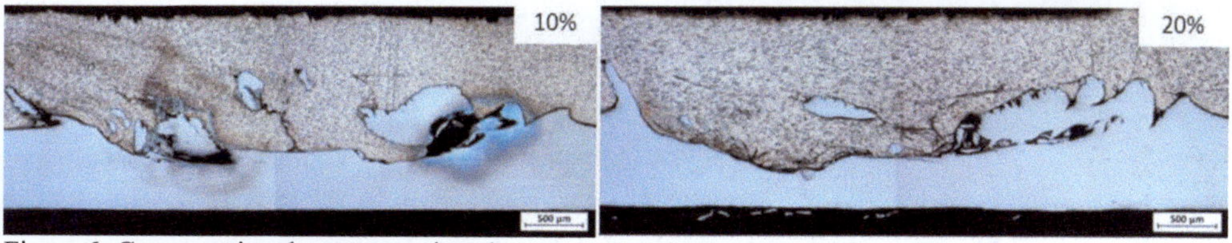

Figure 6: Cross sectional representations between 10 and 20 percent tool configurations

The steel has an electro galvanized coating which adds a layer of zinc to the steel as shown in Figure 7. Zinc coatings have long been added for corrosion protection[10]. This added protection to steel, however, also adds another variable to consider in analyzing the resultant weld microstructure. We see in Figure 4, that the zinc layer appears to stir up into the aluminum, possibly creating a "stringer" feature that could possibly inhibit the strength of the joint. In addition, Figure 8, details a region that exists at the interface between the aluminum, zinc, and steel materials. We see from this image that there is possible melting that is occurring with the zinc layer. Further temperature measurements during welding will be necessary to validate the possibility of the zinc melting. In addition to the interlocking features created from the pin tool itself, the possible melting of the zinc may contribute to a brazing condition to add to the strength of the joint. However, Jana and Hovanski reported that the behavior of the zinc coating when friction stir welding magnesium to steel, was displaced from the joined region of the weld and therefore it was unlikely that the zinc could create a brazing condition[3]. Further examination of the joint interface is necessary to correlate any possible melting and to identify any intermetallics formed at the interface.

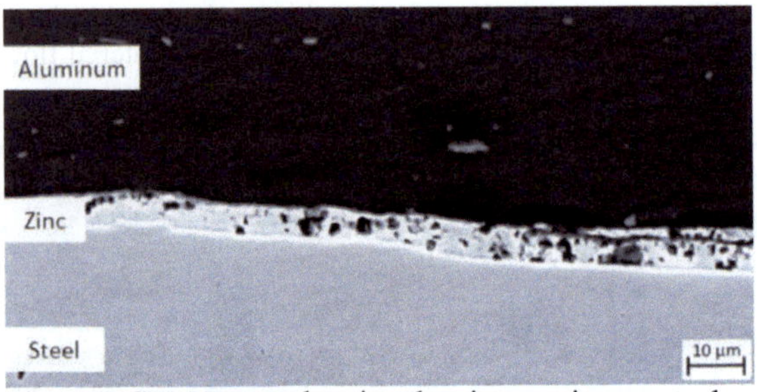

Figure 7: SEM Image showing the zinc coating on steel

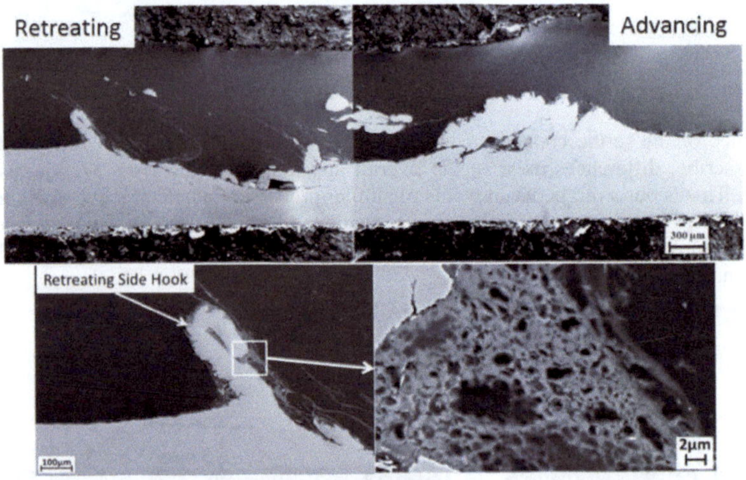

Figure 8: SEM images detailing the zinc layer at the interface features

Conclusions

Friction stir scribe (FSS) welding of dissimilar aluminum to steel lap joints was investigated and the resulting mechanical properties and metallurgical evaluation were reported. Structural joints achieving lap shear strengths as high as 5.8 kN (1300 lbs) are shown to be possible. Because of the nature of the processing conditions, a high level of variability exists in the strengths of the weld and cross sectional differences throughout the weld length can contribute to this variation in strength. Welds were tested to vary placement of loading in either the retreating or advancing sides of the weld; slight though not statistically significant differences were noted, and as such it is not possible to say placement of the advancing or retreating sides affected the overall strength of the joint. No intermetallics were identified at the interface between Al/Zn and Al/Fe. Temperature measurements during welding should be completed to investigate whether or not processing conditions achieve temperatures that allow the zinc to melt.

Acknowledgements

This work was supported by General Motors and was performed under the NSF I/UCRC Center for Friction Stir Processing.

References

1. Bullis, K. *Automakers Shed the Pounds to Meet Fuel Efficiency Standards*. 2013 [cited 2014 9/29/2014]; Available from: http://www.technologyreview.com/news/510611/automakers-shed-the-pounds-to-meet-fuel-efficiency-standards/.
2. Steve Shi, S.W., Stephen Mulligan. *Joining steel to aluminum and magnesium alloys....a new way forward.* TWI Bulletin June 2007; Available from: www.twi-global.com/.../2007/may-june/joining-steel-to-aluminium-and-magnesium-alloysa-new-way-forward/
3. Jana, S., Y. Hovanski, and G.J. Grant, *Friction stir lap welding of magnesium alloy to steel: a preliminary investigation.* Metallurgical and Materials Transactions A, 2010. **41**(12): p. 3173-3182.
4. Chen, Y., et al., *Role of zinc coat in friction stir lap welding Al and zinc coated steel.* Materials Science and Technology, 2008. **24**(1): p. 33-39.
5. Elrefaey, A., et al., *Characterization of aluminum/steel lap joint by friction stir welding.* Journal of materials engineering and performance, 2005. **14**(1): p. 10-17.
6. Movahedi, M., et al., *Effect of tool travel and rotation speeds on weld zone defects and joint strength of aluminium steel lap joints made by friction stir welding.* Science and Technology of Welding and Joining, 2012. **17**(2): p. 162-167.
7. Wei, Y., et al., *Effect of Tool Pin Insertion Depth on Friction Stir Lap Welding of Aluminum to Stainless Steel.* Journal of materials engineering and performance, 2013. **22**(10): p. 3005-3013.
8. Xiong, J., et al., *High strength lap joint of aluminium and stainless steels fabricated by friction stir welding with cutting pin.* Science and Technology of Welding and Joining, 2012. **17**(3): p. 196-201.
9. Hovanski, Y., et al., *Friction stir welding tool and process for welding dissimilar materials*, 2011, Google Patents.
10. Marder, A., *The metallurgy of zinc-coated steel.* Progress in materials science, 2000. **45**(3): p. 191-271.

Friction Stir Welding and Processing VIII
Edited by: Rajiv S. Mishra, Murray W. Mahoney, Yutaka Sato, and Yuri Hovanski
TMS (The Minerals, Metals & Materials Society), 2015

COATING DESIGN FOR CONTROLLING β PHASE IMC FORMATION IN DISSIMILAR Al-Mg METAL WELDING

Yin Wang, Li Wang, Joseph Robson, Basem Mohysen Al-Zubaidy, Philip Prangnell

School of Materials, University of Manchester, Manchester M13 9PL, UK

Keywords: Dissimilar joining; FSSW, Coatings; Intermetallic control

Abstract

The brittle intermetallic compound (IMC), β-Al_3Mg_2, grows very rapidly at the interface during dissimilar welding of Al to Mg. This currently prevents acceptable joints being produced between these two important light alloys. In the present study, the use of Al-alloy inter-layer coatings has been investigated to address this problem, for application to friction welding processes. Thermodynamic principles were employed to aid design of the coatings. Two types of coatings were selected: i) a Si-rich coating, which promoted the slower growing and more stable Mg_2Si phase; ii) an Al-Zn-rich solid solution coating, which replaced the undesired Al_3Mg_2 phase with a ternary $(Al,Zn)_{49}Mg_{32}$ intermetallic compound. In statically annealed diffusion couples both coatings were found to inhibit the growth of Al_3Mg_2. Refill friction stir spot welding (RFSSW) experiments also showed that both coatings were effective in reducing interfacial reaction and can improve the mechanical properties of Al - Mg dissimilar joints.

1. Introduction

Today's automobile industry is facing ever more demanding emissions targets together with consumer pressure for greater safety and performance [1]. The estimated effect of a reduction in vehicle mass on CO_2 emissions is ~ 4.7 g CO_2 per 100 kg's. When combined with the rising cost of fuel, this has aroused wide interest in replacing traditional car structural materials, like steel, with lighter metals, such as Al and Mg [2,3]. In addition, in the most efficient new vehicles designs are based on multi-material bodies, where the best attributes of different materials can be exploited in a single structure. This new direction of the automotive industry presents major challenges in joining dissimilar metals as wells as in joining metals to composites.

Direct joining of aluminum to magnesium alloys has been widely reported to be very difficult, with virtually all welding methods producing poor weld properties that suffer from low toughness [4].This is even true of lower temperature solid state joining techniques, like friction spot welding (FSW) [5]. The problem mainly lies in the rapid formation of several Al - Mg intermetallic compounds (IMCs) at elevated temperature, which can severely embrittle the joining interface. In particular, the β-Al_3Mg_2 phase is reported to have the highest growth rate among the common Al - Mg compounds [6,7] and is the phase in which cracks are most prone to initiate, owing to its low strength and highly brittle nature [8]. Therefore, the β-Al_3Mg_2 compound is the phase in the Al - Mg system that it is most desirable to avoid to improve joint performance. Previous studies have suggested that thin barrier coating inter-layers, such as PVD Mn, can be effective in controlling the Al-Mg interfacial reaction, but unfortunately such coatings are brittle themselves and can readily become damaged by friction welding processes [9]. Hence, better coatings are still required to improve the mechanical properties of Al - Mg dissimilar welds.

In this work, to address the intermetallic reaction problem, thermodynamic calculations have been used to select two ductile Al-alloy coatings; i) a Si-rich coating; and ii) an Al-Zn-rich solid solution coating. Both of these coatings were chosen as they are predicted to change the reaction path and avoid the formation of the β phase. Long-time annealing and refill friction stir spot welding (RFSSW) experiments were carried out in each case to study the intermetallic growth behavior in both static and welding conditions.

2. Experimental

An AZ31 Mg alloy and several different Al-alloy sheets were used in this research, the compositions of which are shown in Table I. All the sheets had a similar thickness of around 1 mm. To test the concept of a Si-rich coating, a commercial Al sheet was used, produced by the 'Fusion™' process, which had an AA6016 core with a 0.1 mm thick ~ 10% Si brazing alloy (AA4045) cladding layer. To investigate the concept of an Al-Zn-rich solid solution coating, sheets of the correct predicted composition were prepared by casting and hot rolling, and then homogenized at 460°C. The effect of the two coating materials on the growth behavior of the intermetallic compounds was compared by using isothermal heat treatments on diffusion couples made with the same AZ31 Mg alloy. The samples were produced by first lightly welding them together using a 2kW Sonobond ultrasonic welder, with a very short weld time of 0.2 sec. Owing to their different melting points, the couple produced between the Al-Si alloy and Mg sheet was heat treated at 400°C, whereas the Al-Zn alloy and Mg couple was annealed at 360°C, before for microstructure analysis. The Refil-Friction Stir Spot welding (RFSSW) technique was adopted to test the performance of the coating materials under previously optimized welding condition: rotation speed of 2500 rpm, welding duration 1 sec, with the AZ 31 Mg alloy placed on the top, and a plunge depth of 1 mm. Baseline uncoated AA6111 - AZ31 samples were also prepared under the same conditions for comparison purposes.

Design of the Al-Si and Al-Zn coatings was aided by the CALPHAD software PANDAT, with the database PanAl2008, and the diffusion simulation package DICTRA. Imaging of the joint interfaces was carried out with an FEI Quanta 650 SEM. Phase identification of the intermetallics was performed by high resolution EBSD with a step size of 0.1 μm, and TEM micro-diffraction on sections removed from the interface by FIB milling. Lap shear tests were carried out on an Instron testing machine to characterize the joining performance of the friction welds.

Table I. Compositions of the Al and Mg alloys employed this work (wt. %)

	Si	Zn	Fe	Cu	Mn	Ti	Cr	Ni	Mg	Al
AA6016	1-1.5	0.2	0.5	0.2	0.2	0.15	0.1	-	0.25-0.6	Balance
AA4045	9.88	-	0.15	-	-	0.02	-	-	-	Balance
Al-Zn alloy	-	52	-	-	-	-	-	-	-	Balance
AA6111	0.63	-	0.05	0.75	0.05	0.06	0.06	-	0.75	Balance
AZ31	0.017	1.03	0.0045	0.0036	0.2	-	-	0.0003	Balance	2.51

3. Results

3.1 Interfacial Characterization of the Si-Rich Coating

The Mg_2Si phase has a low density, high hardness and has high stability in both Al and Mg and has previously been investigated as a candidate reinforcement for forming in-situ composites in both Al and Mg matrices [10,11]. Thermodynamic equilibrium calculations of the Al-Mg-Si system confirmed that the Mg_2Si should be preferably formed over any of the other possible Al-Mg IMC binary phases [12]. Therefore, a Si rich coating was predicted to be a strong candidate for changing the normal reaction path seen in Al - Mg dissimilar welds away from the undesirable rapidly-growing β-Al_3Mg_2 phase.

The typical microstructure of the AA4045 Al-Si clad layer on the Al sheet used in this study, prior to welding, is illustrated in Fig.1. Small globular silicon particles of around 2-4 μm in diameter can be seen to be distributed uniformly in the coating layer, which is about 100 μm thick. The cladding layer also contains some Fe rich primary particles.

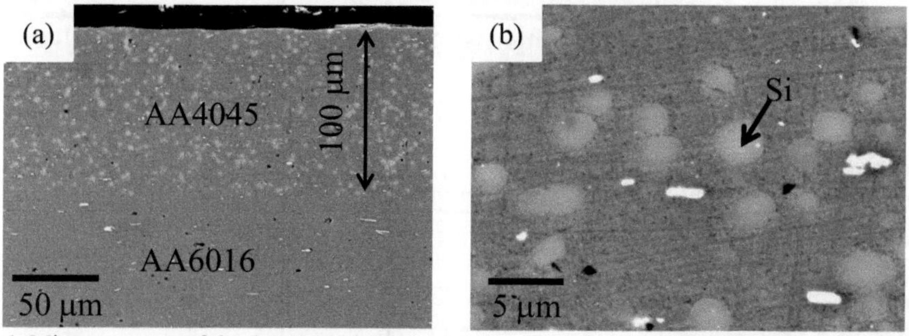

Fig.1. Microstructure of the AA4045 clad layer on the as-received AA6016 aluminum 'Fusion™' sheet product; (a) the whole layer, (b) Si particles at a higher manificaiton.

Fig.2 compares SEM images of the interface region in the Si rich coated Al sheet, after isothermal treatment at 400°C for 3hrs, when joined to the Mg sheet with that of the uncoated AA6111 Al – Mg baseline alloy combination. It can be seen that in the joint with the Si coating the overall reaction layer appears about two thirds as thick, (50 μm, compared to 75 μm) which indicates that the concept of introducing a Si rich coating has been successful. In addition, while the IMC reaction layer between the uncoated standard Al alloy and the Mg alloy is comprised of two continuous sub-layers of different compounds, the thinner reaction layer in the coated Al - Mg joint appears to be made up of a single phase on the Mg side of the joint, above a more complex irregular sub layer, containing second phase particles.

In the baseline couple the reaction behavior is similar to that already reported [13]. The two phases present were readily identified as a thinner layer of γ, which forms first in the reaction sequence on the Mg side of the joint, and β, which forms later, but grows more rapidly after nucleating on the Al side of the joint (see [13,14]). In comparison with the Si rich coating, EDX and TEM micro diffraction analysis indicated that the single phase layer adjacent to the substrate was also γ-$Al_{12}Mg_{17}$, but this was thicker than in the baseline couple and extended down into the region where second phase particles were present (Fig.2a). EDX mapping (Fig. 3) and TEM micro-diffraction analysis (Fig. 2c) showed that the second phase particles in the IMC layer that had grown into the Al-Si cladding were Mg_2Si particles that had formed by reaction between Mg diffusing into the surface and the Si particles distributed in the cladding layer. The lower part of this irregular multi-phase region also contained residual Si particles (Fig. 3). Although analysis is currently incomplete, the β phase has not so far been detected in the IMC reaction between the

Si-rich cladding material and the Mg alloy, with the matrix in the irregular multi-phase region still being predominantly γ phase.

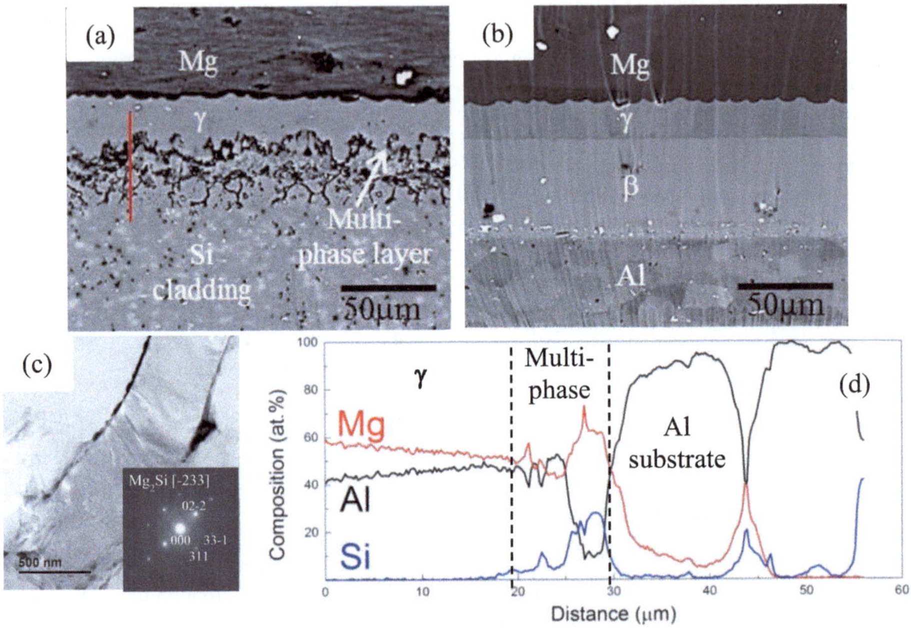

Fig.2. Interface characterization of the Al(Si) clad material (a) and unclad Al (b) joined to Mg AZ31 after annealing at 400°C for 3 hrs; (c) identificaiton of the 2nd phase in the Al(Si)-Mg IMC layer, (c) EDX line profile along the red line shown in (a).

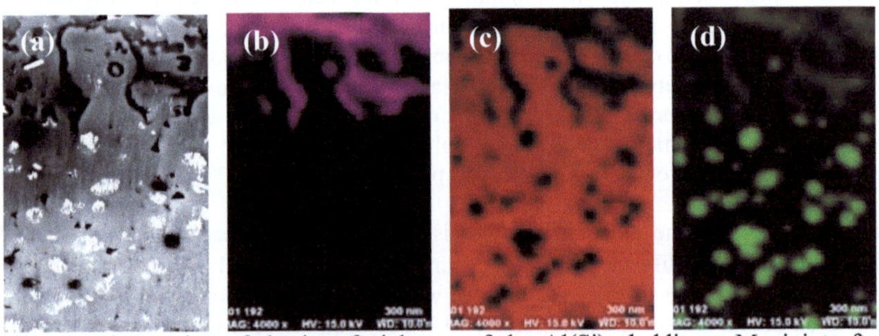

Fig.3. Elemental mapping of the interfacial area of the Al(Si) cladding - Mg joint after heat treatment at 400°C for 3h; (a) BSE image; (b) Mg; (c) Al; (d) Si.

3.2 Interfacial Characterization of the Al-Zn-Rich Coating

Zn is a common alloy element in both Al and Mg alloys. It has the advantage over Si that at elevated temperatures it is highly soluble in Al (~ 83 wt. % Zn at 381°C) and therefore will be homogenously distributed in a coating. Moreover, the presence of Zn can help enhance the corrosion resistance of Mg alloys. Based on this reasoning, an Al-Zn solid solution coating was designed by CALPHAD aided simulation to avoid the formation of the β phase, by again changing the reaction product to an alternative more stable compound. Fig.4 shows a calculated Al-Zn-Mg isothermal section at 360°C. The gray zone on the diagram corresponds to the composition range where the β-Al_3Mg_2 phase is stable. It can also be seen that, with a high enough Zn content one binary phase, $MgZn_2$, and two ternary phases, τ-$(Al,Zn)_{49}Mg_{23}$ and ϕ-$Al_2Mg_5Zn_2$, can also be produced. Diffusion profiles between Al-Zn alloys with various Zn contents and the Mg alloy AZ31 were simulated using DICTRA at 360°C, which is in the range of typical peak temperatures reached during friction welding Al and Mg alloys. The result of such an analysis for an Al(52 wt. %Zn) - Mg couple is marked with a red line in Fig.4. It can be seen that this profile touches the tip of the phase fields in which Al_3Mg_2 exists, implying that at this temperature this is the minimum Zn content required to eliminate β. Moreover, the line goes through phase fields containing several binary and ternary phases (namely the γ-$Al_{12}Mg_{17}$, $MgZn_2$, τ and ϕ phase), indicating that these intermetallics would be expected to be preferentially formed during any inter-diffusion across a weld interface.

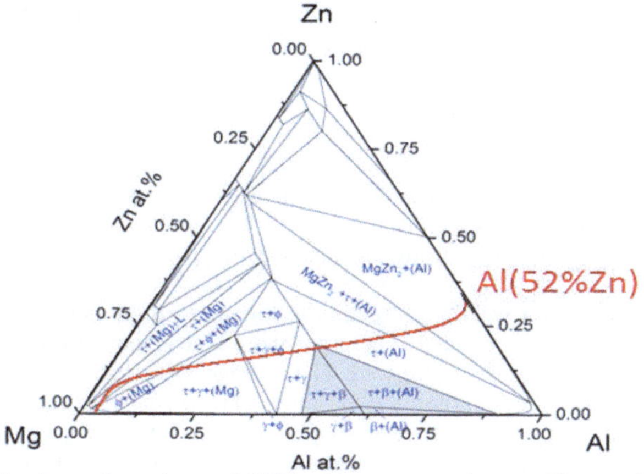

Fig.4. Al-Zn-Mg isothermal section at 360°C with concentration profiles predicted by DICTRA for a diffusion couple between Al(52%Zn) and AZ31.

In Fig.5 the microstructures of the joint interface between the Al(52%Zn) coating to Mg and the baseline Al - Mg couple are depicted after annealing at 360°C for 24 hrs. It can be seen that both interfaces exhibit a three-layer IMC structure. In both cases the IMC layer adjacent to the Mg was indexed as γ-$Al_{12}Mg_{17}$. As before, in Fig. 2b in the standard Al to Mg joint, a uniform thicker layer of β-Al_3Mg_2 phase was found to have formed next to the Al side of the joint, (Fig.5b). In comparison, at the same location in the Al(Zn) - Mg joint (Fig.5a), EBSD and TEM micro-diffraction confirmed the presence of the ternary phase τ-$(Al,Zn)_{49}Mg_{32}$, and no Al_3Mg_2 was detected, confirming the predictions from the diffusion simulations. This means that the less favored intermetallic phases can be successfully prevented during welding, or diffusion bonding, Al to Mg by using an Al-Zn-rich coating based on the composition predicted above. A third IMC layer was also observed in both samples that has yet to be identified, which appears dark in the phase maps. This phase is likely to be $Al_{30}Mg_{23}$ which is known to be stable in the temperature range 250-410°C and is difficult to identify owing to its complex rhombohedra crystal structure and the stoichiometry of $Al_{30}Mg_{23}$ [14,15].

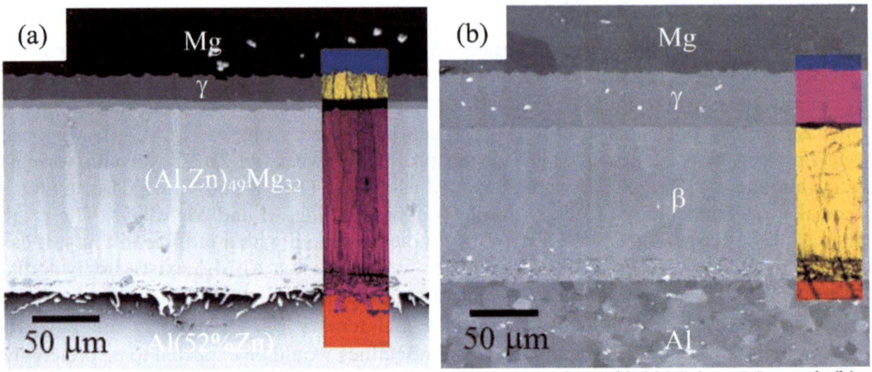

Fig.5. Comparison of the interface microstructures in (a) the Al(52%Zn) – Mg and (b) the standard Al - Mg control joint following annealing at 360°C for 24 hrs. The inserts are from EBSD phase ID maps, where each phase has been indexed from its back scattered diffraction pattern.

3.3 RFSSW Tests

Following these promising initial studies the two coatings were tested under welding conditions, using the RFSSW method. For comparison purposes, baseline uncoated AA6111 - AZ31 welds were also prepared at the same time using identical welding parameters.

The typical morphology of a weld cross section following RFSSW is shown in Fig. 6 with the sleeve and pin plunging areas marked. These are the effective bonding regions where stirring took place and an IMC layers were observed. The average IMC thicknesses was measured near the weld center and near the edge (sites A & B in Fig. 6) to characterize the level of intermetallic growth under the pin and the sleeve, respectively, and the results are illustrated in Table **II**. Among all three combinations, the Al(Zn) coating produced the thinnest IMC layer, which was less than 30% of that in the baseline couple. For the Al(Si) - Mg couple, a similar IMC layer thickness was found at the weld centre (site A) comparing to that seen with the uncoated baseline; however, there was a 50% drop in the intermetallic thickness under the sleeve positon site B, indicating that the coating layer had been effective in retarding the detrimental interfacial reactions. Finally, Fig. 7 shows the peak loads that the three weld combinations achieved in lap shear tests. It can be seen that both coatings have been beneficial in improving mechanical properties of the weld, although the improvement is quite modest.

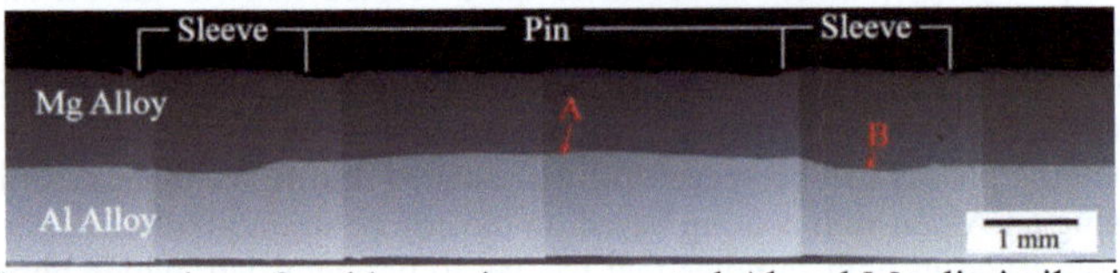

Fig. 6. Typical cross section of weld zone in an uncoated Al and Mg dissimilar joints produced by RFSSW.

Table II. Average IMC thickness at site A and B marked in Fig. 6

Material	Average IMC thickness (μm)	
	Site A	Site B
Al - Mg	3.5	8.0
Al(Si) - Mg	3.4	3.9
Al(Zn) - Mg	0.7	2.1

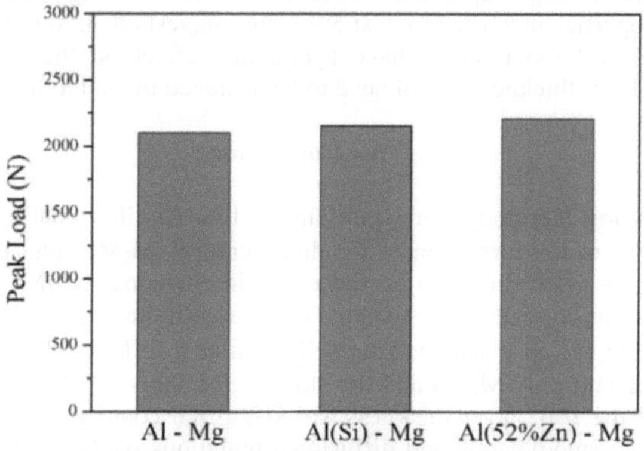

Fig. 7. Peak load Measurements of the Al and Mg RFSSWs during lap shear tests

4. Discussion

The above results have shown that the approach of using thermodynamic predictions to select suitable coatings has some potential for reducing the detrimental effects of IMC formation, in dissimilar metal welds between Al and Mg. In both cases the systems selected have changed the reaction path avoiding the formation of the undesirable β-Al_3Mg_2 phase. Both coatings have also been shown to have had some effect on inhibiting the IMC layer growth in refil-FSSWs.

During the Al - Mg reaction, γ-$Al_{12}Mg_{17}$ is the first intermetallic layer to grow, and β-Al_3Mg_2 only appears after a continuous layer of γ is formed [6]. In the Al-Mg-Si system, the inhibition of the more rapid growing β-Al_3Mg_2 phase was realized via a two-fold mechanism: i) in the initial stage of the IMC reaction, due to the Mg - Si reaction that produces Mg_2Si, Mg atoms diffusing into the aluminum matrix are consumed by the Si particles present in the coating, and their concentration is therefore reduced where β would normally nucleate near the Al–γ interface; ii) at a later stage, as $Al_{12}Mg_{17}$ and Mg_2Si grow simultaneously, when the Mg_2Si particles became more connected with each other they form a partial barrier layer which helps prevent more Al and Mg interdiffusion. Thus, the nucleation of the β phase was restricted which, owing to the lower growth rate of γ, also reduced the overall layer thickness.

In the Al-Mg-Zn system the diffusion simulation calculated the concentration profile between an Al-Zn coating alloy, of a given composition, and the AZ31 composition of the Mg alloy sheet. In Fig.4, β cannot form outside its phase field, because its Gibbs free energy of formation becomes higher than that of alternative compounds. Therefore, by selecting an Al-Zn alloy, where the interdiffusion line was tangential to the β phase field, in the calculated ternary diagram, the Al_3Mg_2 phase was replaced by the ternary compound $(Al,Zn)_{49}Mg_{32}$. In the static annealing tests this phase nucleated and spread once the γ-$Al_{12}Mg_{17}$ reached a critical thickness.

It is known that when welding dissimilar metals, after complete bonding is achieved, the toughness of a joint decreases rapidly with growth of an IMC layer at the interface [9]. Both the Si-rich and Al-Zn-rich coatings have been shown to have some effect on reducing the IMC thickness in the welding zone. As a result, the peak loads that the coated samples achieved in lap shear tests were found to have improved, although this was to a relatively modest level. For example, for the Al(Zn) - Mg couple, the IMC layer thickness was reduced by 70%, although the joint mechanical properties increased by just 5%. This suggests that, although greatly improved, the IMC layer was still too thick to have a dramatic effect on the joint strength and that submicron reaction layer thicknesses will need to be achieved to lead to high performance joints.

5. Conclusions

In this study, following thermodynamic assessment, two Al-alloy coatings (Al-Si and Al-Zn) were selected to prevent the formation of the detrimental β-Al_3Mg_2 phase in Al-Mg dissimilar welding. Both coatings were shown to be successful in changing the IMC reaction path. When using an Al-Si cladding layer, the Mg substrate reacted with the Si particles during annealing to produce a more stable Mg_2Si phase, which itself acted as a diffusion barrier. As a result, the β phase was prevented and only Mg_2Si and the slower growing γ-$Al_{12}Mg_{17}$ phase were observed, which also reduced the IMC layer thickness.An Al-Zn-rich coating composition was predicted by computer aided thermodynamic and diffusion simulations of the Al-Mg-Zn ternary system. This completely suppressed the Al_3Mg_2 phase, which was replaced by the ternary IMC, τ-$(Al,Zn)_{49}Mg_{32}$.

Welding experiments showed that both coatings were effective in reducing the IMC layer thickness produced at the joining interface in Al-Mg Refil-FSSWs, however this only had a modest effect on improving the joint the mechanical properties.

Acknowledgement

The authors are grateful to the EPSRC for funding this work through LATEST2 (EP/G022402/1) as well as to Novelis and MEL-Luxfer for the provision of materials for this research.

References

1. D. Carle and G. Blount, "The suitability of aluminium as an alternative material for car bodies," *Materials & Design*, 20 (5) (1999), 267-272.
2. E. Aghion et al., " The role of the magnesium industry in protecting the environment," *Journal of Materials Processing Technology*, 117 (3) (2001), 381-385.
3. E. Schubert et al., "Light-weight structures produced by laser beam joining for future applications in automobile and aerospace industry," *Journal of Materials Processing Technology*, 115 (1) (2001), 2-8.
4. J. Wang et al., "Microstructure of Al-Mg dissimilar weld made by cold metal transfer MIG welding," *Materials Science and Technology*, 24 (7) 827-831.
5. R. Zettler et al., "Dissmilar Al to Mg alloy friction stir welds," *Advanced Engineering Materials*, 8 (5) (2006), 415-421.

6. A. Panteli et al., "The effectiveness of surface coatings on preventing interfacial reaction during ultrasonic welding of aluminum to magnesium," *Matallurgical and Materials Transactions A*, 44 (13) (2013), 5773-5781.
7. L. Wang et al., "Controlling interfacial reaction during dissimilar metal welding of aluminium alloys," *Materials Science Forum*, 794-796 (2014), 416-421.
8. H. Marsumoto et al., "Fabrication of pure Al/Mg-Li cald plate and its mechanical properties," *Journal of Materials Processing Technology*, 169 (1) (2005), 9-15.
9. A. Panteli et al., "The effectiveness of surface coatins on preventing interfacial reaction during ultrasonic welding of aluminum to magnesium," *Metallurgical and Materials Transactions A*, 44 (13) (2013), 5773-5781.
10. J. Zhang et al., "Microstructural refinement in Al-Mg_2Si in situ composites," *Journal of Materials Science Letters*, 18 (10) (1999), 783-784.
11. M. Mabuchis and K. Higashi, "Strengthen mechanisms of Mg-Si Alloy," *Acta Materialia*, 44 (11) (1995), 4611-4618.
12. Y. Tang et al., "Thermodynamic description of the Al–Mg–Si system using a new formulation for the temperature dependence of the excess Gibbs energy," *Thermochimica Acta*, 527 (2011), 131-142.
13. D. Dietrich et al., "Formation of intermetallic phases in diffusion-welded joints of aluminium and magnesium alloys," *Journal of Materials Science*, 46 (2) (2010), 357-364.
14. J. Murray, "The Al-Mg (aluminum-magnesium) system," *Bulletin of Alloy Phase Diagrams*, 3 (2) (1982), 60-74.
15. S. Shakhshir and M. Medraj, "Computational thermodynamic model for the Mg-Al-Y system," *journal of Phase Equilibria and Diffusion*, 27 (3) (2006), 231-244.

Friction Stir Welding and Processing VIII
Edited by: Rajiv S. Mishra, Murray W. Mahoney, Yutaka Sato, and Yuri Hovanski
TMS (The Minerals, Metals & Materials Society), 2015

FRICTION STIR WELDING OF AUSTENITIC STAINLESS STEEL TO AN ALUMINUM-COPPER ALLOY

S. Babu[1], S.K. Panigrahi[1], G.D. Janaki Ram[2], P.V. Venkitakrishnan[3], R. Suresh Kumar[3]

[1]Dept. of Mechanical Engineering, Indian Institute of Technology Madras, Chennai
[2]Dept. of Metallurgical and Materials Engineering, Indian Institute of Technology Madras, Chennai
[3]Liquid Propulsion Systems Centre, Thiruvananthapuram

Keywords: Stainless steel, Aluminum alloy, Dissimilar friction stir welding

Abstract

Friction stir welding of austenitic stainless steel AISI 321 to an Al-Cu alloy AA 2219-T87 (3 mm thick sheets) was investigated for a specific aerospace application. Welding experiments were carried out using WC-Co tools with different pin profiles. The effects of process parameters, including tool positioning were studied. After careful process optimization, welds with a tensile strength of close to 250 MPa were successfully produced. The welds showed a very rugged stainless steel/aluminum interface as well as some fragments of stainless steel in the stir zone, confirming complete removal of the oxide film on the stainless steel faying surface due to the wearing action of the stirring aluminum. A very thin intermetallic layer (consisting of FeAl, Fe_3Al and $AlCrFe_2$ phases) was observed in the SS/Al interface.

Introduction

Dissimilar joining of steel and aluminum has gathered wide acceptance in automotive and aerospace industries, because it can lead to significant weight reduction in the structure and meet the functional requirements. However, it is difficult to weld aluminum to steel due to their metallurgical incompatibility, large differences in their coefficient of thermal expansion and the formation of brittle intermetallic phases. At present, austenitic stainless steel and an aluminum-copper alloy are widely used in the fabrication of bimetallic adaptors in satellite launch vehicles. These adaptors are used as transition joints between aluminum tanks and stainless steel piping. Though these adaptors are traditionally fabricated by explosive welding, proper bonding at the transition zone is very difficult to achieve. Moreover, this process is very expensive and it also requires a number of safety measures [1]. The difficulty in joining is mainly due to the formation of the brittle intermetallic layer in the bond region, which results in poor joint strength. In order to reduce the thickness of intermetallic layer and to achieve high strength, the heat input near the bond region should be controlled.

Any low heat input process, which does not involve fusion, would be ideal for joining these dissimilar materials. In this regard, one has to consider the efficient processes like friction stir welding to realize satisfactory joints between austenitic stainless steel and an aluminum-copper alloy. Friction stir welding (FSW) is a relatively new and innovative solid state joining process, capable of generating reproducible high quality welds in similar and dissimilar materials. In comparison to fusion welding processes, FSW offers several advantages such as low energy input, short welding time, low distortion and relatively low welding temperatures [2]. Moreover, it was observed that the fatigue properties of stainless steel/AA 6013-T4 dissimilar friction stir welds were approximately 70% of AA 6013-T6 base metal [3]. In another study, a shear strength of nearly 90 MPa was reported in stainless steel/aluminum dissimilar friction stir lap welds, which was higher than that of the pure aluminum base metal [4]. Though FSW appears attractive, extensive research on joining austenitic stainless steel to aluminum-copper alloys is very limited. In the current study, an attempt has been made to join AISI 321 austenitic stainless steel with AA 2219 aluminum-copper alloy using FSW.

Experimental Work

Friction stir welding of 3 mm thick sheets of AISI 321 stainless steel and AA 2219-T87 aluminum alloy (in butt configuration) was carried out. Suitable FSW tools made of a tungsten carbide alloy (WC-Co) were purchased and machined into the required size and shape as per our design. Butt joints were made with variations in rotational speed (1000-3000 rpm), welding speed (15-150 mm/min), plunge depth (2.1-2.6 mm), tool tilt angle (0-5°), tool offset (0-2 mm toward the Al side) and plate position (advancing side and retreating side). Initially visual inspection revealed excess stirring on the stainless steel side, which in turn overheats the aluminum side resulting in large tunnel defects. These trials showed excessive heat developed on the tool due to the equal contact of pin with respective materials.

In order to reduce the heat generated on the tool and to achieve uniform stirring in the weld region, trials were made by offsetting the tool toward the aluminum side. With a tool offset of 1 mm, some improvement in the bead appearance and stirring was observed in the welds. Visual inspection revealed the formation of ripples on the top surface of the weld, which are a common feature in FSW. However, the welds were uniformly observed to exhibit tunnel defects in the offset region. It was therefore decided to increase the tool offset further to improve the joint formation. Defect-free friction stir welded joints between SS and Al (with SS on advancing side) were obtained at a tool offset of 2 mm (Fig. 1).

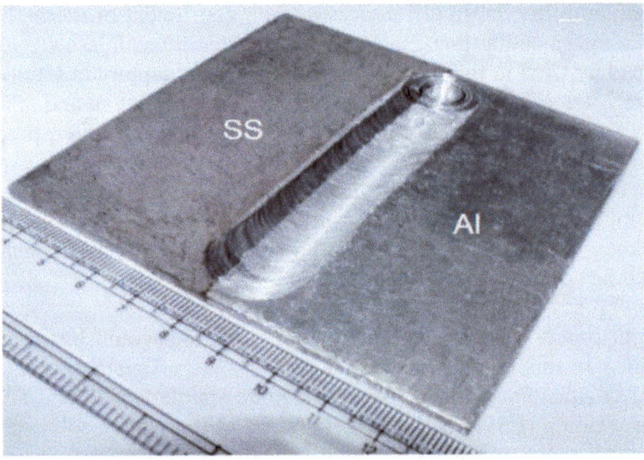

Figure 1. Photograph of a defect-free butt joint produced in AISI 321 and AA 2219 sheets.

The welded joints were sectioned and prepared for microstructural examination following standard metallographic practices. Samples were examined both in as-polished and in etched conditions (Keller's reagent) using an optical microscope. Detailed microstructural analyses of the welds were also made using scanning electron microscopy (FEI Quanta 400) equipped with an energy dispersive spectrometer. As per ASTM-E8M standards, three tensile specimens were extracted from the defect-free welds and tested at a constant load rate of 1 mm/min using a 30 kN Instron Universal Testing Machine.

Results and Discussion

Friction Stir Welding

To obtain reasonable welds, a suitable combination of rotational speed, welding speed, plunge depth, tool tilt angle and tool offset is essential. For the SS/Al combination used in this study, weld parameters of 800-1200 rpm rotational speed, 40-70 mm/min welding speed, 2.1-2.6 mm plunge depth, 1-2° tool tilt angle and 1-2 mm tool offset were found to be appropriate.

Rubbing and jamming of the tool was observed below 800 rpm due to insufficient frictional heating. Above 1200 rpm, abnormal stirring and asymmetric rotation of the tool were noticed due to the high heat input. In the range of 800 to 1200 rpm, the rotational speed did not affect the welding significantly. Incomplete root penetration was observed at smaller plunge depths (2.1 mm), but the welds were found to be good due to reasonable material displacement. At a plunge depth of 2.6 mm, sound welds were produced due to effective stirring in the weld root. Increasing the rotational speed and pin length causes a reduction in shear strength [5]. Welding speed plays a vital role in the welding characteristics of the SS/Al dissimilar materials. No welds were produced below 40 mm/min due to considerable heating of the tool for an excessive duration. At 70 mm/min, a significant reduction in joint formation (material filling) was found due to the rapid translation of tool. However, the joint strength increased, because of the reduction in the intermetallic layer thickness, at high travel speeds ranging from 50-100 mm/min [6].

No defect free welds were produced without a tool tilt. At an angle of 1°, the weld formation was improved by the smooth translation of the tool. However the joint was found to contain worm holes throughout the length. Sound welds were obtained when the tool tilt angle was increased to 2°. Kittipong *et al.,* [7] reported that the increase in tool tilt angle above 2°, decreases joint strength due to the formation of an Al-rich intermetallic compound at the joint interface. In addition, tool offset is a significant parameter in joining steel with aluminum [8]. When the offset was maintained at less than 1 mm, the aluminum side overheats. If the offset was more than 2 mm, no stirring occurs on the stainless side. It was therefore decided to keep a small portion of the pin in the steel side to stir both materials. From our observations, a tool offset in the range of 1 - 2 mm with a 2° tilt can improve the joint formation at the abutting interfaces of SS and Al.

Microstructural Examination

The macro- and micro- structures from a friction stir weld with a 2.1 mm plunge showed good bonding of the stainless steel to aluminum (Fig. 2). In the weld nugget, a small number of stainless steel fragments were scattered throughout surrounded by fine equiaxed grains of the aluminum alloy. All the welds showed distinct boundaries at the stainless steel-aluminum interface. The interface was irregular and wavy at the top (Fig. 3a) and more regular and straight at the bottom (Fig. 3b). This suggests there was considerable stirring by the pin in the aluminum on the top, which was noticeably less at the bottom. From the microstructures, it is believed that no harmful thick intermetallic layer was present at the stainless steel-aluminum faying interface.

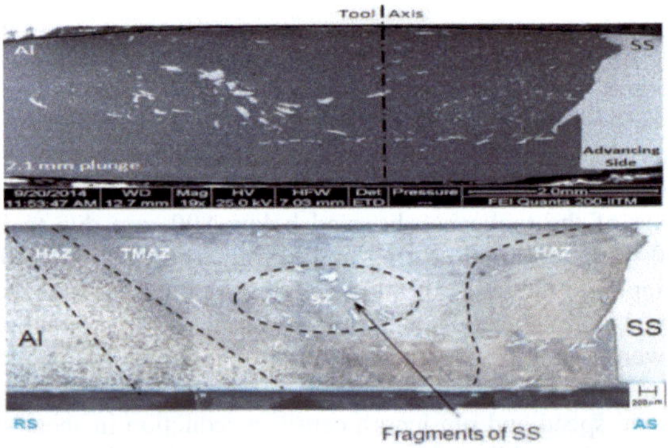

Figure 2. Micrograph of a weld cross section showing bonding at the SS/Al interface.

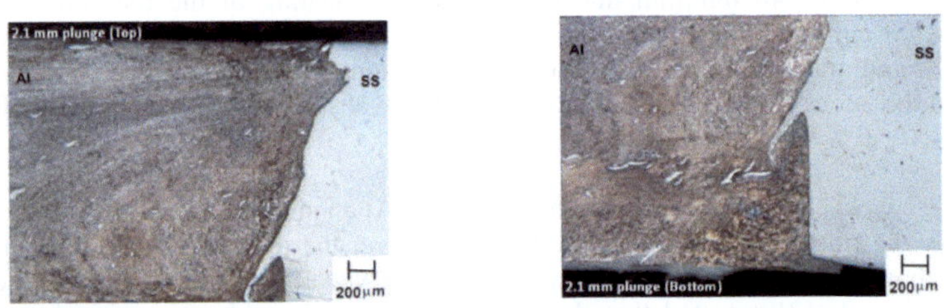

Figure. 3 Magnified view at the top (left) and bottom (right) region of the SS/Al interface (2.1 mm plunge).

Macrographs of friction stir welds with a 2.6 mm plunge depth showed significant improvement in the bonding characteristics of the stainless steel and aluminum interface as a result of additional plunging (Fig. 4). Due to this extra plunge, the SS/Al interface was found to be wavy and rugged (Fig. 5) throughout the entire thickness. Very fine fragments of stainless steel were scattered throughout the weld region (nugget zone and thermo-mechanically affected zone). The quantity of fragments is relatively high when compared with welds made at the lower plunge depth.

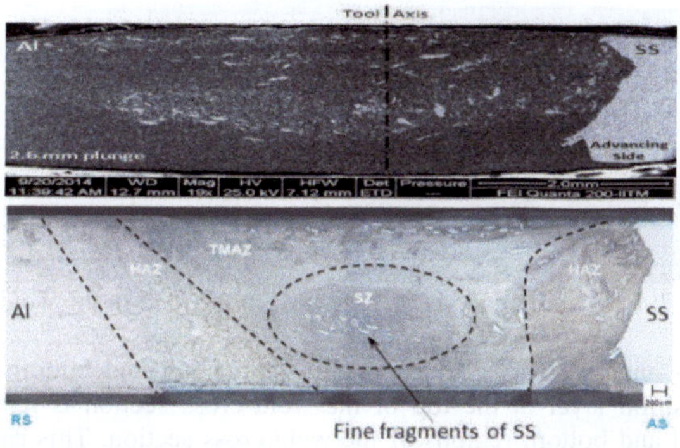

Figure. 4 Macrographs of a weld cross section showing bonding at the SS/Al interface.

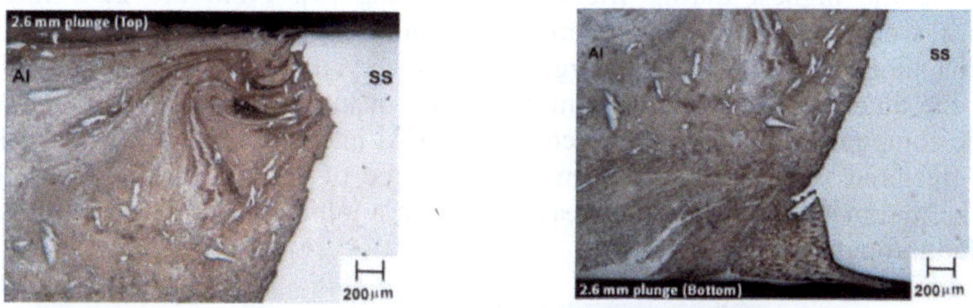

Figure. 5 Magnified view at the top (left) and bottom (right) region of the SS/Al interface (2.6 mm plunge).

In order to quantify the elemental distribution across the SS/Al interface, an energy dispersive spectroscopy (EDS) analysis was performed. The element distribution in the middle region of the weld cross section is shown in figure 6. A sharp variation in the elemental distribution of Fe (or Cr or Ni) and Al in the transition region, suggested that there is no intermetallic layer present in this region. EDS spot analysis confirmed the scattering of stainless steel fragments in the weld nugget (Fig. 7). No trace of oxide films around the SS fragments was observed.

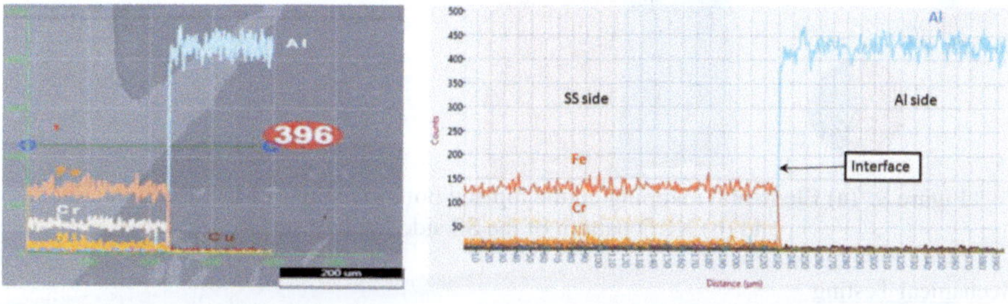

Figure 6. EDS line scan across the middle of the SS/Al interface.

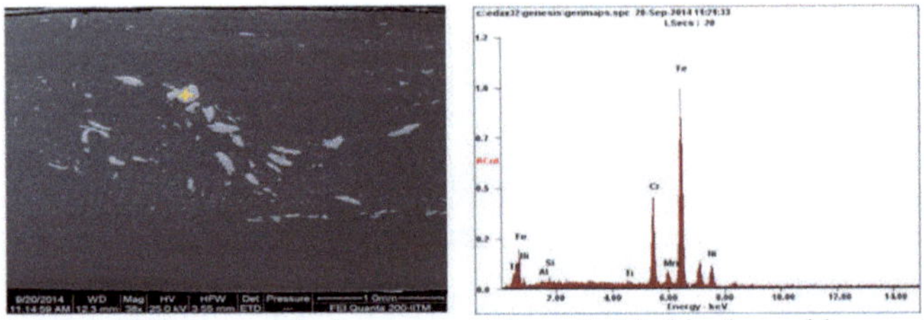

Figure 7. EDS spot analysis on a particle embedded in the weld nugget.

Elemental mapping analysis was also conducted at the top and bottom regions of the joint interface. The transition layer at the top of the weld cross section is relatively thicker, when compared to central and bottom regions of the weld cross section. This can be due to the large frictional heat generated by the shoulder at the top surface where the temperature has been reported to be the highest during friction stir welding [9]. From the EDS line scan, no intermetallics were formed at the central and bottom regions of the interface. However, elemental mapping on these regions suggests that a very thin layer of intermetallics can be present at the interface (Fig. 8a). From the EDS quantitative analysis, FeAl and Fe_3Al intermetallic compounds might be expected at the SS/Al interface [10]. XRD analysis (Fig. 8b) on the SS fractured surface also confirms the presence of phases such as FeAl, Fe_3Al and $AlCrFe_2$. The intermetallic layer is the weakest zone in a joint and it might be believed that the welds can fail through these regions.

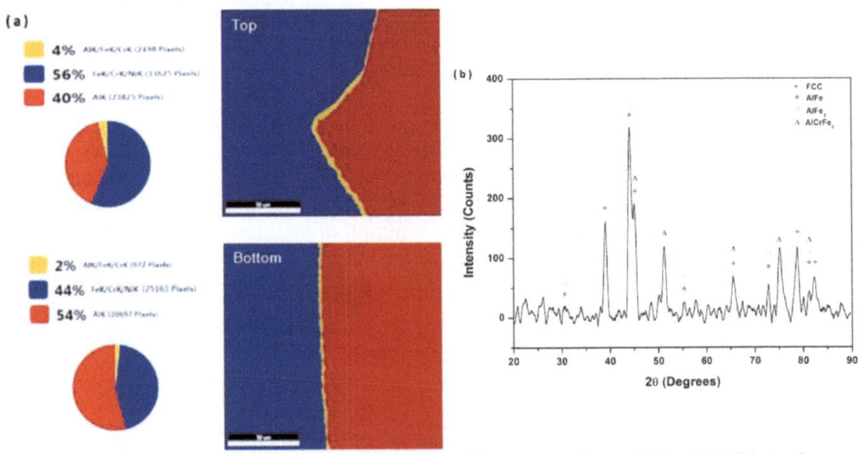

Figure 8. (a) Elemental mapping on the top and bottom regions of the SS/Al interface, and (b) XRD pattern of the SS side fracture surface.

Mechanical Testing

Transverse tensile tests showed that the ultimate tensile strength of the SS/Al friction stir welds (average of 5 tests) was found to be 246 MPa (Fig. 9) which is approximately 60% of the aluminum alloy base material strength. All the welds failed at the SS/Al interface in a brittle manner. No necking was observed near the fractured regions. The observed microstructural features for the two conditions presented in figures 3 and 5 correlates well with the tensile strength. Higher strength in the welds produced at higher plunge depths can be attributed to the improved bonding characteristics at the weld root (Fig. 5). Lack of stirred material at the bottom resulted in poor strength of the joints (Fig. 3). It appears that the combined stirring action of the pin and aluminum in the weld root region can lead to a more rugged SS/Al interface throughout the thickness, which in turn increases the joint strength significantly. Further improvements in the joint strength are likely achieved by using different pin profiles and tool designs.

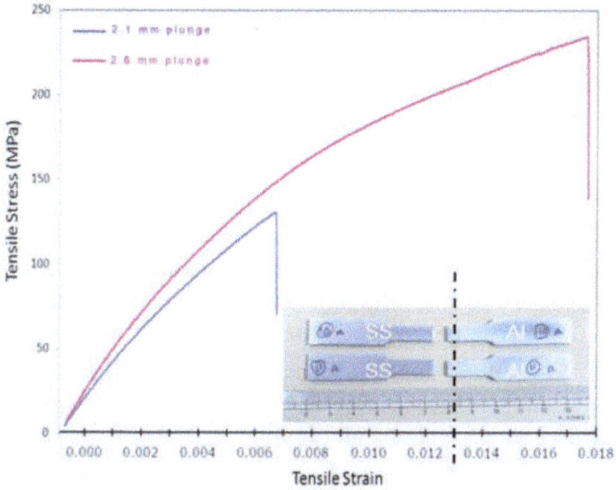

Figure 9. Stress-strain curves of friction stir welds produced at different plunge depths.

Failure Analysis

From the fractured surfaces of SS/Al (Fig. 10), the cracks clearly initiate from the weld root on the advancing side and fail in a brittle manner. This is mainly due to a lack of stirred material at the weld root, which initiates premature failure. No coincidence between the fracture line and the weld centre line also suggests that no stirring has taken place in this region. The fracture surface of the welds produced at higher plunge depths reveals the presence of dimples in the aluminum fracture surface indicating yielding of the alloy prior to fracture, resulting in the higher strength of the joints.

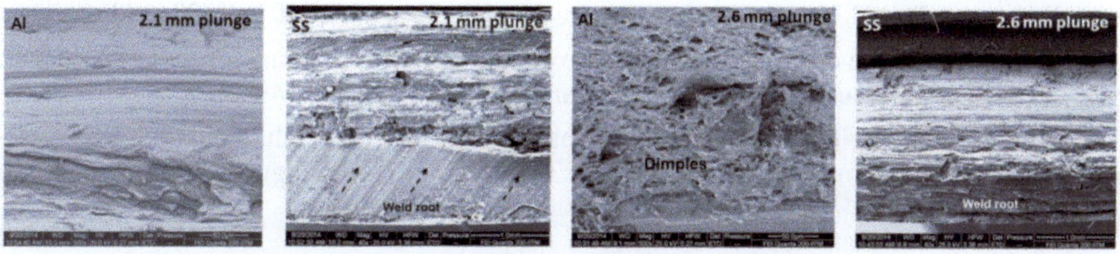

Figure 10. Fracture surfaces of SS and Al sides.

Conclusions

Defect-free welds between austenitic stainless steel AISI 321 and an Al-Cu alloy AA 2219-T87 were produced successfully using FSW. A maximum tensile strength of about 246 MPa (60% of the Al-Cu alloy) was obtained. Metallographic studies revealed a very thin intermetallic layer consisting FeAl, Fe_3Al and $AlCrFe_2$ phases at the weld interface.

Acknowledgements

Financial assistance from the Indian Space Research Organization (ISRO) is gratefully acknowledged.

References

1. F. Findik, "Recent developments in explosive welding", *Materials and Design*, 32 (2011) 1081-1093.
2. R.S. Mishra and Z.Y. Ma, "Friction stir welding and processing," *Materials Science and Engineering R*, 50 (2005), 1-78.
3. H. Uzun, C. D. Donne, A. Argagnotto, T. Ghidini, and C. Gambaro, "Friction stir welding of dissimilar Al 6013-T4 to X5CrNi18-10 stainless steel," *Materials and Design*, 26 (2005) 41-46.
4. J. T. Xiong, J. L. Li, J. W. Qian, F. S. Zhang, and W. D. Huang, "High strength lap joint of aluminium and stainless steels fabricated by friction stir welding with cutting pin," Science and Technology of Welding and Joining, 17 (3) (2012), 196-201.
5. W.B. Lee, M. Schmuecker, U. A. Mercardo, G. Biallas, and Seung-Boo Jung, "Interfacial reaction in steel-aluminum joints made by friction stir welding," *Scripta Materialia*, 55 (2006) 355-358.
6. T. Tanaka, T. Morishige, and T. Hirata, "Comprehensive analysis of joint strength for dissimilar friction stir welds of mild steel to aluminum alloys," *Scripta Materialia*, 61 (2009) 756-759.
7. K. Kimapong and T. Watanabe, "Effect of Welding Process Parameters on Mechanical Property of FSW Lap Joint between Aluminum Alloy and Steel," *Materials Transactions*, 46 (10) (2005) 2211-2217.
8. W. H. Jiang and R. Kovacevic, "Feasibility study of friction stir welding of 6061-T6 aluminium alloy with AISI 1018 steel," Proceedings of the Institution of Mechanical Engineers, Part B: *Journal of Engineering Manufacture*, 218 (2004), 1323- 1331.
9. P. Ulysse, "Three dimensional modeling of the friction stir welding process," *International Journal of Machine Tools and Manufacture"*, 42 (14) (2002), 1549-1557.
10. T. Watanabe, H. Takayama, and A. Yanagisawa, "Joining of aluminum alloy to steel by friction stir welding," *Journal of Materials Processing Technology*, 178 (2006) 342-349.

FRICTION STIR WELDING AND PROCESSING VIII

Friction Stir Processing

Friction Stir Welding and Processing VIII
Edited by: Rajiv S. Mishra, Murray W. Mahoney, Yutaka Sato, and Yuri Hovanski
TMS (The Minerals, Metals & Materials Society), 2015

FRICTION STIR PROCESSING OF DIRECT-METAL-DEPOSITED 4340 STEEL

Bharat K. Jasthi[1], Todd Curtis[1], Christian A. Widener[1], Michael West[1], Matthew Carriker[1], Ashish Dasgupta[2], and Robert B. Ruokolainen[3]

[1] Arbegast Materials Processing and Joining Laboratory
Materials and Metallurgical Engineering
South Dakota School of Mines and Technology
501 E. St. Joseph St., Rapid City, SD 57701, USA

[2] Center for Advanced Technologies, Focus: HOPE
1400 Oakman Blvd, Detroit MI 48238, USA

[3] Chrysler Group LLC,
4800 Chrysler Drive, Auburn Hills, MI 48326, USA

Keywords: Direct metal deposition, Friction stir processing, and 4340 steel

Abstract

The main objective of this work is to investigate the effect of friction stir processing on a multi-layered 4340 steel, deposited on a 4340 steel substrate via laser-assisted direct metal deposition (DMD). FSP was performed using W-25%Re-4%HfC and W-25%Re-6%HfC pin tools. X-Ray micro computed tomography (CT) was performed on the as-deposited and processed material to evaluate porosity present in the material. Mechanical properties were also evaluated before and after FSP. FSP effectively eliminated porosity in the as-deposited material and improved the yield and ultimate tensile strengths in the processed region. No significant improvement in the ductility with FSP was observed and the ductility of the processed material is comparable with the as-deposited material. Pin tool wear and microstructural changes associated with FSP were also analyzed and reported.

Introduction

4340 alloy steel is a heat treatable, high strength low alloy (HSLA) steel used for many critical structural applications by the Navy due to its toughness, high strength and ability to retain good fatigue strength at elevated temperatures. This alloy is also used in many other structural applications such as automobile components, high pressure vessels, nuclear power plants, and aircraft components. Naval support equipment and aging weapon systems are often subjected to wear and corrosion. Replacement of these old components, which are low in quantity but high in value, can be difficult and very expensive. Therefore, research regarding repair, refurbish and return to service of these components is extremely important.

Recent studies have indicated that laser-assisted DMD can be used for repairing turbine blades, torpedo shells, tools with surface cracks, wear and other defects in many metallic components [1, 2]. However DMD is often limited by porosity, cracking, and segregation of alloying elements in the deposit which can adversely affect corrosion resistance and mechanical properties [3-5]. Recently, FSP has been employed to refine the microstructure and eliminate porosity in many high temperature cast alloys including Alloy 718 and NiAl bronze [6, 7]. Since FSP is a solid state microstructural modification technique, most of the problems associated with DMD can be avoided. Therefore, the main objective of this work is to perform FSP of a DMD 4340 deposit, as a means to repair damaged areas of expensive Naval components.

Experimental

DMD-deposited 4340 on 4340 substrates were supplied by Focus: HOPE, Detroit, MI, USA. The substrate dimensions were 125 mm x 125 mm x 25.4 mm and the DMD layer dimensions were 100 mm x 100 mm x 5 mm. The DMD plates have a very rough surface finish in the as-deposited condition. A rough surface could impact the quality of the FSP. Therefore, the surfaces of the plates were surface ground such that the final thickness of the DMD layer is 4.4 mm. A W-25%Re-4%HfC pin tool was used for the initial FSP parameter development. Optimized FSP passes for mechanical properties evaluation were performed using W-25%Re-6%HfC tools. These tools have a 25.4 mm shoulder diameter and a convex shoulder with a pin length of 5.8 mm. The FSP setup is shown in Figure 1a. FSP was performed using an iSTIR-10 FSW system in a force control mode. The FSW system was equipped with a Megastir water-cooled, high-temperature pin tool adapter. Argon gas was used to protect the tools and the FSP surface from oxidation.

Repair of a large area often requires multiple overlapping FSP passes. Therefore, overlapping FSP was performed with FSP passes running parallel to each other in the same direction with an offset of 7.6 mm between the passes as shown in Figure 1b. The overlaps were made in such a way that the advancing side of one pass was overlapped by the retreating side of the next pass. A total of 11 multiple side-by-side passes were made with the optimized process parameters (200 rev min^{-1}, 102 mm min^{-1} and 35.5 kN forge force) using the W-25%Re-6%HfC pin tools. Optical microscopy was used to analyze the microstructures of the DMD and FSP 4340 specimens. Tensile testing was performed on the 4340 clad, substrate and friction stir processed specimens in accordance with ASTM E 8. Vickers microhardness measurements were made on the cross-sections of as deposited and friction stir processed specimens. X-ray Micro CT was performed to estimate the porosity present in the material before and after FSP.

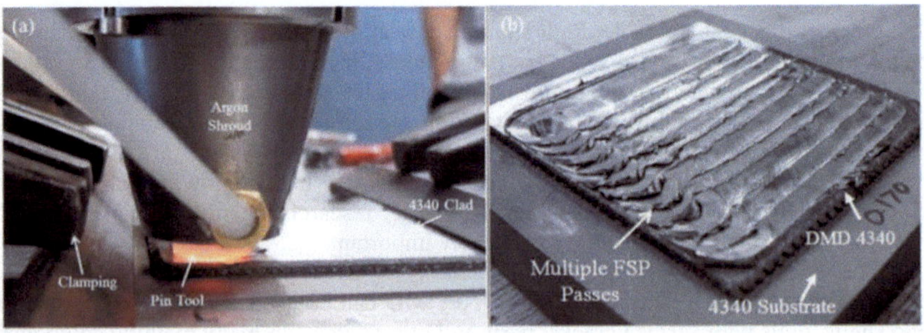

Figure 1: (a) Friction stir processing setup; (b) Photograph showing multiple overlapping FSP Passes

Results and Discussion

An optical micrograph of the as-deposited 4340 specimen is shown in Figure 2a. Porosity was identified in the clad region in both longitudinal and transverse directions. No delamination or any interfacial defects were identified at the interface between the clad and the substrate. Successful defect free FSP passes were made using both 4% and 6% HfC (W-Re-HfC) pin tools. A macrograph showing a cross-sectional view of the friction stir processed specimen is shown in Figure 2b. A macrograph showing the cross sectional view of the overlapping FSP passes is shown in Figure 3. All the passes showed good consolidation and no defects identified in any of the FSP passes. Defect free FSP passes were made with both 4% and 6% HfC tools. Both tools exhibited flattening after the initial FSP pass as shown in Figure 4. Pin tools with 4% HfC flattened to ~0.5 mm, whereas the tools with 6%HfC were only flattened to ~0.25 mm. The increase in HfC content in the pin tool material definitely showed improvement in terms of pin tool flattening. No significant wear or flattening was observed for subsequent FSP passes (after 11–plunges and 44-inches of total weld length) on 6% HfC tools, which suggests a need for a change in pin tool design.

Figure 2: Optical micrographs showing cross-sectional views of (a) as-deposited 4340 DMD Material; (b) Friction stir processed 4340 material.

Figure 3: Macrograph showing the cross sectional view of overlapping FSP passes

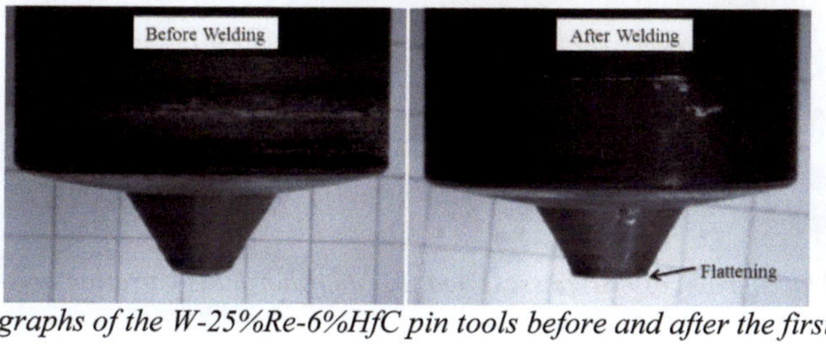

Figure 4: Photographs of the W-25%Re-6%HfC pin tools before and after the first FSP pass.

Microhardness evaluations were performed on the as-deposited and also on the friction stir processed specimens. Figure 5a shows that the hardness profiles in both longitudinal and transverse directions were similar for the as-deposited 4340 DMD specimens. The top layer of the clad has the highest hardness (~600 HV). The hardness decreased for the bottom layers. The average hardness for the parent material is ~ 320 Vickers. DMD of 4340 steel powders generally produce a fully martensitic microstructure with very high hardness in the deposited layer [8]. This explains the very high hardness observed on the top layers of the DMD material. The bottom layers in the DMD material generally experience thermal cycles from the top layers during the deposition process. So, the decrease in the microhardness from top to bottom layers of DMD material is most likely caused by the transformation of martensite to tempered-martensite in the bottom layers [8]. Figure 5b shows a microhardness profile across a friction stir processed nugget. The average hardness in the friction stir processed region is ~640 Vickers. The Microhardness is consistent in the processed region and no softening in the heat affected zone (HAZ) region was observed.

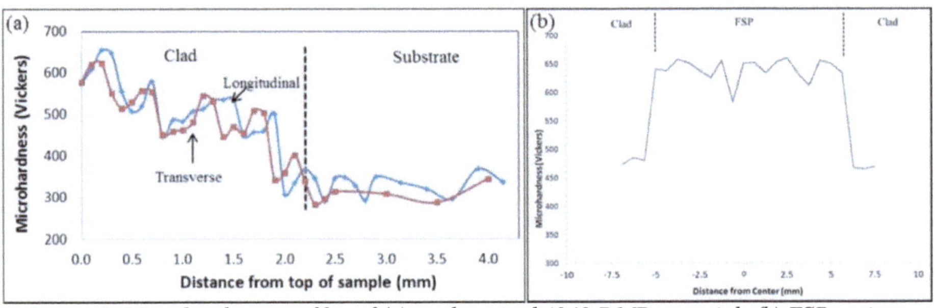

Figure 5: Microhardness profiles of (a) as-deposited 4340 DMD material; (b) FSP region.

Microhardness measurements were also made on the side-by-side passes with results shown in Figure 6. The results show that the final FSP pass had a higher hardness when compared to the preceding FSP passes. The temperatures measured in the heat affected zone (adjacent to the weld nugget) of the FSP pass showed a maximum temperature of ~900 °C. Thus, every time an FSP pass was made, the preceding pass experiences thermal cycles (most likely a tempering process). Therefore, the decrease in the hardness in the preceding FSP passes is most likely because of the thermal cycles they experience during subsequent FSP passes.

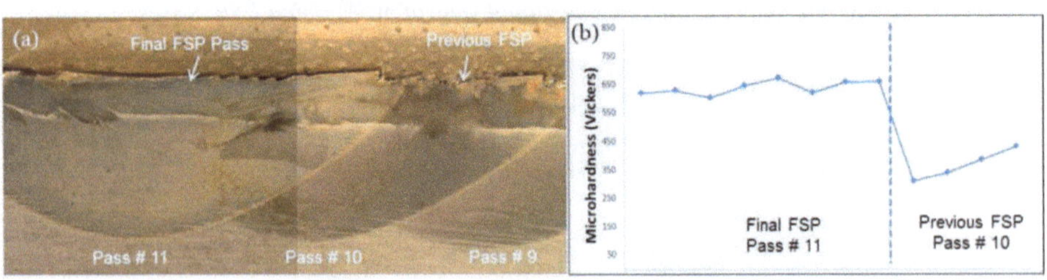

Figure 6: (a) Macrograph of FSP passes; and (b) Microhardness variation across the passes.

Tensile results are tabulated in Table I. The reported tensile results are the average of two specimens per condition. Testing was performed in both longitudinal (along the laser deposition direction- based on top layer) and transverse orientations. Tensile testing was also performed on the overlapping passes in order to evaluate the mechanical properties following FSP. The specimens were extracted across the overlapping passes and the top and bottom layers were machined such that the specimens had only processed material. As shown in Table I, the tensile strengths of FSP specimens are significantly higher compared to the parent and clad materials. However, both the clad and the FSP specimens showed a significant decrease in the ductility of the material. The 4340 substrate is in the annealed condition, and this explains the lower strengths and higher ductility for the substrate material. The decrease in ductility for the clad material is most likely due to the presence of porosity in the material. Friction stir processing is generally expected to improve the ductility of the material along with the improvement in strength. However, the tested FSP specimens only improved the strength of the material but with no improvement in ductility. This is most likely due to the formation of a martensitic microstructure in the friction stir processed region. Though not investigated in this research effort, an additional tempering heat treatment step would most likely improve the ductility for the FSP specimens.

Table I: Tensile properties of 4340 steel

Sample	Yield Strength (Mpa)	Ultimate Tensile Strength (Mpa)	Percent Elongation (%)
4340 Substrate Transverse	481	941	20
4340 Substrate Longitudinal	484	941	20
4340 Clad Transverse	723	1402	2
4340 Clad Longitudinal	669	1307	2
FSP Longitudinal –no surface machining	836	1632	2
FSP Longitudinal – surface machining	871	1557	1.5
FSP – Multiple overlap Passes	731	1389	2.3

The fracture surfaces of the FSP tensile specimens were analyzed and found that the top surface of the weld has a "lip" like feature on the surface (~0.5 mm from the top surface) of the weld as shown in Figure 7. These thin features on the surface are generally caused because of insufficient vertical material flow beneath the pin tool shoulder during FSP.

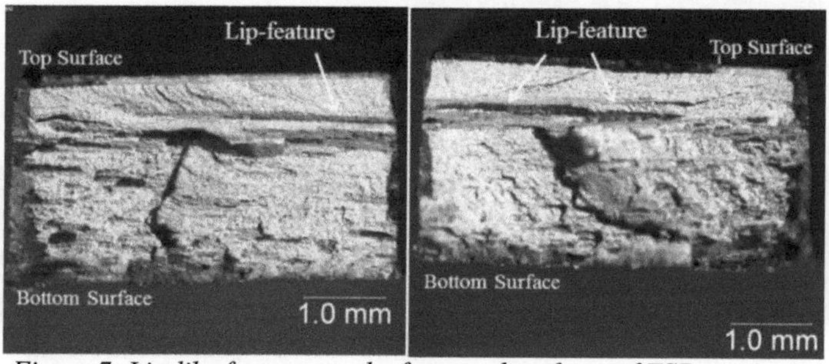

Figure 7: Lip-like features on the fractured surfaces of FSP specimens

The vertical material flow during processing can be improved by modifying the pin tool design or by adding scroll features on the shoulder and spiral features on the pin. However, features in W-based tools are not expected to survive the process forces during FSP due to localized wear. Alternatively, the top surface can be post-weld machined to remove these features. It was assumed initially that the decrease in ductility of the FSP material is due to the presence of these lip-like features on the surface. Therefore, a few tensile specimens were also tested after machining these features from the surface to see the effect of surface machining on the ductility of the material. The results, as shown in Table I, suggest that surface machining (or the presence of lip features in the nugget) had no effect on the tensile properties of the FSP specimens. Multiple overlap FSP specimens showed a slight decrease in yield and tensile strengths. These reductions can be associated with the microstructural changes as a result of overlapping passes.

X-ray Micro CT was performed to understand the location and also the shape and size of the porosity present in the material before and after FSP. Micro CT is a technique to nondestructively image and quantify the internal structure of the specimen in three dimensions. X-ray tomograms were scanned using a MicroXCT-400 system with a 140 kV voltage. The grey levels in a CT scan corresponds to X-ray attenuation, which reflects the proportion of X-rays scattered or absorbed as they pass through each voxel. Therefore, X-ray attenuation is primarily a function of X-ray energy and the density and atomic number of the material being scanned. Figure 8 shows two-dimensional (2D) micrographs showing porosity in the as-deposited clad material.

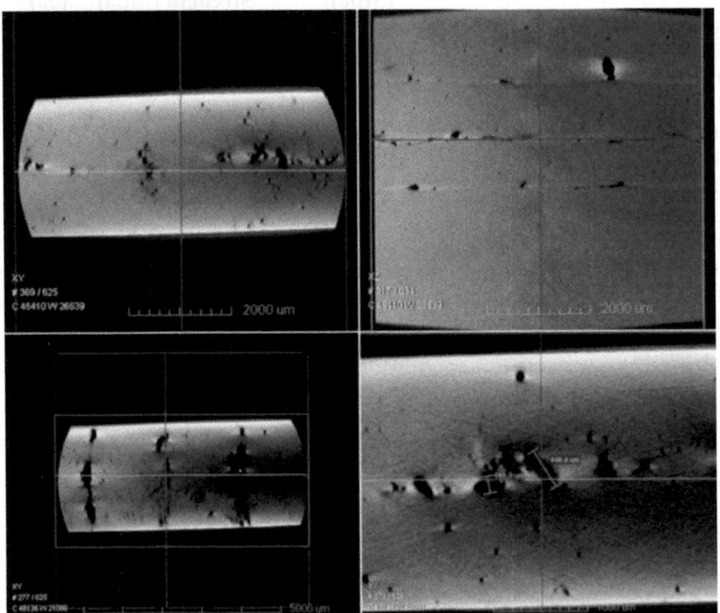

Figure 8: X-ray Micro CT 2-D micrographs showing porosity in the clad material

Significant porosity was identified in the clad region in both longitudinal and transverse directions. Porosity was primarily identified between the deposition layers. This can be clearly seen in the 3D views as shown in Figure 9. A similar image for the friction stir processed material was not possible because the porosity had been completely eliminated. This fact can be more clearly seen by looking at a comparison of Micro CT scans showing the as-deposited clad and friction stir processed material, Figure 10. No porosity or volumetric weld defects were identified in any of the processed material samples that were analyzed. This illustrates that FSP was effective in eliminating porosity in the as-deposited material.

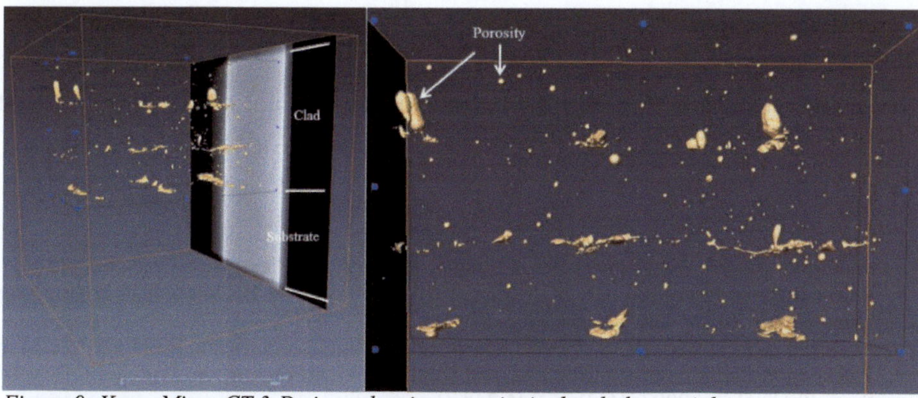

Figure 9: X-ray Micro CT 3-D views showing porosity in the clad material

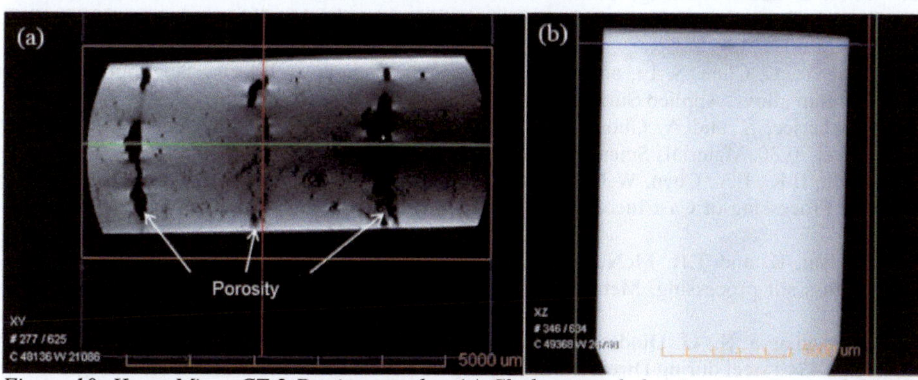

Figure 10: X-ray Micro CT 2-D micrographs: (a) Clad material showing porosity; and (b) FSP material showing no indications of porosity

Conclusions

Friction stir processing was successfully performed on 4340 DMD material. Defect free FSP passes were made using W-Re-HfC pin tools. Tools with 6% HfC performed better than tools with 4% HfC in terms of pin tool wear. Significant porosity was observed in the as-deposited 4340 clad material. FSP effectively eliminated porosity in the as-deposited material and improved the yield and ultimate tensile strengths in the processed region. No significant improvement in ductility with FSP was observed with ductility of the processed material comparable with the as-deposited material.

Acknowledgements

The authors gratefully acknowledge the financial support provided by the Office of Naval Research under the contract number N00014-10-C-0258.

References

1. Pinkerton, A., W. Wang, and L. Li, Component repair using laser direct metal deposition. Proceedings of the Institution of Mechanical Engineers, Part B: Journal of Engineering Manufacture, 2008. 222(7): p. 827-836.
2. Nowotny, S., S. Scharek, E. Beyer, and K.-H. Richter, Laser Beam Build-Up Welding: Precision in Repair, Surface Cladding, and Direct 3D Metal Deposition. Journal of Thermal Spray Technology, 2007. 16(3): p. 344-348.
3. Choi, J. and Y. Chang, Characteristics of laser aided direct metal/material deposition process for tool steel. International Journal of Machine Tools and Manufacture, 2005. 45(4): p. 597-607.
4. Tian, Y., C. Chen, S. Li, and Q. Huo, Research progress on laser surface modification of titanium alloys. Applied Surface Science, 2005. 242(1): p. 177-184.
5. Mukherjee, S. and A. Ghosh, Friction stir processing of direct metal deposited copper–nickel 70/30. Materials Science and Engineering: A, 2011. 528(9): p. 3289-3294.
6. Jasthi, B.K., E.Y. Chen, W.J. Arbegast, M. Heringer, D.R. Bice, and S.M. Howard, Friction Stir Processing of Cast Inconel 718. Friction Stir Welding and Processing VI, 2011: p. 25-32.
7. Oh-Ishi, K. and T.R. McNelley, Microstructural modification of as-cast NiAl bronze by friction stir processing. Metallurgical and Materials Transactions A, 2004. 35(9): p. 2951-2961.
8. Bhattacharya, S., G. Dinda, A. Dasgupta, and J. Mazumder, Microstructural evolution of AISI 4340 steel during Direct Metal Deposition process. Materials Science and Engineering: A, 2011. 528(6): p. 2309-2318.

Friction Stir Welding and Processing VIII
Edited by: Rajiv S. Mishra, Murray W. Mahoney, Yutaka Sato, and Yuri Hovanski
TMS (The Minerals, Metals & Materials Society), 2015

MANUFACTURING A SURFACE COMPOSITE MATERIAL MADE OF NANOCERAMIC PARTICLES OF TiC AND ALUMINUM ALLOY 7075 BY MEANS OF FRICTION STIR PROCESSING

David Verdera[1], Pilar Rey[1], Felipe García[2], Rocío Saldaña[2]

[1] AIMEN Technology Centre, Relva 27A, 36410, Porriño, Pontevedra, Spain
[2] COMIMSA, Calle Ciencia y Tecnología No 790 Colonia Saltillo 400
Saltillo, Coahuila, Mexico

Keywords: friction stir processing, surface composite, nanoparticles

Abstract

Friction Stir Processing is a useful tool to produce composite material. By friction stirring ceramic particles into an aluminum matrix it is possible to obtain functional surfaces with an improved performance compared with monolithic alloy. In this work, nano-particles of TiC (2% in weight) were added to aluminum alloy AA7075 to produce functional surface composite material. Different processing conditions were tested (number of passes, direction of each pass, sealed or unsealed groove, etc.) in order to find out the most suitable way to uniformly disperse the particles inside the matrix. Two different kinds of samples were compared: 7075 matrix plus an addition of 7075 powder, and 7075 matrix plus an addition of 7075 powder mixed with nano-particles of TiC. Micrographs, microhardness and wear resistant tests show the properties of the composite material made by means of friction stir processing.

Introduction

Metal Matrix Composites are a kind of material, formed by two different materials (matrix and reinforcement), with in general improved properties in comparison with monolithic alloy. Among these improved properties, it is possible to highlight wear and erosion resistance, stiffness or hardness. However, these special materials have also drawbacks, for instance an important decrease in ductility and toughness. In this sense it may however be noted that there are many applications in which only surface properties play an important role, as for instance, wear resistance. In these situations, only the surface layer needs to be reinforced by ceramic phases, while the bulk of the component should retain the original composition and structure with higher toughness. Hence, instead of bulk reinforcement, if ceramic particles are only added to the surface, wear and erosion resistance could be improved without sacrificing bulk properties [1,2].

There are several methods to fabricate particulate reinforced Al based composites to create functional surfaces. On one hand, methods based on applying a coating such as plasma, thermal spraying or laser surface treatment. On the other hand, methods based on using an insert, such as infiltration or squezze casting. In many cases these methods alter the component thickness or increase design complexity. In plasma/thermal spraying, the bond between the coating and the substrate is primarily mechanical and not metallurgical, so surface preparation of the substrate is extremely important. In processes involving liquid phase, important problems come up: interfacial reactions between reinforcement and matrix, segregation of the reinforcement; high viscosity of the melt compared with that of the un-reinforced alloy; porosity derived from the occluded gas and additional problems associated with the different CTEs and conductivities between metal and ceramic particles, etc. Furthermore, critical control of processing parameters is necessary to obtain ideal solidified microstructure in surface layer. Obviously, if processing of surface composite is carried out at temperatures below melting point of substrate, the problems mentioned above can be avoided [3-6]. For all these reasons, FSP has emerged as a promising alternative to create functional surfaces based on metal matrix composites reinforced with particles. Different methods have been reported on literature to create a composite surface by means of FSP [7-11]. One of them is by mechanizing a groove in the plate surface, deposit the powder over the surface or inside on a groove and then friction stir process on top of the groove.

In this work, aluminum metal matrix composites surfaces, using a solid state friction stir techniques were obtained. Pre-alloyed 7075 powder and 7075 powder reinforced with TiC nano particles in a 2% in weight were deposited into a surface grooves under different conditions. Microscope techniques were used to characterize the surfaces in order to determine the best way to incorporate the TiC particles: number of passes, direction of each pass, sealed or unsealed groove, etc. On the other side, a comparison between samples processed with unreinforced powder (without TiC particles) and reinforced (2% weight particles) was done. Microhardness and wear resistance tests were carried out in order to see differences derived from the addition of TiC particles

Experimental Procedure

AW 7075 T651 plates of dimensions 300x150x10mm were used as substrate, with nominal chemical composition as indicated in table 1.

Table 1. Chemical composition of 7075 T651 plates

Mg	Si	Fe	Cr	Cu	Zn	Mn	Ti	Al
2,80	0.05	0.15	0.01	1,93	5,92	0.01	0.02	Balanced

Two different powders added by FSP to the substrate: pre-alloyed 7075 (as received) and mechanical alloyed 7075 with 2% nano-TiC. Pre-alloyed 7075 powder aluminum alloy (90,53 Al; 5,14 Zn; 2,51 Mg; 1,46 Cu; 0,21 Cr; 0,071 Fe y 0,024 Si) with an average size of 30 μm was used. TiC particles with purity of 99% and average particle size of 20 nm were used as reinforcement. The as-received AA7075 and TiC powders were supplied by the Aluminium Powder Company Limited (APC) and Iolitec (nano) respectively.
A cyclic process of mechanical alloying, typical in ductile materials, was carried out with an attritor mill. This cyclic mode allows modification of the process kinetics to avoid sticking and agglomeration.

The mechanical alloying cyclic operation mode was performed with a 48s high rotary speed of 1300 rpm and a 12s low rotary speed of 1000 rpm (1 min cycle) during 480 min. The discharging cycle is the opposite of the operation cycle (12s at 1300 rpm and 48s at 1000 rpm). Powder oxidation was prevented using an argon atmosphere and 0.5% of process control agent (Licowax C from Clariant) which, furthermore, reduces the excessive cold welding between AA7075 particles at the beginning of the mechanical alloying. Initial crystallite size for the pre-alloyed 7075 is 80 nm while crystallite size for the mechanical alloyed 7075 with 2% nano-TiC after 480 min is almost 22 nm. Microhardness for pre-alloyed 7075 is 105 Hv approximately and 280Hv after 480 min of mechanical alloying with the 2% of nano-TiC reinforcement.

All FSP tests were performed on a friction stir welding machine PDS-4 Intelligent-Stir (MTS). A FSW tool with tapered pin of 5mm in its widest part and 4mm in its shortest part; pin length of 2.8mm and shoulder diameter of 12mm was used. Pin material is MP159 alloy and shoulder material is H13 steel. For the sealed of the groove, a pinless tool made of H13 was employed.
In previous work of this research group, it was found that a good combination of parameters for processing this alloy (AW7075 T651) was 1000rpm, 300mm/min, 1.5° tilt angle and 8kN of forge force. Based on this set of basic parameters, changes in other parameters such as groove condition, type of powder, number and direction of passes were analyzed. Table 2 with some of the samples tested is presented:

Table 2. FSP samples and conditions

Sample	Groove condition	Type of powder	Number of passes	Direction of 2nd pass
S6	Sealed	7075	1	----
S8	Sealed	7075	2	Same
S11	Sealed	7075 + TiC	1	----
S13	Sealed	7075 + TiC	2	Same
S17	Sealed	7075 + TiC	2	Opposite
S19	Unsealed	7075 + TiC	2	Opposite
S21	Sealed	7075	2	Opposite
S23	Unsealed	7075	2	Opposite
S24	Unsealed	7075 + TiC	1	----
S26	Unsealed	7075 + TiC	2	Same

In order to check if the powder was incorporated, macrographs in the transverse direction were obtained. Grinding, polishing and NaOH etching was used. Micrographs were performed by optical and scanning electron microscope. Specimens were mounted, polished up to 0.05 μm with colloidal alumina and finally etched with HF 0.5%. Pin on disk tests were performed on the stirred surface with a high temperature tribometer Microtest MT2/60/SCM. Wear tests were performed for 32 minutes against a 4 mm rotating steel ball (AISI 52100) at the constant load of 10 N and a sliding speed of 0.026 m/s. Hardness tests on the transverse section were carried out in a microhardness tester SHIMADZU using 1kg load. The plates were stored at room temperature and ambient atmosphere after FSP.

Results and Discussion

Basically, the visual aspect of the samples after FSP was very good, with no defects such as voids or cracks observed on the surface. The top surface shows very smooth quality and there are almost no prominences or depressions, due to the tool stirring.

Macrostructure for all samples is very similar. No gross defects such as voids or tunnel defect are detected and a homogeneous processed zone is visible in all samples. Figure 1 shows optical macrographs of all the cross section of the samples processed with different conditions. For samples S17, 19, 21 and 23, advancing and retreating side correspond to the second pass. The most significant difference in the stirred zone is due to the incorporation of AA7075 mechanical alloyed with TiC powder. S6 and S11 correspond with samples processed in the same conditions with pre-alloyed AA7075 (Sample 6) and mechanical alloyed with TiC AA7075 (Sample 11). Differences are caused due to different flow of the material. Mechanical alloyed powder with ceramic TiC particles are less plastic and cause a wider nugget than when original 7075 powder is used. In addition, differences can be found between samples processed when mechanical alloyed reinforced AA7075 powder is used with 1 or 2 passes (Sample 11 and 13 respectively). Again, more material around the pin is plasticized when 2 passes are carried out. Finally, in samples S11, S13, S17, S19, S24 y S26 processed with TiC powder reinforcement, the distribution of TiC particles is not uniform. The sealing of the groove results in more TiC incorporated (S11 compared with S24 and S17 with S19).

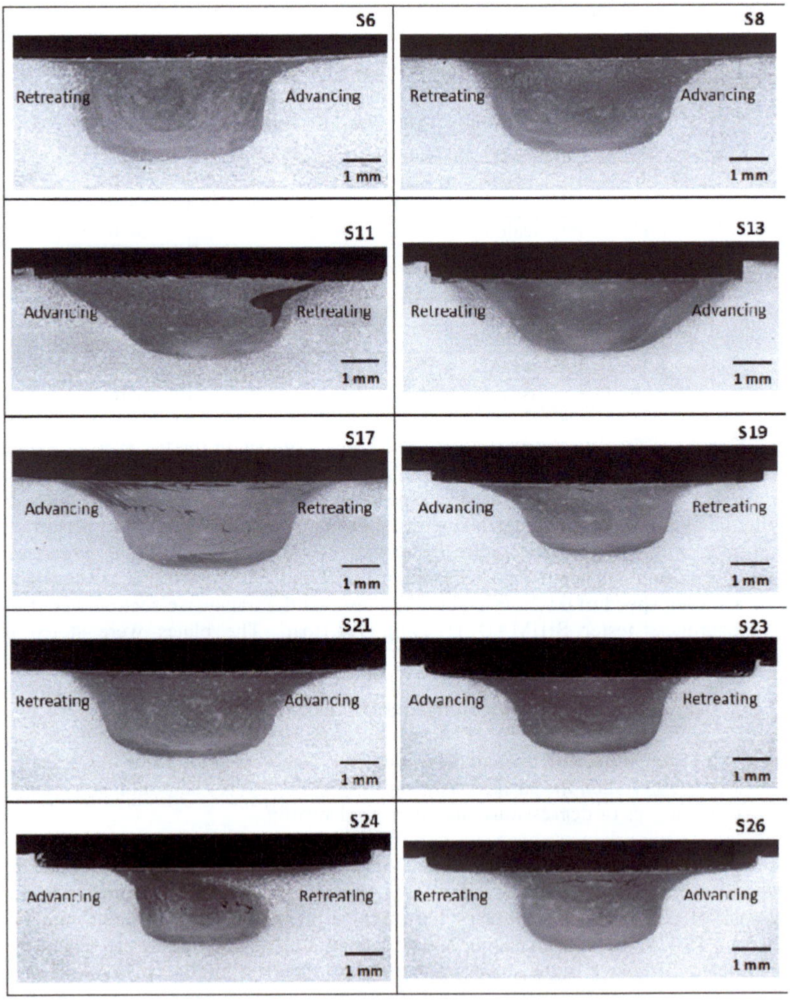

Figure 1. Optical macrographs of the cross section of all samples with different FSPed conditions

S17 is the sample that presented more homogeneous distribution of TiC particles at the surface. This sample was processed with sealed groove and two passes on opposite direction. The results with this double pass are consistent with previous research about multi-pass FSP [12]. Figure 2 shows the cross section and amplification of TiC distribution zones. FSP causes intense plastic deformation and high strain rates in the processed material resulting in precise control of the microstructure through material mixing and densification. It was observed that FSP refines the microstructure and the grain size, resulting in closure of porosities and provides a convenient method to improve the surface properties of aluminum alloy by forming surface composites.

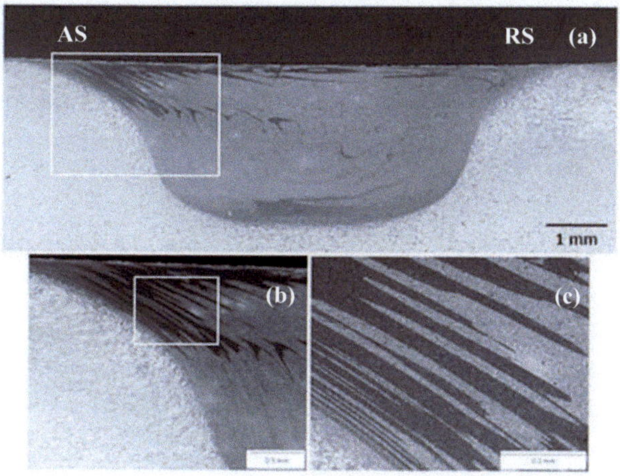

Figure 2. (a) Cross section of the sample S17. (b) and (c) Amplified zones of the TiC particle distribution.

Darker zones in Figure 2 correspond to TiC reinforced AA7075 zones. Figure 3 shows a SEM micrograph corresponding with this dark zone and the EDS analysis. Very small quantity of TiC is observed due to the low percentage of reinforcement in the AA7075 mechanical alloyed powders. It is not possible to differentiate the TiC particles because of their nanometric size.

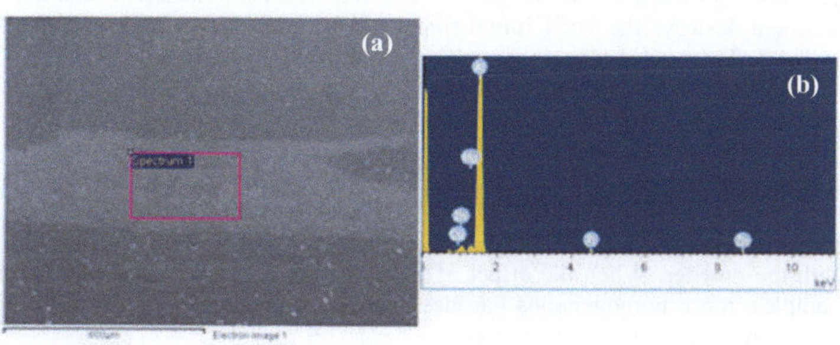

Figure 3. (a) SEM image of the particle distribution on sample S17. (b) EDS spectrum of the rectangle indicated zone.

Microhardness tests were carried out in the polished surface of the cross section of all samples according to Figure 4. Values of microhardness measurements are also illustrated in the graph. Advancing side of all samples is situated on the right (for samples S17, 19, 21 and 23 advancing side correspond to the second pass). Results showed in figure 4 correspond to microhardness tests performed one week after processing the samples by FSP. These tests were repeated one and two months after processing the samples. Several authors [13, 14] established the increase in mechanical properties for friction stir welded 7xxx aluminum alloys for very long natural aging times. However, no increasing was detected in the hardness values due to natural aging in the FSP samples reinforced with 7075 powder and 7075+TiC powder. Incorporation of powder and ceramic particles could have affected the natural aging of the alloy after FSP. Further research should be conducted to investigate why hardness values keep constant in time.

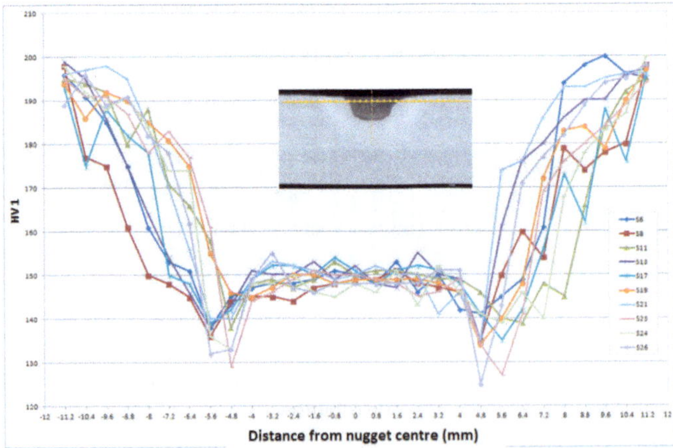

Figure 4. Distribution of the micro hardness values along the cross section of FSPed samples with image illustrating the measurement zones.

The hardness of the unprocessed 7075 aluminum alloy was about 195 HV. As a result of FSP, the average hardness measured on the stirred zone was 150 HV. Hardness falls due to the FSP thermal treatment destroys the T651 initial treatment resulting in lower hardness in the nugget, TMAZ and HAZ. The microhardness pattern of FSPed AA7075 samples exhibit the "W" shape for all samples. The two troughs of the W are located in the TMAZ and the crest occurs in the nugget zone. Hardness is more decreased on the advancing side and this effect is more noticeable with two passes processed samples. On the other hand, comparison between samples processed with the reinforced powder and with 2 passes, show that the hardness is higher in the nugget for the samples with the sealed groove (S13 and S17) compared with the unsealed ones (S19 and S26). In sealed samples, the powder remains into the groove while the pin tool stirs the base material with the powder. When the effect of the direction of the second pass is compared, for unsealed samples, more homogeneous hardness values are observed throughout the nugget for S19 (Opposite direction for the second pass).

Figure 5 shows the specific wear rate coefficient, K, for all FSPed samples and the base metal. All FSPed samples presented a decreased wear resistance that can be attributed to the softening of the aluminum. Probably, the small quantity of nano-TiC is not enough to increase the wear resistance on the surface, counteracting the softening due to the elimination of the T651 thermal treatment. On the other hand, the best wear rate of the samples corresponded to S17, which presented a homogeneous distribution of TiC particles on the surface of the stir zone. Despite sample S11 was processed with TiC particles and one pass, it presented a higher wear rate value. This higher wear rate is related with not homogeneous distribution of the TiC reinforced powder.

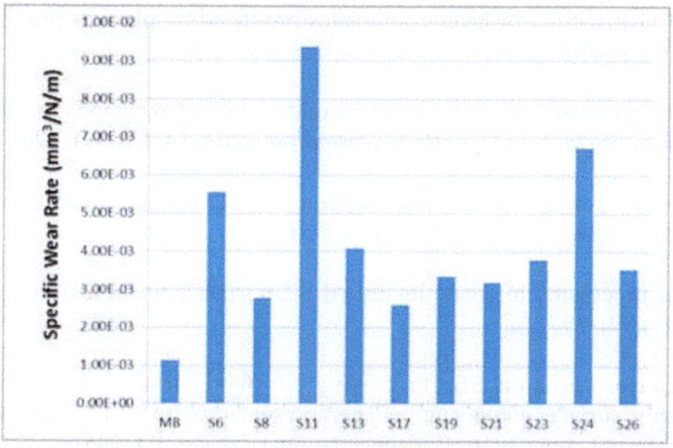

Figure 5. Specific wear rate coefficient, K, for the base alloy and for the FSPed samples

Conclusions

- All the processed samples with unreinforced and reinforced AA7075 powder exhibited a homogeneous stir zone without defects. However differences in the nugget can be observed as function of the groove conditions, the type of powder incorporated (with or without TiC) and the number of FSP passes.
- Hardness falls in the stirred zone in all FSPed samples as a consequence of the elimination of the T651 thermal treatment.
- Sample S17 processed with sealed groove and two passes on opposite direction presented more homogeneous distribution of TiC particles at the surface. S17 also presented the lower wear rate due to a homogeneous distribution of TiC particles on the surface of its stir zone.
- FSP is a good technique to incorporate reinforcement on the surface of metal alloys. However, additional tests with more quantity of TiC reinforcement and best distribution have to be made in the future to counteracting the softening due to the elimination of the T651 thermal treatment.

Acknowledgements

This work has been developed under the project: *Strengthen International Research Collaborations on the Development of Functional Surfaces between the European Union, Brazil and Mexico* (Grant Agreement nº 295254), supported by the European Commission under the FP7-People Programme Marie Curie's International Research Staff Exchange Scheme (IRSES).

References

1. Aleksandar Rac, Aleksandar Vencl, Tribological coatings on aluminium alloys used in engine cylinder bore-piston system, MVM– International Journal for Vehicle Mechanics, Engines and Transportation System, 30, Special Edition, December 2004, 105-113.

2. Reconditioning aluminium engine blocks, published by MSI Motor Services International GmbH, http://www.ms-motor-service.com/ximages/ks_50003804-02_web_leseprobe.pdf

3. Gui et al., 6061Al/Al-SiC_p bi-layer composites produced by plasma-spraying process, *Mater. Lett.* **46** (2000) 296-302.

4. B. Torres et al. Properties and microstructure of Al-11Si/SiCp composite coatings fabricated by thermal spray, *Surface and Coatings Technology*, 203, 14, (2009), 1947–1955.

5. D. Pantelis et al., Formation of wear resistant Al-SiC surface composite by laser meeting injection process, *Mater. Sci. Technol.* **11** (1995) 299.

6. Hu et al. Laser processing to create in-situ Al SiCp surface metal matrix composites', *J. Ma.t Sci.*, **30**, (1995) 891-897

7. P. Rey, D. Gesto. "Friction Stir Processing for fabricating a zirconium nitride/AW5083 aluminium alloy surface composite" (Paper presented at the 14th European Conference on Composite Materials, June 2010, Budapest, Hungary. Paper ID: 145)

8. A. Shafiei-Zarghani et al. "Microstructures and mechanical properties of Al/Al2O3 surface nano-composite layer produced by friction stir processing", *Materials Science and Engineering A*, 500 (2009) 84-91.

9. J. C. Huang et al. "Mg based nano-composites fabricated by friction stir processing", *Scripta Materialia* 54 (2006) 1415-1420.

10. J. C. Huang et al. "On the hardeming of friction stir processed Mg-AZ31 Based Composites with 5-20% Nano-ZrO2 and Nano-SiO2 particles", *Materials Transactions*, Vol 47, No 12 (2006) 2942-49.

11. A. Sert and O. N. Celik, "Wear behavior of SiC reinforced surface composite Al7075-T651 aluminum alloy produced using friction stir processing", Indian J. Eng. Mater. Sci., 21 (2014) 35-43.

12. Z.Y. Ma et al., Effect of Multiple-Pass Friction Stir Processing on Microstructure and Tensile Properties of a Cast Aluminum–Silicon Alloy. Scripta Mater. 54, (2006) 1623-26.

13. T.W. Nelson et al., In situ thermal studies and post-weld mechanical properties of friction stir welds in age hardenable aluminium alloys. Sci. Tech. Weld. Join. 8 [4] (2003) 283-288.

14. Christian B. Fuller et al. Evolution of microstructure and mechanical properties in naturally aged 7050 and 7075 Al friction stir welds, Mat. Sci. Eng. A 527 (2010) 2233–40.

Friction Stir Welding and Processing VIII
Edited by: Rajiv S. Mishra, Murray W. Mahoney, Yutaka Sato, and Yuri Hovanski
TMS (The Minerals, Metals & Materials Society), 2015

MICROSTRUCTURAL EVALUATION OF COLD SPRAY DEPOSITED WC WITH SUBSEQUENT FRICTION STIR PROCESSING

Tom Peat[1], Alexander Galloway[1], Tiziana Marrocco[2], Naveed Iqbal[2]

[1]University of Strathclyde, 16 Richmond St, Glasgow, G1 1XQ, United Kingdom
[2]TWI, Wallis Way, Catcliffe, Rotherham, South Yorkshire S60 5TZ, United Kingdom

Keywords: Cold Spray, Friction Stir Processing, Aluminium, Hardness, Microstructure

Abstract

Friction Stir Processing (FSP) has been shown to improve the strength, ductility and toughness of both aluminium and steel materials through grain refinement and the even distribution of precipitates within the substrate matrix. This article presents the application of FSP of Cold Spray deposited Tungsten Carbide – Cobalt (WC-Co) of two distinct types, on a series of aluminium substrates. Microstructural investigations of FSP processed samples exhibit interaction between the deposited WC-Co particles and aluminium alloy and show the homogeneous dispersion of deposited particles through the metal matrix. Results show that the dispersion of these particles varies with powder type, FSP parameters and substrate characteristics. A parallel study focusing on the hardness of the FSP generated Metal Matrix Composite (MMC) further demonstrate the potential of combining Cold Spray and FSP technologies to tailor surface properties for specific applications.

Introduction

Friction Stir Processing (FSP) is a variant of the widely adopted Friction Stir Welding (FSW) process used to carry out solid state welding of a variety of metals and alloys. FSP makes use of a rotating tool that is plunged into a chosen substrate. The heat generated as a result of friction, causes the substrate to plastically deform with the rotation of the tool forcing material from the leading edge round to the rear of the tool. Details of the FSP process can be found elsewhere, **[1] [2] [3] [4]**, but the beneficial features of the process include substantial grain refinement and homogeneous distribution of impurities and precipitates throughout the metal matrix, both of which lead to improvements in the mechanical properties of the material **[5] [6]**.

Cold Spray (CS) Deposition is a surface coating technique that was developed in the mid 1980's at the Institute for Theoretical and Applied Mechanics, Russian Academy of Science in Novosibirsk **[7]**. The process accelerates micro particles to high velocities by the use of compressed gas through a supersonic nozzle. Particles are fired at the material surface at a temperature lower than that of their melting point, with the bonding mechanism occurring as a result of plastic deformation due to the high velocities involved **[8]**. This subsequently leads to the coating layer being formed from solid-state particles. Given that the deposition is achieved without the melting of the powder particles, any undesirable phase transformations, common to high temperature deposition methods, are avoided **[9]**.

The present study is a highly novel preliminary assessment that considers the effect of CS deposited WC-Co, combined with subsequent FSP. The microstructure and hardness are evaluated for a variety of substrate and coating combinations. Limited data exists in the area of combined CS deposition and FSP. Previously, Hodder et al **[10]** have reported the co-deposition of Al and Al_2O_3 onto AA6061 Al alloy with the aim of investigating the influence of Al_2O_3 content on the coating following stirring. Hodder found that homogenous distribution of the hard Al_2O_3 particles within the MMC lead to increased hardness, with a reduction in the mean free particle distance of the Al_2O_3 particles below a distance of 5µm resulting in the greatest hardness increase. The aim of the present study is to investigate the effect of varying FSP tool geometry, coating type and substrate material on the microstructure and mechanical properties of the modified surface. The results from this study will form an initial insight to evaluate the interaction between deposited coating and substrate following FSP and develop a pathway for further research in this area.

Experimental Procedure

CS deposition was carried out using a CGT Kinetics 4000/47 CS system, connected to a PF 4000 Comfort series powder feeder operating at a powder feed rate of 3.5rpm. The carrier gas used to accelerate the powder particles was nitrogen and operated at a pressure of 30Bar for all WC-Co coatings. The CS apparatus was mounted to an OTC 6 axis robotic arm to increase repeatability and reproducibility of results. The spray gun was fitted with a long pre-chamber to increase the nozzle exit temperature of the powder to around 700°C. An iterative approach was used to determine the most appropriate stand-off distance and traverse speed, with a series of test coupons being sprayed with varying parameters. The stand-off distance between the substrate and nozzle was set to 50mm, with the nozzle traversing horizontally across the substrate at a speed of 50mm/s. It was found that further reducing the stand-off distance resulted in a shot blasting affect, with very little powder adhering to the substrate. In order to provide a suitably large coated area for subsequent FSP, eight passes were carried out, with an overlap of 2mm. This gave a total track width of 50mm. The two WC-Co powders used in the study have distinct internal structures. WC-17Co powder is comprised of nano scale Tungsten Carbides, evenly distributed throughout a Cobalt matrix **[11]**. WC-25Co is comprised of a Tungsten Carbide core surrounded by a Cobalt binder. The diameter of powder particles lies between 5µm and 30µm. The CS system was used to deposit the two powder types onto four aluminium substrates. Each test plate measured 200mm by 120mm by 3mm. Test plates were degreased using acetone prior to CS deposition. Table 1 outlines the FSP parameters that were applied to the coated specimens and the specific alloy grades assessed.

Substrate	Coating	Tool Type	Rotation Speed (RPM)	Traverse Speed (mm/min)	Plunge Depth (mm)	Dwell Time (s)
AA2024-T3	WC-25Co	Pinless Tool	300	50	0.25	0
AA2024-T3	WC-17Co	Pinless Tool	300	50	0.25	5
AA5083-O	WC-25Co	M44185	600	150	5.05	0
AA5083-O	WC-17Co	M44185	600	150	5.05	0
AA6082-T6	WC-17Co	M44185	600	300	5.05	0
AA6082-T6	WC-25Co	M44185	600	300	5.05	0
AA6N01-	WC-25Co	M44185	600	250	5.05	0
AA6N01-	WC-17Co	M44185	600	250	5.05	0

Table 1 – Test specimens with associated FSP tool type and process parameters.

As FSP tool design is crucial to producing a high quality stirred zone, two tool types were investigated. The first being a pinless tool with fixed shoulder comprised of concentric circular features, manufactured from tool steel, the second being an MS-M-005 PCBN tool with 5mm pin length and a fixed shoulder with concentric circular features. The 5mm pin also featured a spiral pattern to force material from the surface down to the root of the stir zone.

The following testing programme was conducted to achieve the outcomes of this study:

- Optical Microscopy: to establish the interaction, (if any), between deposited coating and substrate. Evaluate the distribution of deposited coating through the substrate. Images were taken using an Olympus G51X series optical microscope.
- Scanning Electron Microscopy (SEM): To verify images obtained from optical microscopy and to identify the elements present within the MMC. WC-Co content was quantified by taking SEM images at prescribed locations and carrying out Energy-dispersive X-ray spectroscopy (EDX) analysis. Hitachi S-3000N VP-SEM series Scanning Electron Microscope (SEM) with EDX used to carry out analysis.
- Micro-hardness measurement: To establish the hardness at various locations within the stir zone. Mitutoyo MVK-G1 micro-hardness tester was used to evaluate hardness with a 200gf load.
- The areas assessed are outlined in Figure 1 with the locations kept consistent throughout all analysis.

Figure 1 displays the cross section of a friction stir processed region. Numbers 1-4 signify the locations in which optical and electron microscopy were carried out with number 5 denoting the friction stirred root region. The advancing side and retreating side of the stir zone have also been indicated. It should be noted that the advancing side and retreating side on samples W30 and W31 are reversed due to the rotation of the pinless tool being opposite to that of the PCBN tool.

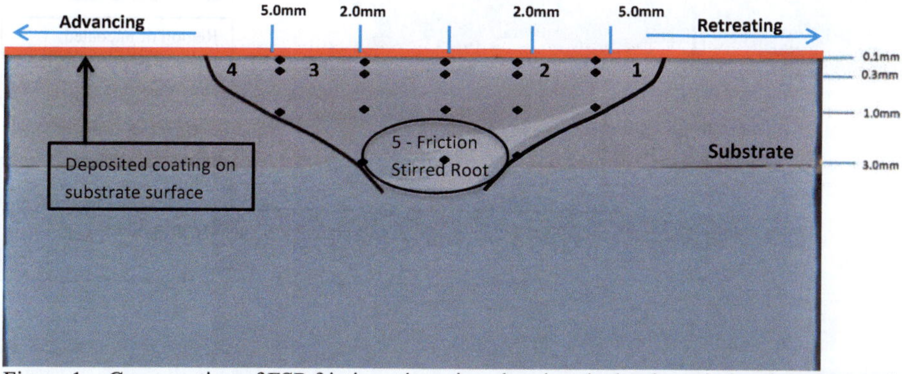

Figure 1 – Cross section of FSP friction stir region showing the hardness indent locations and locations of optical and SEM images.

Results

Table 2 outlines the nomenclature adopted to describe individual specimen corresponding to a particular coating and substrate combination.

Specimen Code	Substrate	Coating
W30	AA2024-T3	WC-25Co
W31	AA2024-T3	WC-17Co
W41	AA5083-O	WC-25Co
W44	AA5083-O	WC-17Co

Specimen Code	Substrate	Coating
W50a	AA6082-T6	WC-17Co
W50b	AA6082-T6	WC-25Co
W54a	AA6N01-T6	WC-25Co
W54b	AA6N01-T6	WC-17Co

Table 2 – Sample coding system.

Optical Microscopy

Optical microscopy was employed for initial assessment to determine whether CS deposited particles had been successfully embedded into the substrate during FSP processing, or simply removed by the FSP tool. Figure 3 exhibits a magnified region of the top surface of sample W31, showing a larger concentration of WC-Co particles, compared to that seen in W30. This suggests that the amount of WC-Co coating present in W31 is greater than that found in W30. A comparison of both images reveals that W31 exhibits a more homogeneous layer of embedded particles, with the WC-Co regions present in W30 being relatively inconsistent in size and shape.

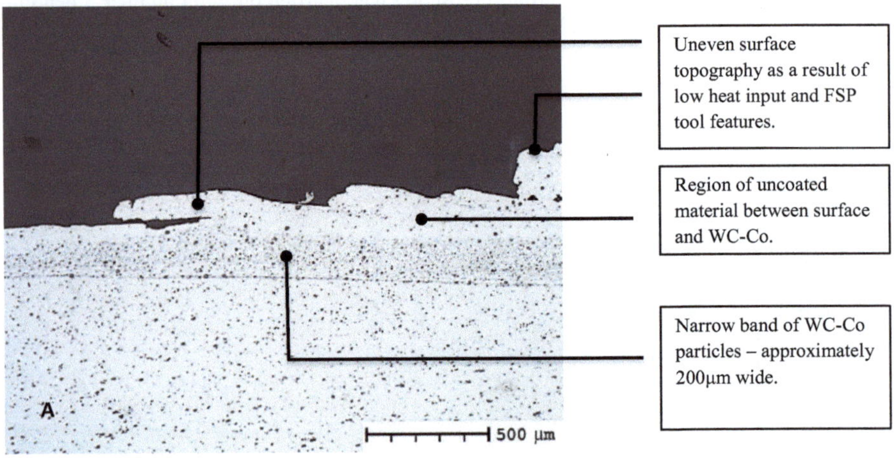

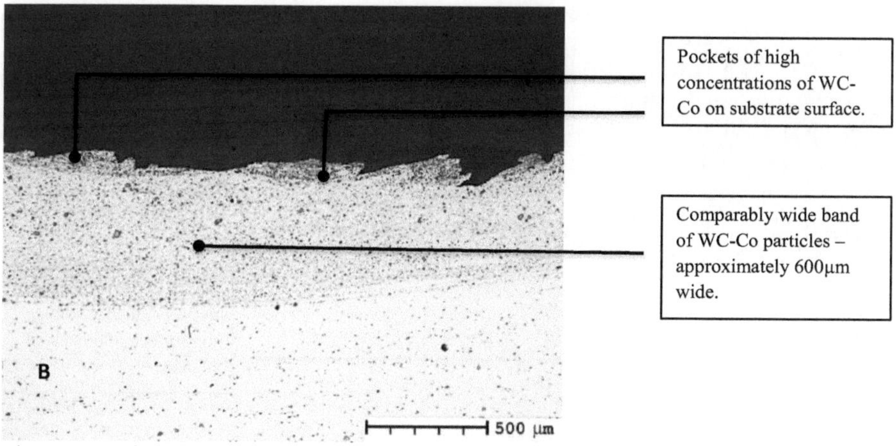

Figure 3 – Micrographs displaying advancing side of A) Specimen W30 B) W31 located in region 2 from Figure 1, prepared to a 1μm finish. [x50 Unetched]

Previous studies have shown that the heat input to the stir zone relates to the plasticity of the substrate material **[12]**. Based on this, the 5s dwell time during FSP of sample W31 has increased the heat in the FSP region, thereby increasing the plasticity of the substrate alloy. This, in turn has permitted a greater quantity of the coating to be embedded into the softer substrate. It is likely that particle morphology has also influenced the particle flow behaviour during FSP, and has affected the distribution pattern seen in the substrate material. The nano scale carbides within the Cobalt matrix in WC-17Co could permit the larger (≈200μm) particles to break down more easily and embed within the substrate. It is important to highlight that despite the differences between the two samples, both exhibit significant interaction and mixing between the coating and substrate. The next series of images refer to the specimens processed using the 5mm PCBN tool. Figure 4 shows magnified images of region 3 in specimens W50a and W50b, with Figure 5 exhibiting the stir root for specimens W54a and W54b.

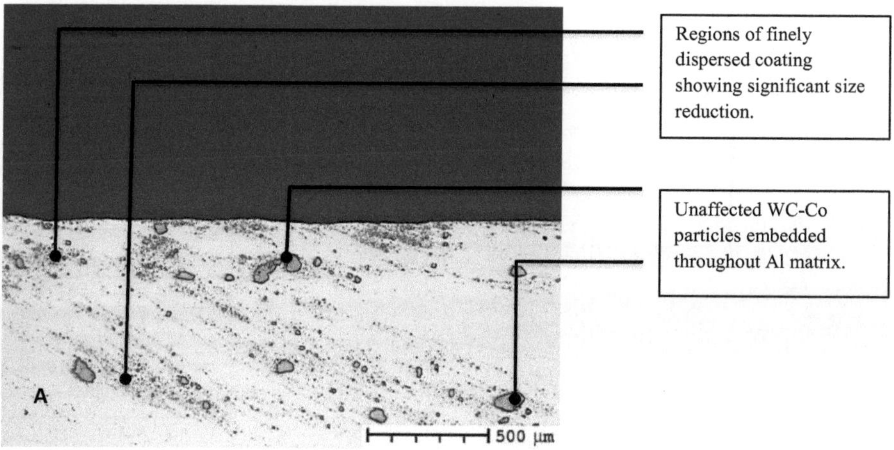

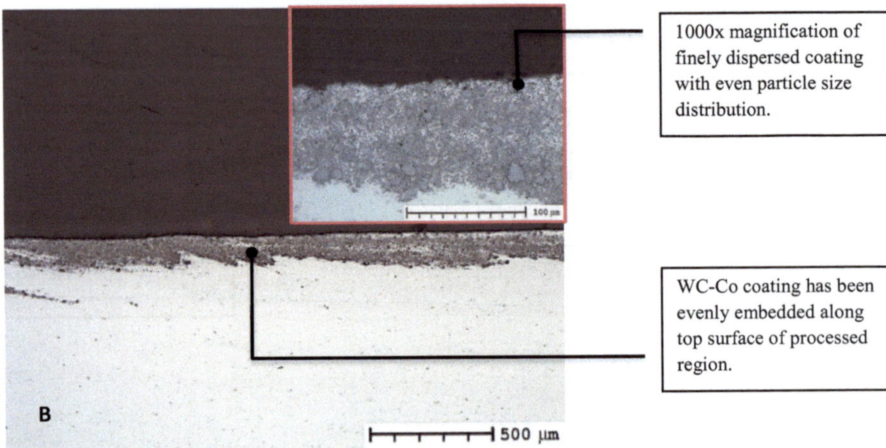

Figure 4 – Micrographs of A) Specimen W50a, B) Specimen W50b taken at Location 3 from Figure 1, prepared to a 1µm finish. [x50 Unetched]

Specimen W50b exhibits a homogeneous distribution of particles along the top surface of the substrate, whereas W50a contains large, (≈200µm) particles that are present throughout the stir zone. Given that both samples were stirred using the same tool, and identical FSP parameters, it is likely the variation in particle morphology and their distinct flow behaviours in plasticized aluminium during processing, has resulted in this distinct difference in microstructure.

Tool traverse speed was reduced to 250mm/min for samples W54a and W54b, with all other parameters kept consistent with previous samples. In doing so, the heat input into the sample is increased. A by-product of this increased heat is increased plasticity in the stir region, which can potentially permit greater flow of the deposited coating particles. This feature of material flow was investigated by Thangarasu et al; and demonstrated that an increase in the traverse speed leads to an increase in micro hardness **[13]**.

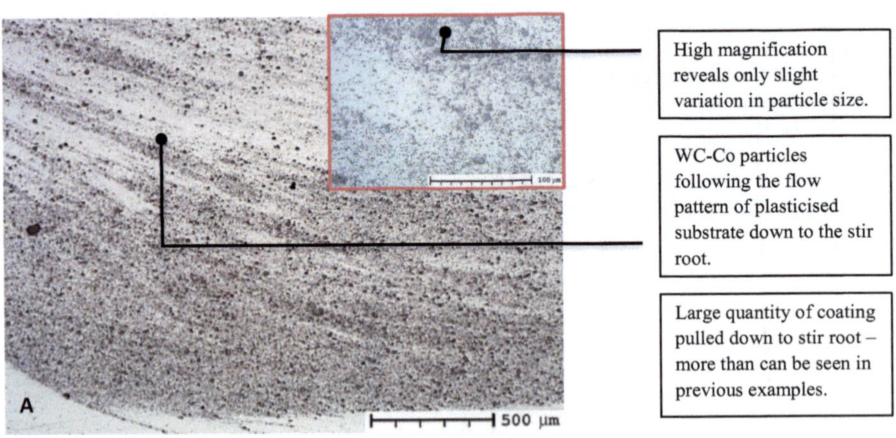

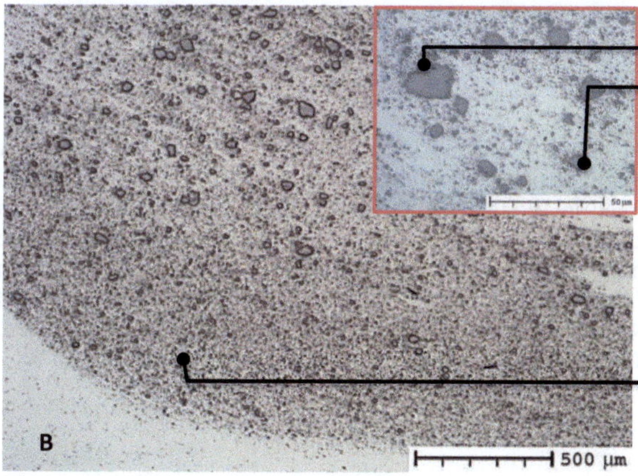

Figure 5 – Friction Stir root (Figure 1) comparison for A) Specimen W54a, B) Specimen W54b. [x50 Unetched]

Assessment of microstructure at locations 2 and 3 of specimens W54a and W54b showed significantly less WC-Co particles at the surface when compared to that seen in Figure 4, with the quantity of coating particles varying significantly between the advancing and the retreating sides. However, as shown in Figure 5, large quantities of WC-Co particles were present at the root of the stir zone in both specimens. Though heat input has potentially caused more of the coating particles to flow down towards the root of the stir zone, the varied material flow patterns of different aluminium alloy substrates, even with similar FSP tool and process conditions, needs to be considered and may well determine the distribution pattern of coating particles within the substrate. Moreover, particle size also reduces towards the base of the pin, which aligns with the area exposed to the highest shear forces **[14]**. The results from specimens W54a and W54b also corroborate those found in W50a and W50b in relation to the breakup of powder particles. Although not shown within this report, similar studies on specimens W41 and W44 again showed that coating WC-25Co is more easily broken down and dispersed through the substrate.

EDX Analysis

Using SEM, EDX analysis was performed to assess the elements present within the stir zone. Table 3 illustrates the results from the EDX analysis of various locations within the specimens. A diagram showing the assessed locations can be seen in Figure 1.

Spectrum		**O**	**Mg**	**Al**	**Mn**	**Co**	**Cu**	**W**	**Total**
Sample I.D.	**Location**								
W31	1	0	1.21	83.25	0.63	**1.07**	5.29	**8.56**	100
W31	2	0	1.25	85.73	0.55	**0.95**	5.17	**6.35**	100
W31	3	0	1.15	80.83	0.63	**1.67**	4.52	**11.18**	100
W31	4	1.81	1.2	81.43	0.55	**1.19**	4.92	**8.89**	100
W50b	2	0	0.52	63.27	0	**5.44**	0.48	**30.29**	100
W50b	3	0	0.5	62.9	0	**4.67**	0.56	**31.31**	100
W50b	5	0	0.48	89.08	0.54	**0.93**	0.7	**8.28**	100
W54a	2	0	0.53	85.56	0	**1.89**	0	**12.02**	100

W54a	3	0	0.53	69.63	0	**4.84**	0	**25**	100
W54a	5	0	0.48	75.91	0	**2.88**	0.63	**20.1**	100

Table 3 – EDX analysis of FSP regions.

The results indicate a distribution pattern of W and Co particles across the width of each stir zone, with a slight reduction in weight % observed in the retreating sides of W31 and W50b, with a significant reduction observed in W54a. This variation in elemental distribution is a potential consequence of reduced frictional heating, observed in the retreating side of the tool. The impact of tool design on the distribution of coating particles is also evident through relative comparison of sample W31, produced using a pinless tool, with samples W50b and W54a, produced using the PCBN tool. Higher heat input due to higher tool rotation speed, resulting in increased plasticity is also an important factor in increasing the quantity of WC-Co in the Al matrix.

Hardness

Micro hardness measurements were taken for the FSP processed region with and without deposited coating, along with unaltered substrate. The results of these measurements can be seen in Table 4, with the location of hardness intends being outlined in Figure 1.

				<Advancing		Retreating>	
Sample I.D.			Distance from Centre (mm)				
	Un-modified Substrate (Hv)	**Distance from Surface (mm)**	-5	-2	0	2	5
			Hardness (Hv)				
W31-1	130	0.1	226	201	214	124	166
		0.3	166	148	150	148	158
W44 (coated + stirred)	77	0.1	233	171	90.1	82.8	87
		0.3	217	84.6	161	79.9	84
		1	90	82.2	91.3	81.5	102
		3	85.3	86.3	84.5	80.9	87.2
W50b (stirred)	85.5	0.1	70.2	67.3	67.2	75.8	73.4
		0.3	77.8	76.8	76.7	75.1	76.3
		1	74.6	75.1	75.2	72.6	75.6
		3	N/A	57.8	71.3	72.2	N/A
W50b (coated + stirred)	85.5	0.1	76.4	91	86.7	84.2	77.2
		0.3	77.8	79.4	79	79.6	81.4
		1	77.7	76.8	72.6	79.5	80
		3	N/A	60.6	75.5	71.3	N/A
W54a (stirred)	74	0.1	63.4	51.3	66	53.2	65
		0.3	66.7	65.9	67.7	67.7	68.1
		1	66.5	66.4	63.9	65.7	66
		3	N/A	64.9	69.7	66.5	N/A
W54a (coated + stirred)	74	0.1	62.4	55.8	47.6	59.2	70.2
		0.3	74.6	72.1	71.2	69.4	69.1
		1	72.4	70.4	72.4	68.1	70.3
		3	N/A	74.3	85.5	82.4	N/A

Table 4 – Hardness variation throughout stirred region.

Results indicate that samples W31 and W44 exhibit a significant increase to hardness post FSP processing, in regions with a high weight % of WC-Co. Specimens W50 and W54, however, exhibit a drop in hardness when compared to the parent substrate. Given the elevated temperature associated with the high rotation speeds in these samples, it is likely this has had a softening effect on the hardened, T6, condition. This softened matrix is incapable of supporting the hard WC-Co particles under load and as a result, the hard phase particles have little impact on the hardness of the matrix, even in regions of high coating density, i.e. the stir zone root. Despite this, the non-heat treated alloy, AA5083-O responded well to the process – showing an increase in hardness of over 200% in one location.

Conclusion

- The hardness of AA6XXX-T6 alloys decreases with FSP. The high heat input results in an annealing effect of the hardened condition.
- The AA2024-T3 alloy shows an increase to hardness of around 43%, after FSP of CS deposited WC-Co coating.
- The AA5083-O alloy demonstrated significant hardness increase of over 200% after FSP of CS deposited WC-Co coatings
- When compared to pinless tool geometry, the PCBN tool promotes greater mixing and increases weight % of WC-Co embedded within the Al matrix due to the higher rotation speeds and addition of the 5mm spiral pin.
- Specimens produced using the PCBN tool demonstrate that WC-25Co experiences a higher level of particle size reduction compared with WC-17Co. This confirms that powder morphology has a significant effect on the resultant MMC.

References

[1] H. Krohn, R.M. Miranda, P. Vilca, L. Quintino, J.F. dos Santos J. Gandra, "Friction Surfacing - A Review," *Journal of Materials Processing Technology*, (2013).

[2] Devinder Yadav, G. Suhas Ranjit Bauri, "Effect of friction stir processing (FSP) on microstructure and properties of Al–TiC in situ composite," *Materials Science and Engineering* , vol. 528, no. 13-14, (May 2011), pp. 4732–4739.

[3] M.L. Santella et al, "Effects of friction stir processing on mechanical properties of the cast aluminium alloys A319 and A356," *Scripta Materialia* vol. 53, (2005).

[4] João Gandra, "Friction stir processing," *Surface Modification by Solid State Processing*, pp. 73-111, (2014).

[5] C.J. Lee, J.C. Huang C.I. Chang, "Relationship between grain size and Zener–Holloman parameter during friction stir processing in AZ31 Mg alloys ," *Scripta Materialia* , vol. 51, (May 2004), pp. 509-514.

[6] Huijie Liua, Sudarsanam Suresh Babu Xiuli Feng, "Effect of grain size refinement and precipitation reactions on strengthening in friction stir processed Al–Cu alloys," *Scripta Materialia*, 65, (September 2011), pp. 1057-1060.

[7] V. Kosarev, K.V. Klinkov, A. Alkhimov, V.M.Formin A. Papyrin, *Cold Spray Technology*. Oxford: Elsevier, (2006).

[8] E.R. Nivinski B.A. Kushner, "Thermal Spray Coatings," vol. 18, (1992).

[9] V K Champagne, *The Cold Spray Materials Deposition Process*.: Woodhead Publishing, (2007).

[10] H Izadi, AG McDonald, A.P. Gerlich K.J. Hodder, "Fabrication of aluminium-alumin metal matrix composites via cold gas dynamic spraying at low pressure followed by friction stir processing," *Materials Science & Engineering, vol. 556,* (2012), pp. 114-121.

[11] , Xiao Chen a,b, Xiao-Bo Bai a,b, Gang-Chang Ji a,b,□, Zeng-Xiang Dong a,b, Deng-Liang Yi a Hong-Tao Wang a, "Microstructure and properties of cold sprayed multimodal WC–17Co deposits," *Int. Journal of Refractory Metals and Hard Materials*, vol. 45, (July 2014), pp. 196-203.

[12] Y. Miyamoto, Y. Watanabe, Y. Natio, T. Nose H. Sato, "Fragmentation behavior of Al3Ti particles in Al-Al3Ti composite deformed by friction stir processing," *Proceedings of the 1st International Joint Symposium on Joining and Welding* , vol. 1, (November 2013), pp. 453-457.

[13] N. Muruganb, I. Dinaharanc, S.J. Vijayd A. Thangarasua, "Influence of Traverse Speed on Microstructure and Mechanical Properties of AA6082-TiC Surface Composite Fabricated by Friction Stir Processing," *Procedia Materials Science*, vol. 5, (2014), pp. 2115-2121.

[14] Ehab A. El-Danafa Magdy M. El-Rayesa, "The influence of multi-pass friction stir processing on the microstructural and mechanical properties of Aluminum Alloy 6082," *Journal of Materials Processing Technology*, vol. 212, no. 5, (May 2012), pp. 1157-1168.

[15] M. Saadatmand, J. Aghazadeh Mohandesi, M. Salehi, "Optimization of process parameters for producing AA6061/SiC nanocomposites by friction stir processing," *Transactions of Nonferrous Metals Society*, vol. 22, no. 5, (May 2011), pp. 1055-1063.

FRICTION STIR WELDING AND PROCESSING VIII

Friction Stir Related Technologies

Friction Stir Welding and Processing VIII
Edited by: Rajiv S. Mishra, Murray W. Mahoney, Yutaka Sato, and Yuri Hovanski
TMS (The Minerals, Metals & Materials Society), 2015

FRICTION STIR WELDING TECHNOLOGY FOR MARINE APPLICATIONS

Jonathan Martin, Sam Wei

TWI Technology Centre (Yorkshire)
Wallis Way, Catcliffe, Rotherham, S60 5TZ, UK

Keywords: FSW, Bobbin, Marine, Mobile

Abstract

Friction Stir Welding (FSW) is a very attractive alternative to MIG welding for joining aluminium alloys and is an established joining technique in the marine industry. One of the limitations of FSW is that the process requires relatively high levels of force to be applied, primarily to maintain the correct tool position, during welding. This requirement typically results in large, heavy, expensive FSW machines with limited scope for making very large panels. A new FSW technique called “floating-bobbin” which allows welds to be made with almost zero vertical force on the weld, offers the potential for smaller, less expensive, and critically mobile equipment to be developed. Floating Bobbin Friction Stir Welding (FBFSW) is accomplished with a "fixed" bobbin tool where the two shoulders and pin features are constrained rigidly to each other however the whole tool can float in a holder in the vertical axis. This arrangement has advantages in that the tool automatically compensates its position relative to the component eliminating the need for accurate set up procedures and sophisticated position or force control systems. This approach is designed to provide uniform heat input, minimising distortion and eliminate the potential for root defects. This paper describes the development and assessment of joining aluminium alloys using the FBFSW technique and its suitability for use on a mobile FSW system designed for use in a shipyard. The principles of this technique are explained including the assessment of welding forces, tool life, weld properties and panel distortion when compared to conventional welding techniques. The development and exploitation of FBFSW offers the potential for a low cost, reliable solid-phase joining technique and the possibility of mobile FSW equipment to be developed.

Introduction

Aluminium is used extensively in shipbuilding worldwide. It is an ideal metal being light, strong, and corrosion-resistant. FSW was adopted as a suitable fabrication technique within 5 years of its invention by TWI in 1991 [1]. FSW offers many benefits including:

- Minimal distortion and good aesthetic appearance – thus reducing the need and cost for post-welding remedial work such as plate straightening and filling.
- User and environmentally friendly – no UV radiation or fumes.
- Elimination of flaws associated with conventional fusion welding (hot cracking, porosity, lack of fusion).
- Wider availability of operators, as the process is assured through automation.
- Ability to weld in any plane or orientation (unaffected by gravity as the weld is not molten).

The first commercial application of FSW was the production of on-board-ship fish freezing panels by Swedish extruder Sapa [2]. However since that time many other applications have been put into production such as deck panels, floors, hulls, superstructures and platforms.

Friction stir welding has been used extensively in the manufacture of pre-fabricated panels, away from the shipyard before final assembly on site using fusion welding. These panels are manufactured on high capital, high process force machines which can fabricate panels of a finite size. The panel size is also often limited by the panel transportation method to the shipyard.

A number of portable FSW systems have been developed for use in shipyards. Examples of such systems include a machine manufactured by the University of Adelaide and a prototype low-cost, portable FSW machine built by the US Navy Metalworking Centre for production of large panels [3]. However currently there is no commercially available mobile FSW system.

One of the major limitations of FSW which prevents its use on a mobile machine is that the process requires relatively high levels of force to be applied, primarily to maintain the correct tool position, during welding. The challenge was therefore to develop a mobile FSW system incorporating a crawler system which would be low cost, transportable and have no limit on panel and weld length. This system must be capable of resisting the welding forces and provide a precise, smooth FSW tool path movement. A project called Mobi-Weld was therefore funded by the European Union Seventh Framework Programme (FP7) with the objective of developing a prototype mobile FSW system for use in final fabrication/assembly in a shipyard.

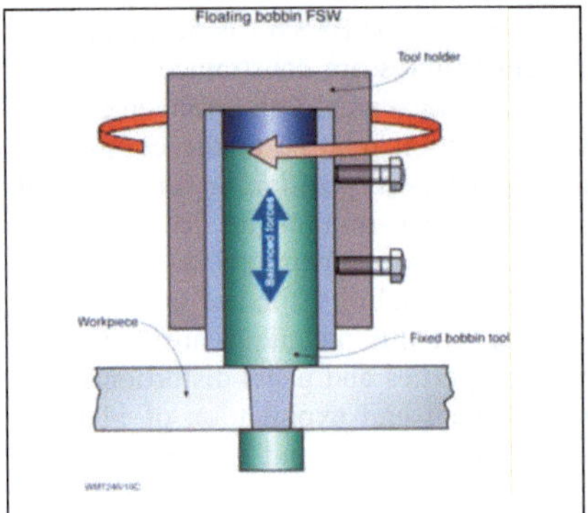

Figure 1. Basic principle of the floating-bobbin FSW tool which can find its own operating position

Reduction of process forces was one of the key requirements allowing the mobile application of FSW and this was thought to be achievable using a variation of the fixed gap bobbin technique. Fixed gap bobbin tools were considered very early in the development of FSW by researchers at TWI based on the perceived benefit of internalizing the forging force on the welding tool, elimination of potential lack of penetration welding and eliminating the welding fixture backing bar. More recent work by CTC has developed the technique further using tapered shoulders [4]. TWI has taken this a step further by allowing the fixed bobbin tool to ‘float’ axially in a sleeve which allows the bobbin tool to find the position of least resistance during welding by balancing the force on each of the shoulders automatically resulting in reduced process forces and offering very simple control required by the FSW machine. This technique is called Floating-Bobbin Friction Stir Welding (FBFSW) and a schematic diagram is shown in Figure 1.

This paper presents some of the development work involved in designing and testing FBFSW tools suitable for use on the Mobi-Weld system and assessing tool life, effect of plate conditions on weld quality and panel distortion when compared with conventional welding techniques.

FBFSW Development

Initially 4 and 6mm thick AA6082-T6 and AA5083-H111 components were selected to represent the range of typical aluminium alloys and plate thicknesses used in the fabrication of marine hulls and decks.

Two floating-bobbin FSW tools were designed and manufactured to weld the two aluminium alloy thicknesses which featured balanced tapered scrolled shoulders and a four flat probe. A picture of a typical tool is shown in its individual components and assembled in Figures 2a and b respectively.

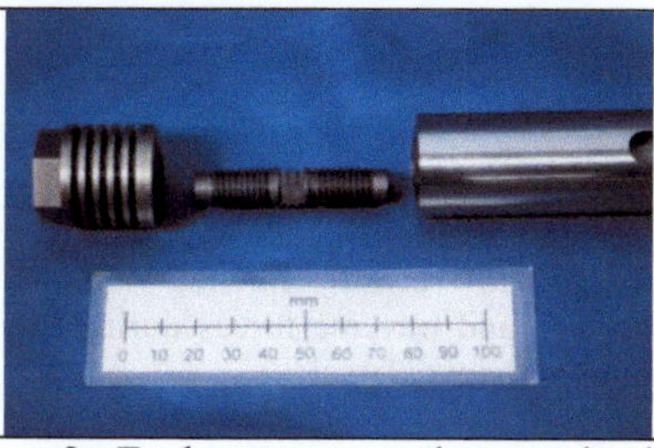

Figure 2a Tool components: bottom shoulder, probe and top shoulder

Figure 2b Bobbin tool as assembled

Bobbin welds do not have an initial phase where the welding tool probe is plunged axially into the workpiece as is done in conventional FSW. Bobbin welds are typically started by driving onto the edge of the plate with an initially slow travel speed until plastic deformation starts (engage phase), followed by acceleration of the travel speed (ramp phase) to the final steady state value. The typical weld surface quality produced by the FBFSW tool is shown in Figure 3. The tapered shoulders produce an almost 'flashless' weld surface on either side of the component.

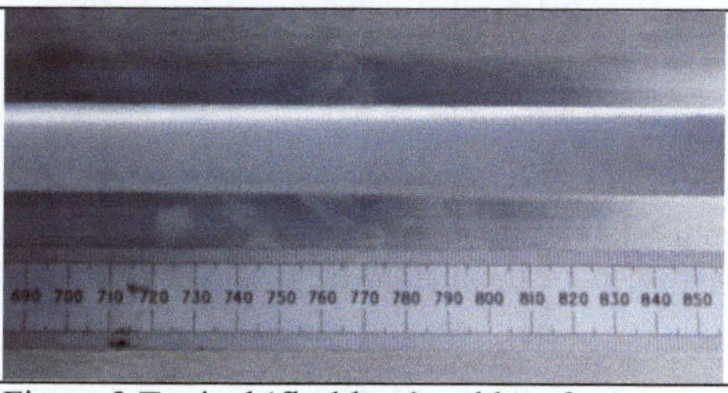

Figure 3 Typical 'flashless' weld surface

The initial weld development work focused on minimising the forces experienced during the weld cycle. Low force welding was very important to the design and capability of the Mobi-Weld mobile crawling system as vacuum pads were the chosen technique for clamping it to the plates to be welded.

The engage and ramp phase of the weld cycle were found to be the most unstable and welding 6mm thickness AA5083-H111 produced the highest forces. Figure 4 shows the first 100mm of a typical weld record. Peak welding forces for both the traverse force (X-Force) and the force normal to this (Y-Force) were observed during the engage and ramp phases. These forces were reduced by using an increased tool rotation speed during these phases, increasing the heat generated; softening the component material which reduced the forces as shown in Figure 5.

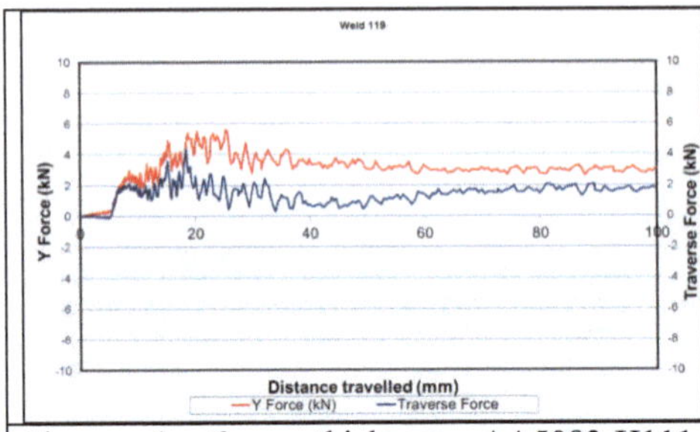

Figure 4. 6mm thickness AA5083-H111 bobbin weld record showing the Y and traverse forces during the first 100mm of the weld cycle.

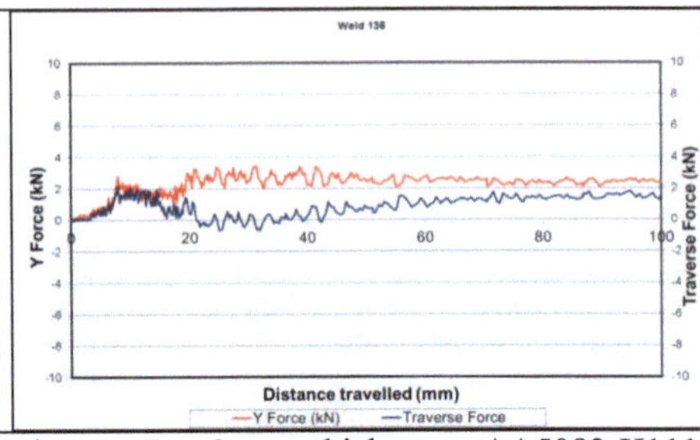

Figure 5. 6mm thickness AA5083-H111 bobbin weld record showing significantly reduced Y and traverse forces during the first 100mm of the weld cycle

These initial trials provided the general design data for the crawler system, as set out in Table I.

Table I Design Data for the Crawler System

Alloy grade (thickness mm)	Rotation speed	Traverse rate	Traverse force	Y force	Z force	Spindle Torque
	rpm	mm/min	kN	kN	kN	Nm
AA5083 (6)	250	240	2.0	3.9	0.0	80
AA6082 (6)	600	600	1.6	2.1	0.0	60
AA5083 (4)	400	400	0.8	1.7	0.0	40
AA6082 (4)	800	750	1.0	0.9	0.1	30

These parameters produced welds exceeding the ISO 25239 quality specification as assessed by one cross tensile and two bend tests extracted from a fabricated sample test plate at the start of the weld, and similar tests with an additional macroscopic examination from samples extracted from the end portion of a test plate.

Typical macrosections taken from the range of aluminium alloys and thicknesses are shown in Figure 6.

a 6mm Thickness AA6082-T6

b 4mm Thickness AA6082-T6

c 6mm Thickness AA5083-H111

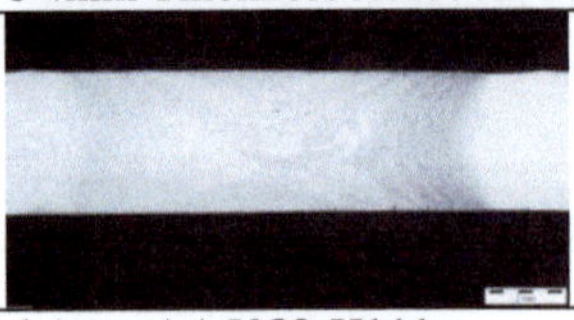

d 4mm AA5083-H111

Figure 6 Typical macrosections taken from the range of aluminium alloys and thicknesses

The typical welding efficiency obtained when welding AA6082-T6 was 75-80% and failure occurred in the parent material when testing AA5083-H111.

The Effect of Plate Conditions on Weld Quality

An assessment of the FBFSW techniques capability to produce welds to the ISO25239 quality standard when joining plates with gaps between the butting faces and different plate thicknesses was then made.

Firstly plates were machined as shown in Figure 7 and then a standard ISO25239 test plate was fabricated. The plates were assessed as previously described. It was found that the FBFSW tools could produce welds which exceeded the ISO standard between plates with a gap in the joint equivalent to 5% of the plate thickness.

To assess the technique's capability to weld plates of different thicknesses, four different joint configurations were produced and welded as shown in Table III.

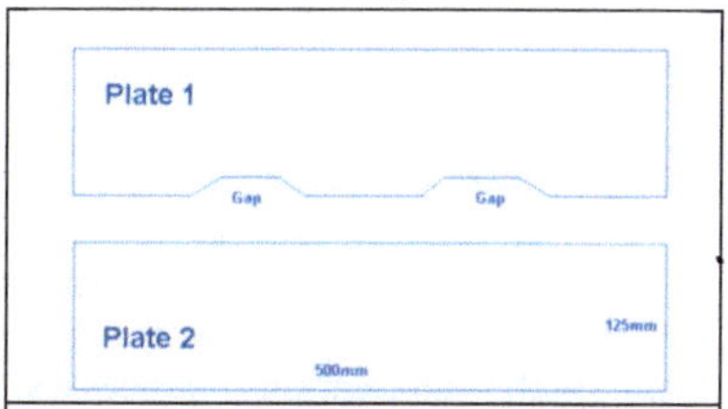

Figure 7. Schematic drawing of areas of machining to introduce a gap between the plates to be welded.

Table III Plate Thickness and Position Relative to Weld Joint Line

	Plate thickness tolerance	
Weld no	Advancing side of weld	Retreating side of weld
1	Lower	Lower
2	Lower	Upper
3	Upper	Lower
4	Upper	Upper

A plate thickness tolerance of 0.3mm was the limit of the FBFSW tools before voids appeared (Figure 8). Voids were typically found in joints which did not have sufficient material to provide the necessary reaction with the tool shoulders to generate the required heat to adequately soften the workpiece, leading to poor material flow. Welds having small voids typically passed the ISO25239 assessment as they were typically on the neutral axis of the specimen for bend testing (Figure 9).

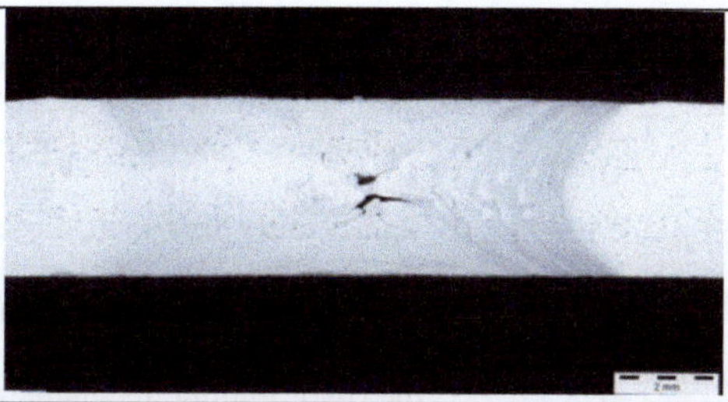

Figure 8. Typical weld flaw

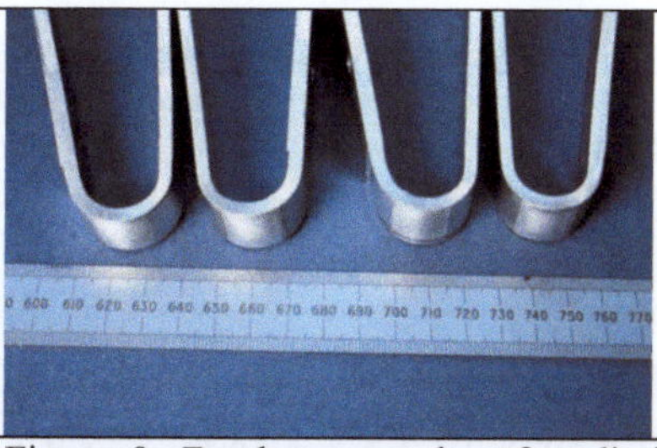

Figure 9. Bend test results of a dissimilar thickness weld showing 180° passes on both top and bottom

Tool Life Assessment

An assessment of minimum tool life was made for each alloy grade and thickness. Each tool life trial consisted of 5 identical cycles, each cycle including nine 1m bead on plate welds followed by a 1m long butt weld. Four coupons were taken from each butt weld for bend testing to assess weld quality.

Figure 10 Tool life trials when welding 6mm thickness AA5083-H111

All the welds made from each assessment of the two different alloy grades and thickness successfully passed top and bottom face bend tests concluding that the tools can consistently produce high quality welds to a minimum total length of 50 metres. No tool wear was noticed and it is believed that the tool life could be much longer than 50 metres. One set of welded test plates and resulting bend tests are shown in Figure 10.

Distortion Assessment

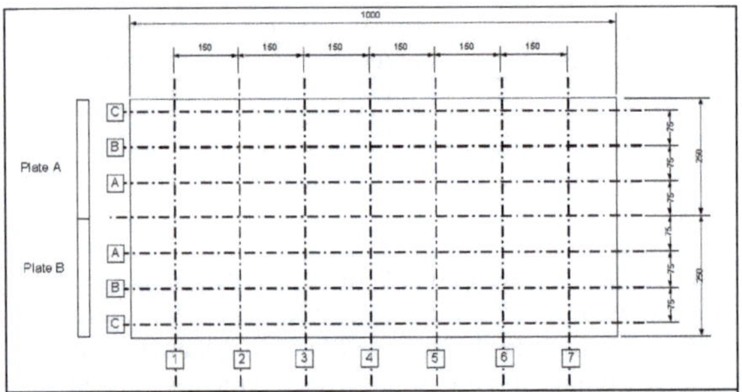

Figure 11 Location of measurement points used to assess distortion pre and post welded plates

To assess typical plate distortion produced by FSW when compared to traditional fabrication processes, pairs of test plates were measured pre weld and then again after welding using the FBFSW and a single-sided MIG technique. A total of 21 measurement points per plate and 42 per fabricated plate were measured and recorded using a coordinate measuring machine as detailed in Figure 11. The results of this assessment are shown in Table II and visually in Figure 12.

Table II Results of the distortion test

Thickness and alloy grade		Distortion (Peak-valley value; mm)			
		Before FSW welding	After FSW welding	Before MIG welding	After MIG welding
4mm	AA5083	0.2	19.9	0.7	32.3
	AA6082	0.2	15.5	0.1	25.4
6mm	AA5083	0.4	14.3	0.3	28.3
	AA6082	2.0	8.2	0.6	26.0

Although the plates fabricated using the FBFSW technique showed some distortion it was significantly less than those made using MIG. This test gave a good comparison of the distortion produced by the two joining techniques however it is expected that the form and extent of the distortion would be different when welding larger components typical of commercial marine fabrications (i.e. those that the Mobi-Weld system was designed to fabricate).

		Welding Technique	
		FSW	MIG
4mm thickness	AA5083-H111	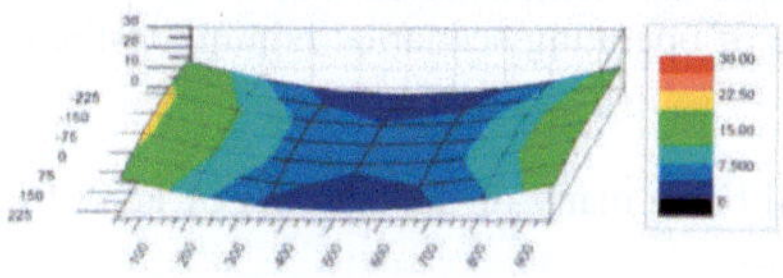	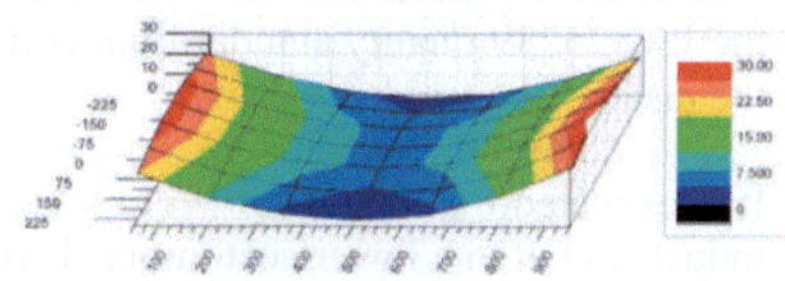
	AA6082-T6	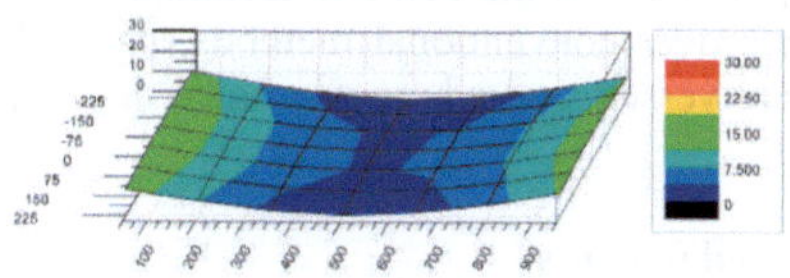	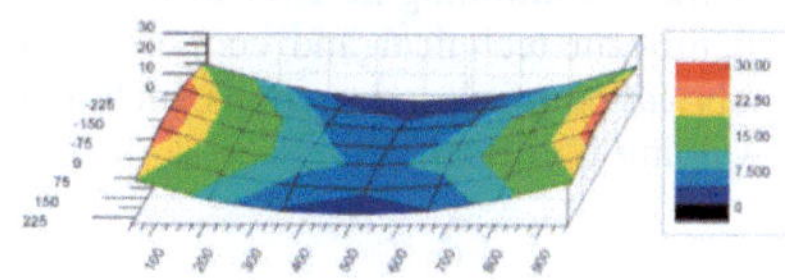
6mm thickness	AA5083-H111	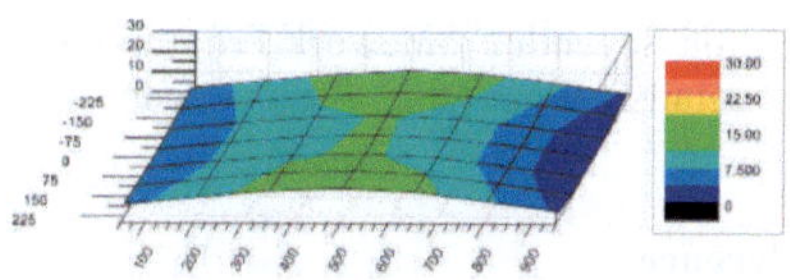	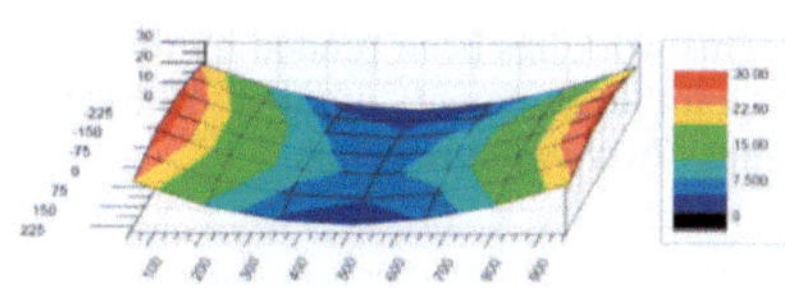
	AA6082-T6	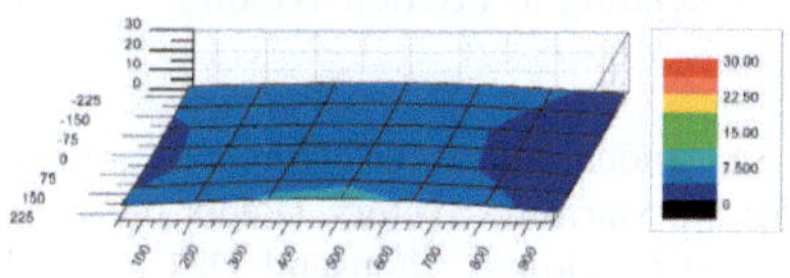	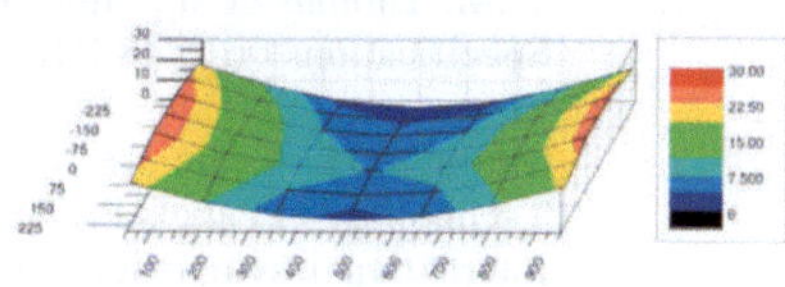

Figure 12. Summary of distortion plates measured after welding by FBFSW or MIG welding techniques. All numbers are in mm.

Mobi-Weld System

Figure 13. The Mobi-Weld System

The Mobi-Weld system, at the time of writing this paper, has been built and is being commissioned. A picture of the system is shown in Figure 13.

The system has the capability of unlimited continuous motion through sequencing of a walking mechanism with integral vacuum cups which provide the reaction against the welding forces. The system also has a seam tracking capability to align the tool to the plate interface. It is anticipated that the system will be completed by the end of 2014.

Summary

This paper provides an overview of recent work at TWI in developing FBFSW tooling for use on a mobile FSW machine. This welding technique has demonstrated that it can produce welds to the ISO 25239 quality standard under a range of operating scenarios relating to the condition of the plates to be welded.

It is expected that the Mobi-Weld system will bring many of the benefits of FSW to the marine industries but in a lower cost, more flexible way.

Work is continuing at TWI to further develop and assess the FBFSW technique as it may offer significant technical and economic advantages over conventional methods, particularly in the fabrication of closed sections such as tanks and vessels and where guarantee of full penetration welding is critical.

Acknowledgements

This work has been funded by the European Union Seventh Framework Programme (FP7) under Grant Agreement number 606156. More details about the Project can be found on http://www.mobiweld.eu/

References

1. W.M. Thomas et al., "Improvements Relating to Friction Welding". European Patent Specifications 0615 48 B1.

2. Midling OT, Kvale J S and Dahl O, "Industrialisation of the Friction Stir Welding Technology in Panels Production for the Maritime Sector". (Paper presented at the 1st International symposium on friction stir welding, Thousand Oaks, CA, USA, June 1999).

3. K Colligan, "Low-Cost Friction Stir Welding of Aluminum for Littoral Combat Ship Applications" (Paper presented at the 8th International Friction Stir Welding Symposium, 18-20 May 2008, Timmendorfer Strand, Germany).

4. K Colligan, G Hostetter, A O'Donnell, "CTC, Friction Stir Welding of Thin Aluminum Using Fixed Gap Bobbin Tools". (Paper presented at the 9th International Friction Stir Welding Symposium, Huntsville, AL, USA, May 2012).

FRICTION STIR WELDING AND PROCESSING VIII

Simulations and Measurements

Friction Stir Welding and Processing VIII
Edited by: Rajiv S. Mishra, Murray W. Mahoney, Yutaka Sato, and Yuri Hovanski
TMS (The Minerals, Metals & Materials Society), 2015

PREDICTION OF JOINT LINE MOVEMENT AND TEMPERATURES IN FRICTION STIR SPOT WELDING OF DP 980 STEEL

M.P. Miles[1], U. Karki[1], T. Lee[1], Y. Hovanski[2]

[1]Brigham Young University, Provo, UT 84602
[2]Pacific Northwest National Lab, Richland, WA 99352

Keywords: Friction stir spot welding, modeling, advanced high strength steel

Abstract

The friction stir spot welding (FSSW) process is non steady state and therefore lends itself to modeling using a Lagrangian approach. However a 3 dimensional model of this process can be overly time-consuming, particularly when an implicit scheme is used to solve the equilibrium equations that arise from discretization of the continuum body into finite elements. In this paper a novel 2 dimensional, finite element approach was used to model the FSSW process. An updated Lagrangian scheme was employed to predict the flow of the sheet material, subjected to boundary conditions of a descending tool and a fixed backing plate. Material flow was calculated from a velocity field that is two dimensional, but heat generated by friction was computed by an approach where the rotational velocity component from the tool surface was included in the thermal boundary conditions. An isotropic, viscoplastic Norton-Hoff law was used to model the material flow stress as a function of strain, strain rate, and temperature, while a viscoplastic friction law was used to compute the shear stress at the tool/sheet interface. The model predicted welding temperatures to within a few percent of experiment, but welding loads were significantly overpredicted. Comparison with a 3D model of FSSW showed that frictional heating and proportion of total heat generated by friction were similar. It was also shown that the position of the joint interface was reasonably well-predicted by the 2D model, compared to experiment.

Introduction

At the present time resistance spot welding (RSW) is the primary means of joining AHSS, but this has proven to be a challenge, because these steels have high alloy content and have a tendency to form brittle microstructures with small cracks in the weld nugget, especially when galvanized steels are spot welded. Some studies have postulated that shrinkage stresses in the weld nugget, along with the presence of zinc, were responsible for the forming of cracks [1-3]. The presence of microcracks in the weld can be exploited by fatigue cycles, potentially limiting joint durability. As a result, some automakers use adhesive in combination with RSW for joining of AHSS, in order to provide a margin of safety [4]. Another issue that occurs when RSW is used to join high strength steel sheets is the type of failure mode which appears during testing of the weld. Some studies have shown that dual phase (DP) 600 steel exhibited interfacial failures when joined by RSW [2, 5, 6]. It was found that using specific weld currents and weld times could reduce interfacial failures and cause the desired "button pullout" failure [6]. However, while DP 600 steel can be welded by RSW in the laboratory under controlled conditions, welding process control is a challenge for the production environment and more refined control strategies are required [7]. DP 980 alloy has also been studied with mixed results. Failures in lap shear tension were found to occur at the weld interface while other failures were of the button pullout type. In particular, when expulsion of the weld was observed, which is common in an RSW production process, interfacial failures were the most common type [8].

Friction stir spot welding (FSSW) is an alternative to RSW for spot joining of steel sheets. Prior work has shown promising results in FSSW of AHSS, with good joint strength, an absence of cracks in the weld, and a "pullout" type failure mode. These benefits are facilitated by the solid state bonding mechanism of FSSW, avoiding the problems that result from melting and solidification of a weld pool [3, 8].

The modeling of the FSSW process has been aimed at predicting welding temperatures and the effect of tool speeds and feeds on welding temperatures and material flow. Modeling of FSSW is a challenge, because unlike friction stir welding, which can be approximated as a steady-state process, FSSW is non steady state and has a constantly evolving material flow and temperature distribution. An updated Lagrangian or an arbitrary Lagrangian Eulerian (ALE) formulation can be used for modeling of FSSW, with a material law that provides flow stress as a function of strain, strain rate, and temperature. Many models have been developed for friction stir welding, but there has not been a large body of work on modeling of FSSW. Two models were developed in three dimensions with Abaqus/Explicit, and were used for studying FSSW and the plunge phase of friction stir welding of aluminum alloys [9, 10]. The Johnson-Cook material law was used in both cases to calculate flow stresses at different strain rates and temperatures, and an arbitrary Lagrangian Eulerien (ALE) approach was used to minimize mesh distortion over the course of the simulation. These approaches were reasonably successful in predicting welding temperatures, but required up to 14 days in order to complete a simulation, even using an explicit solver. It was found that frictional dissipation contributed about 97% of the heat generated during welding, with plastic deformation contributing the balance [10]. Another investigator used the Forge® software to build a three dimensional model of friction stir welding, including the plunge phase [11]. An ALE approach was developed using an implicit solver and the model was successful at predicting welding temperatures at the end of the plunge phase, as well as material flow phenomena like void formation behind the tool during the translation phase. In the current work a two dimensional, axisymmetric approach was used for modeling FSSW. This limits tool geometry to one where features like threads or scrolls cannot be modeled, but it does allow for modeling of shoulder size and profile, as well as pin size and profile. An approximation of the heat generated by frictional contact between the rotating tool and the workpiece was also employed in order to make reasonable predictions of welding temperatures and material flow in the two dimensional section plane. From a computational cost viewpoint, the two dimensional simulation can be completed within a few hours, versus many days, with equivalent CPU, for a three dimensional simulation.

Modeling Approach

The two dimensional, axisymmetic model of the FSSW process was developed using the Forge® [12] simulation software. An updated Lagrangian scheme was employed to model the flow of the sheet material, subjected to boundary conditions of a rotating tool and a fixed backing plate. The flow stress of the workpiece material was computed as a function of strain, strain rate, and temperature, where the expression for the deviatoric stress tensor is shown below:

$$\mathbf{s} = 2K\left(\sqrt{3}\dot{\bar{\varepsilon}}\right)^{m-1}\dot{\boldsymbol{\varepsilon}} \tag{1}$$

where $\dot{\boldsymbol{\varepsilon}}$ is the strain rate tensor, $\dot{\bar{\varepsilon}}$ is the effective strain rate, K is the material consistency, and m is the strain rate sensitivity. The material consistency K is a function of temperature T and equivalent strain $\bar{\varepsilon}$, where n is the strain hardening exponent and β is a thermal softening parameter:

$$K = K_0(\varepsilon_0 + \bar{\varepsilon})^n e^{\frac{\beta}{T}} \tag{2}$$

This viscoplastic law is capable of modeling material flow stresses in the region of the weld, while also providing the contact stresses with the tool that were used to calculate the friction shear stress at the tool/sheet interface. The data for this law were obtained from JMatPro specifically for DP 980 steel, for the range of strain rates and temperatures that would be encountered during FSSW [13]. Friction at the sliding interface between the sheet and the tool were modeled using a Norton law. The Coulomb friction law has been used for modeling of friction stir welding, but this law is more appropriate for room temperature sliding. For high temperature metal working the standard law does not provide a threshold beyond which the shear stress saturates. Rather than using the normal stress at the interface to calculate a shear stress induced by friction, the Norton law models the shearing of a thin layer of viscoplastic material having the same flow stress as the workpiece, as follows:

$$\tau = -\alpha_f K |v_s|^{p-1} v_s \tag{3}$$

where α_f is the friction coefficient, K is the material consistency given by equation 2, v_s is the relative sliding velocity at the sheet/tool interface, and p is a rate sensitivity parameter. Therefore this friction law is a function of the material properties of the interface, including temperature, strain, and strain rate, while also being a function of the sliding velocity. Calculation of the flow of material was based on a finite element discretization using an enhanced (P1+/P1) 3-noded triangular element [12], where equilibrium equations were solved at each increment using the Newton-Raphson method. The unilateral contact condition was applied to the sheet surfaces by means of a nodal penalty formulation, where the FSSW tool and backing plate were considered rigid. An explicit time integration scheme was used to update the sheet geometry at each increment of calculation:

$$X_{t+\Delta t} = X_t + V_{mesh}\Delta t \tag{4}$$

X is a mesh material coordinate, V_{mesh} is a component of velocity of the mesh at time t, and Δt is a time increment chosen sufficiently small. The evolution of temperature in the tool was modeled in order to provide accurate boundary conditions at the tool/sheet interface. The temperature in both the FSSW tool and the sheet were calculated at the end of each material flow increment. The calculated velocity field allowed for computing strain rates and stresses in the sheet, which were then used to determine the heat dissipated by friction and plastic deformation. The heat dissipated by plastic deformation (derived from the velocity field) is given by:

$$\dot{q}_v = f\bar{\sigma}\dot{\bar{\varepsilon}} \tag{5}$$

where $\bar{\sigma}$ is equivalent stress. The factor f takes into account the fraction of deformation energy converted into heat, taken as 0.9 in this paper. For a Norton-Hoff viscoplastic material the heat generation rate from material deformation is computed as follows:

$$\dot{q}_v = fK\left(\sqrt{3}\dot{\bar{\varepsilon}}\right)^{m+1} \tag{6}$$

Heat from friction was computed as a function of sliding that occurred along the radial direction of the tool surface, as well as sliding that occurred from rotation, as shown below:

$$\dot{q}_f = \tau \cdot (v_{rad} + v_{rot}) \tag{7}$$

where τ is the shear stress calculated from equation 3, v_{rad} is the radial sliding velocity between sheet and tool, calculated from the velocity field in the sheet, and v_{rot} is a virtual rotational sliding velocity between sheet and tool, taken as the surface speed at a given point on the tool, based on its rpm. The approach used here for FSSW is similar to one employed for the modeling of inertia friction welding [14], except that in the case of FSSW the tool is much harder than the workpiece. The heat generated by friction at the sliding interface is shared between the two bodies in contact according to their thermal/physical properties. For example, the heat entering the sheet would be calculated as:

$$\dot{q}_{sheet} = \frac{\dot{q}_f b_{sheet}}{b_{sheet} + b_{tool}} \tag{8}$$

where b is the effusivity of each material, defined as $\sqrt{k\rho c}$, where k is thermal conductivity, ρ is density, and c is heat capacity. The boundary conditions for the backing plate were included in the model, based on thermocouple measurements that were done in prior work [15]. The thermal conductivity, heat capacity, and density of both the Si_3N_4 tool and the DP 980 sheet were modeled as a function of temperature, over a range from 25°C to 1400°C.

Description of Experiments

Welding experiments employed tool speeds of 3000 – 6000 rpm on a Fadal machining center, where the lap shear tension configuration was used to evaluate joint strength. Lap shear specimens were made using coupons of 1.2 mm bare DP 980 steel, with dimensions of 25 mm by 100 mm, and an overlap of 25 mm. The DP 980 steel composition, in weight percent, was 0.15% C, 1.44% Mn, 0.011% P, 0.007% S, 0.32% Si, and 0.02% Cr, where the balance was Fe. The spot weld was positioned in the center of the overlap in each case. Si_3N_4 tools with a shoulder diameter of 10 mm and a smooth pin, with three small flats, were used for all spot welding experiments. The flats tended to wear off very quickly, leaving a completely smooth pin after approximately 10 welds. A constant plunge rate of 25.4 mm/min was used for each experiment. Loads on the machine spindle were measured using a special fixture with four load cells linked to a data acquisition system, using a sampling rate of 60 Hz. Lap shear specimens were tested on an Instron screw-driven frame at a rate of 10 mm per minute. Some welded specimens were sectioned, mounted, and polished for examination by optical microscopy. Optical microscopy of welded specimens was done by first sectioning a weld through its center, using wire electrical discharge machining (EDM), then mounting in Bakelite. Polishing of the specimens was carried out in stages. The process started with 120 grit sand paper, and continued through 240, 400, 600, 800, and 1200 grit. This was followed by polishing with 6 micron diamond paste, then with 3 micron paste, and finally 1 micron paste. Etching was done using 2% Nital for about 15 seconds, followed by a rinse in methanol.

Results

The model employed 12,543 linear triangle elements for the sheet and 4,695 linear triangles for the tool. The number of elements in the sheet mesh varied somewhat during the course of the simulation, because it was automatically remeshed when element distortion reached a prescribed level. The backing plate was assumed to be rigid and at a constant temperature of 25°C, since the mass of the backing plate was relatively large compared to the mass of the sheet, and the welding time was relatively short. The heat transfer coefficient between the tool and the sheet was 20,000 W/m^2-°C, and the coefficient between the sheet and the backing plate was 2,500 W/m^2-°C. The sheets were modeled as one body in order to simplify the contact problem, but some virtual Lagrangian sensors were used to track the interface between the sheets as a function of deformation. An image of the model is shown in Figure 1.

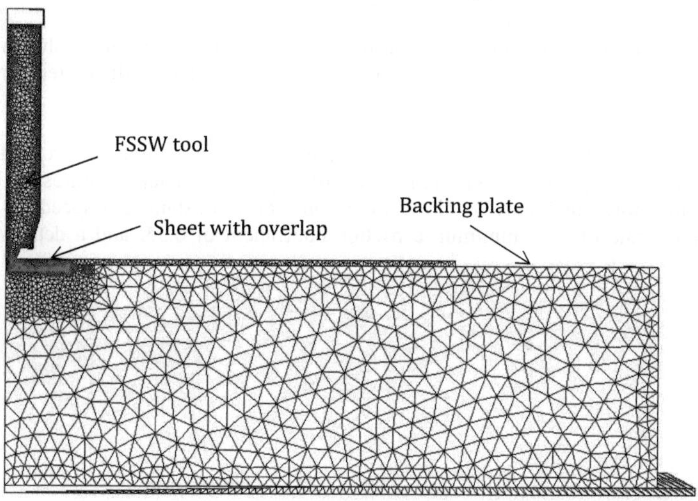

Figure 1. View of 2D axisymmetric finite element model of the FSSW process. One half of the section view is shown.

Temperature measurements for model validation were done by placing thermocouples in the lap shear welding fixture, at 1.2mm beneath the surface of the lower sheet. The thermocouples were positioned along the central axis of the specimen, at 3mm, 4mm, and 6.5mm from the center of the weld, in channels machined out of one of the spacers beneath the specimen. The locations of the thermocouples are shown in Figure 1, for reference, on a magnified image of the model shown in Figure 2.

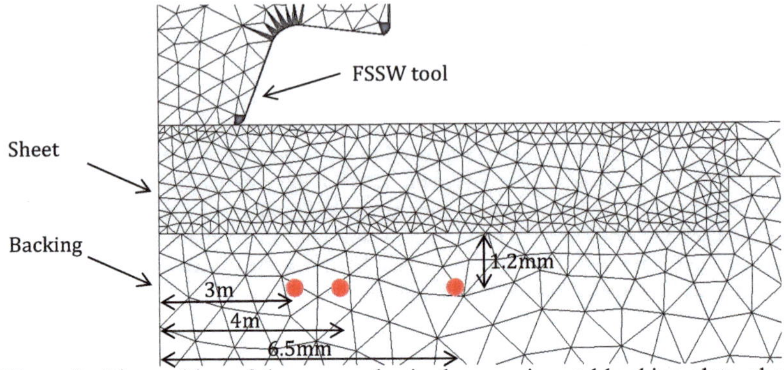

Figure 2. The position of thermocouples in the experimental backing plate, shown at the same locations in the model. The measurements were compared with the predicted temperature curves for model validation.

Three temperature measurement replications were done at the three different thermocouple positions shown in Figure 2. The comparison of experimental temperatures with the model predictions are shown in Figure 3. These were done for a constant tool speed of 5000 rpm, a constant plunge rate of 25.4 mm/min, a friction coefficient of 0.35, and a depth of plunge of 1.9mm.

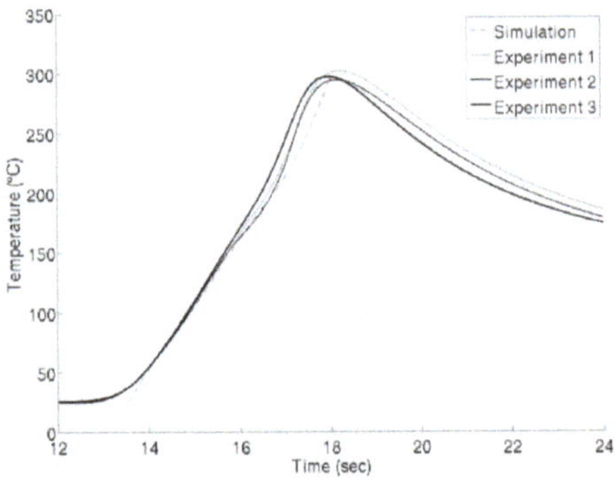

(a)

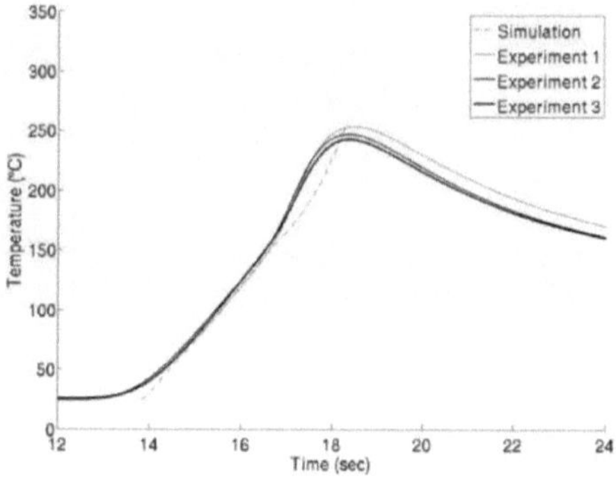

(b)

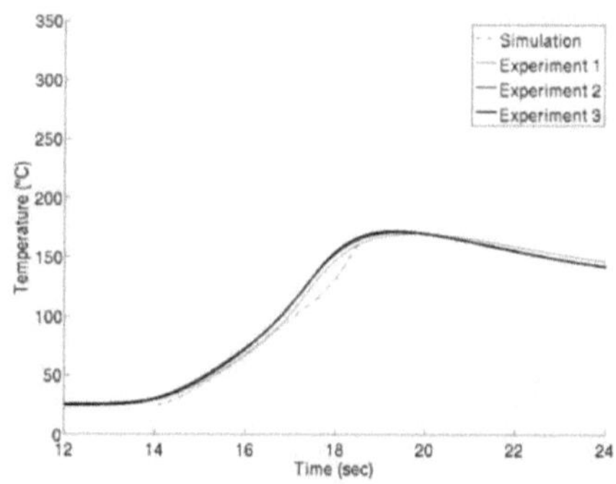

(c)

Figure 3. Experimental temperatures and predicted temperatures at (a) 3mm from center of weld, (b) 4mm from center of weld, and (c) 6.5mm from center of weld.

Peak temperatures were predicted within 4% of the average experimental peaks. The slopes of the temperature-time curves also appear to be predicted with reasonable accuracy [16].

While the temperature predictions appear to be reasonable, the welding load predictions are overestimated, compared to experiment. They are also higher than the loads predicted by a 3D implicit model developed using the Forge® software. The 3D model was created for comparison with the 2D model load predictions. The 3D model took approximately 3 weeks to run (using an Intel i7 2820 QM cpu, with 4 cores), but it did not run to completion. Using a purely Lagrangian framework, it was necessary to remesh the volume of the sheet frequently in order to avoid excessive mesh distortions. At a certain point the flash forming under the shoulder of the tool had element degeneration and contact difficulties that stopped the simulation. It is likely that further development would enable completion of the simulation, but the implicit solver makes the 3D calculation prohibitive, from a practical viewpoint. The stopping point of the 3D simulation is shown in Figure 4.

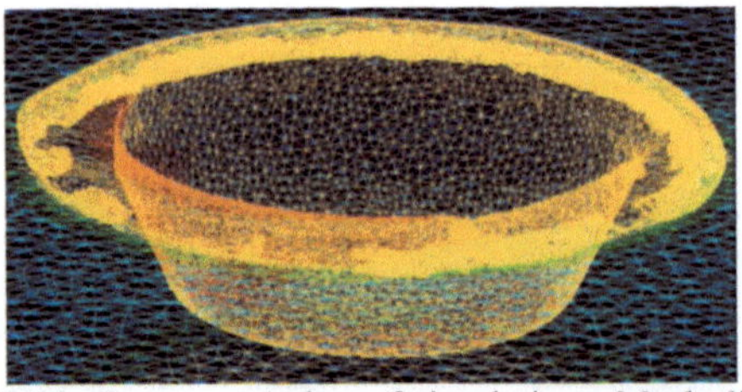

Figure 4. 3D model of FSSW before completion of simulation. Mesh distortion in the flash, as it curled between the shoulder and the sheet, caused numerical difficulties that prevented the simulation from finishing. Shading shows temperature gradients in the part.

While the 3D simulation did not reach completion, it did at least provide a partial comparison of the welding load prediction with the 2D model. The welding load is the vertical load required to plunge the tool to a specified depth of penetration. A comparison of the 2D and 3D loads is shown in Figure 5.

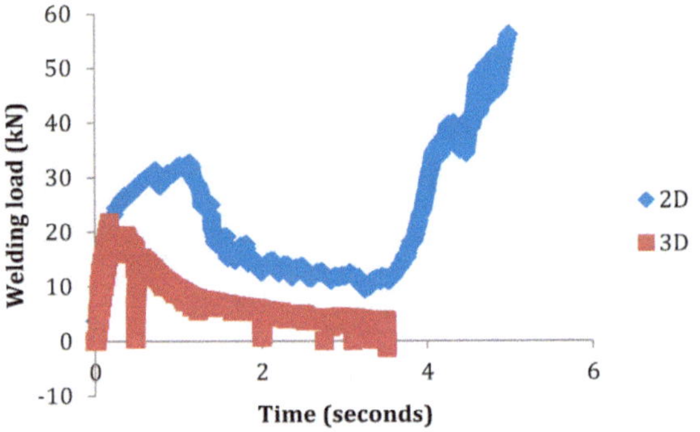

Figure 5. Comparison of welding loads for the 2D and 3D simulations. The tool speed was a constant 5000 rpm, the plunge rate was 25.4 mm/min, the friction coefficient was 0.35, and the plunge depth was 1.9mm.

The experiments done under the conditions that correspond to Figure 5 had peak loads of 9-12 kN. The loads are clearly overpredicted by both models, but especially by the 2D model. The difference in the 2D versus 3D prediction is likely in part due to the lack of a rotational component in the velocity field. The 2D model is essentially simulating the plunge of a non rotating tool that is heating up as though it were rotating at the prescribed speed of 5000 rpm. The 3D model is simulating the literal rotation of the tool at 5000 rpm, and therefore includes the shearing of the material close to the tool surface by friction. Despite the overprediction of the welding load, the 2D model does a reasonable job of predicting the joint line movement at the end of welding. Figure 6 shows the evolution of the joint line, which was tracked by Lagrangian markers whose positions were updated at each increment of the numerical simulation.

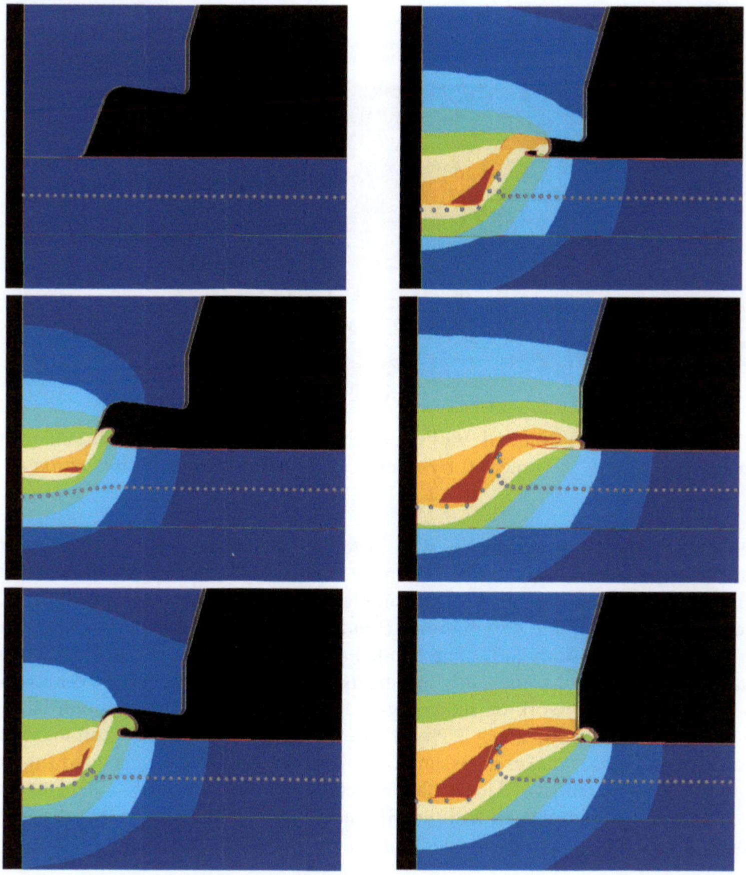

Figure 6. Evolution of the joint line during simulation of FSSW of 1.2mm DP 980 steel. The joint line is shown as a grey dotted line. Temperatures are also shown, from 25°C (dark blue) to 1400°C (red).

The joint line shown in Figure 6 curls up and forms a hook as the welding process progresses. Hook formation is a well-known phenomenon in FSSW. The entire joint line is visible in the simulation, but experimentally there is bonding that takes place, so the joint line vanishes in the weld zone. A micrograph of a lap joint of 1.2mm DP 980 steel, welded under the same process conditions as the simulation, is shown in Figure 7.

Figure 7. Lap joint line in 1.2mm DP 980 steel, with hook formation. The distance from the tip of the hook to the edge of the hole left by retraction of the tool gives an indication of the bond area.

The model predicts the hook reasonably well in terms of it shape and position. The tip of the hook from the experiment from Figure 7 can be related to the predicted joint line from the simulation. At a corresponding point on the simulated joint line the distance to the edge of the hole created by the tool is about 450 microns. Bond area has been shown to be highly correlated to joint strength in a prior study [17]. The bonded area of the joint is a function of temperature, pressure, and time as the tool compresses the sheets together and heats them over a typical cycle time of 4-5 seconds. Prior work on modeling of FSSW has shown that friction is the primary source of heat, accounting for about 97% of the heat, with the balance coming from plastic deformation. The 2D model makes an assumption about the heat that is generated from rotational sliding of the tool on the sheets. The surface speed of the tool is used to calculate the heat, as shown in equation 7. A comparison with the 3D model in Figure 8 shows that the difference in friction power for the same simulation conditions is small.

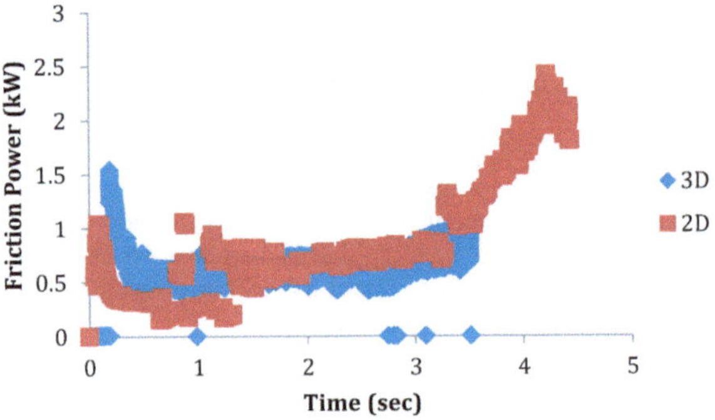

Figure 8. Friction power for the 2D and 3D simulations (note that as stated earlier the 3D simulation did not reach completion).

The data are scattered, partly because contact conditions between the part and the tool evolved with each increment of the simulations, including slight differences in contact after remeshing. But a good comparison can be made in what appears to be a relatively stable period of frictional heat rate between 1.5 and 3 seconds of the simulation (process time), with the 2D model having about 3% more friction power than the 3D model, based on the averages over this time span. The other difference between the 2D and 3D models is the percentage of power that comes from frictional sliding versus plastic deformation, which turns out to be relatively small. The 2D model had 97.9% of the total power coming from friction and 2.1% coming from plastic deformation; the 3D model had 96.2% coming from friction and 3.8% coming from plastic deformation. These values are close to the 97% number cited in [10] for a 3D explicit simulation of FSSW, although the material was aluminum and not steel.

Finally, the effect of tool rpm and plunge rate on predicted welding temperatures at the end of the weld cycle was studied for two different plunge rates: 12.7 mm/min and 6.4 mm/min. The temperatures were taken from the top and bottom sheet surfaces, along the model axis of symmetry. Figure 9 shows the trends tools speeds ranging from 3000 – 6000 rpm. As expected, the temperatures at the end of the welding cycle increase with increasing rpm for a given plunge rate. And for a given rpm a faster plunge rate results in lower temperatures at the end of the welding cycle.

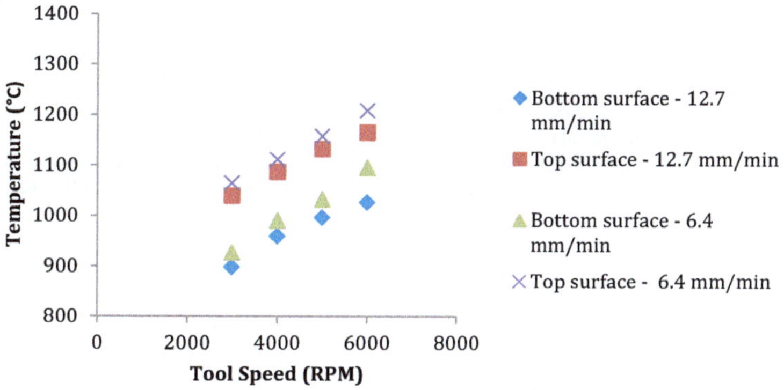

Figure 9. Predicted temperatures in the sheets at the end of the weld cycle, for two different plunge rates and 4 different tool speeds.

Conclusions

The friction stir spot welding process was modeled using a simplified 2D, axisymmetric approach. While the velocity field and temperatures were calculated in the section plane of the model, the thermal boundary conditions included both the heat from frictional sliding in the plane of the calculation, as well as the virtual frictional sliding from tool rotation. As such, the model effectively simulated the plunging of a non rotating tool that was heated progressively as though it were rotating at a prescribed rpm. It was found that peak temperatures were predicted quite accurately, within a few percent of the experimental data. The slopes of the predicted time-temperature curves also agreed reasonably well with the experiments. Welding loads were significantly overpredicted compared to experiments, and were also overpredicted compared to a 3D model. The inability of the 2D model to simulate the shear imparted by the rotating tool is likely one reason for the load overpredictions. However, it was found that the amount, and proportion, of heat generated by friction was very similar for the 2D and 3D models, within a few percent. The evolution of the interface between the sheets was also modeled over the course of the tool plunge, both in terms of shape and position, to within 10% of the experiment.

Acknowledgements

This work was supported by National Science Foundation grant CMMI-1131203 and by funding from Department of Energy, EERE-Vehicle Technologies Office, via PNNL subcontract 116126.

References

1. Cortez, V.H.L., F.A.R. Valdes, and L.T. Trevino, *Weldability of Martensitic Steel by Resistance Spot Welding a Neural Network Optimization in the Automotive Industry*. Materials and Manufacturing Processes, 2009. **24**(12): p. 1412-1417.
2. Khan, M.I., et al., *Resistance and friction stir spot welding of DP600: a comparative study*. Science and Technology of Welding and Joining, 2007. **12**(2): p. 175-182.
3. Santella, M., et al., *Friction stir spot welding of DP780 carbon steel*. Science and Technology of Welding and Joining, 2010. **15**(4): p. 271-278.
4. Hovanski, Y., *Personal communication*. 2010.
5. Marya, M. and X.Q. Gayden, *Development of requirements for resistance spot welding dual-phase (DP600) steels part 2: Statistical analyses and process maps*. Welding Journal, 2005-2. **84**(12): p. 197s-204s.
6. Marya, M. and X.Q. Gayden, *Development of requirements for resistance spot welding dual-phase (DP600) steels - Part 1 - The causes of interfacial fracture*. Welding Journal, 2005-1. **84**(11): p. 172s-182s.
7. Ma, C., et al., *Expulsion monitoring in spot welded advanced high strength automotive steels*. Science and Technology of Welding and Joining, 2006. **11**(4): p. 480-487.
8. Nikoosohbat, F., et al., *Microstructure and failure behaviour of resistance spot welded DP980 dual phase steel*. Materials Science and Technology, 2010. **26**(6): p. 738-744.
9. Mandal, S., J. Rice, and A.A. Elmustafa, *Experimental and numerical investigation of the plunge stage in friction stir welding*. Journal of Materials Processing Technology, 2008. **203**(1-3): p. 411-419.
10. Awang, M. and V.H. Mucino, *Energy Generation during Friction Stir Spot Welding (FSSW) of Al 6061-T6 Plates*. Materials and Manufacturing Processes, 2010. **25**(1-3): p. 167-174.
11. Fourment, L. and S. Guerdoux, *3D numerical simulation of the three stages of Friction Stir Welding based on friction parameters calibration*. International Journal of Material Forming, 2008. **1**: p. 1287-1290.
12. Transvalor, S.A., *Forge 2009*. 2009.
13. JMatPro, *Sente Software Ltd.* 2014, Sente Software Ltd.: Surrey Technology Centre, 40 Occam Road, United Kingdom.
14. Moal, A. and E. Massoni, *Finite Element Simulation of the Inertia Welding of Two Similar Parts*. Engineering Computations, 1995. **12**: p. 16.
15. Miles, M., et al. *Steady-State Simulation of Material Flow and Temperature in Friction Stir Welding of Aluminum Alloy 6061*. in *Materials Science & Technology 2008 - Joining of Advances and Specialty Metals*. 2008. Pittsburgh, PA.
16. Miles, M., U. Karki, and Y. Hovanski, *Temperature and Material Flow Prediction in Friction Stir Spot Welding of Advanced High Strength Steel*. Journal of Metals, 2014. **in press**.
17. Saunders, N., et al., *Joint strength in high speed friction stir spot welded DP 980 steel*. International Journal of Precision Engineering and Manufacturing, 2014. **15**(5): p. 8.

APPLICATION OF ACOUSTIC EMISSION AS AN EFFECTIVE TOOL TO MONITOR FSW OF AA2024-T3 ALUMINUM ALLOY

B M Rajaprakash[1], C N Suresha[2], Sarala Upadhya[1]
[1] Department of Mechanical Engineering, University Visvesvaraya College of Engineering, Bangalore University, Bangalore-560 001, Karnataka, ,India.
[2] Department of Mechanical Engineering, Jyothy Institute of Technology, Bangalore -560 062, Karnataka, India

Abstract

Friction Stir Welding (FSW) is a solid-state welding technique used for joining metals and alloys to avoid problems associated with fusion welding. Acoustic Emission (AE) has been successfully used to monitor processes like metal cutting, grinding, electron beam welding and FSW. In this work, an attempt has been made to study the application of AE to monitor FSWs to produce defect free welds. During welding of aluminum alloy AA 2024-T3 5mm thick plates, AE signals were acquired. Patterns of AE signals produced during welding are helpful in identifying the defects produced. The lower and higher values of AE parameters help to decide the quality of welded joints. In order to have a better understanding of the behavior of AE parameters when defects were supposed to have occurred during welding, a time domain analysis of AE signals was carried out. The time domain analysis has resulted in justifying the behavior of the AE signals at the instant of occurrence of defects. The range of values of AE parameters, derived from AE signals found to be helpful in monitoring FSW, was accomplished by identifying the time of occurrence of the defect during welding followed by suitable corrective action to produce defect free welds.

Keywords: Friction Stir Welding, Acoustic Emission Test, Monitoring, Time Domain Analysis.

Introduction

FSW is a solid–state, hot–shear joining process invented at The Welding Institute (TWI), United Kingdom in 1991. The schematic diagram of the process is shown in Figure.1 in which a rotating tool with a shoulder terminating in a pin with a particular geometry moves along the butting surfaces of two rigidly clamped plates placed on a backing plate [1].

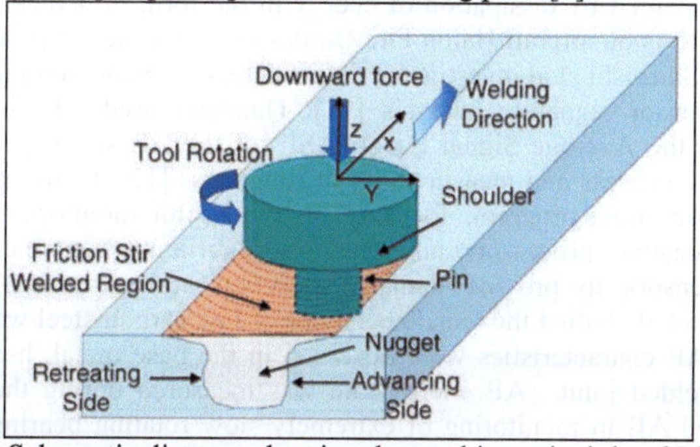

Figure 1. Schematic diagram showing the working principle of FSW [1-3]

The shoulder makes firm contact with the top surface of the work piece. Heat generated by friction at the shoulder and to a lesser extent at the pin surface softens the material being welded. Severe plastic deformation and flow of this plasticized material occurs as the tool is translated along the welding direction [2-3]. Material is transported from the front of the tool to the trailing edge where it is forged into a joint due to a small tilt provided to the rotating tool. The half–plate where the direction of rotation is same as that of welding is called the advancing side, while the other side is designated as the retreating side. Since its discovery, FSW has evolved as a technique of choice in the routine joining of aluminum components and its applications for joining different metals other than aluminum are also growing, although at a slower pace [4]. In FSW, the important process parameters which affect the quality of weld are tool rotational speed, weld traverse speed, axial force resulting from shoulder contact with the surface of the base metal (plunge depth), and tool tilt angle. However, there is lot of scope for effective monitoring of FSW due to the varied combination of tool material and its geometry, work material and thickness etc. Further, it is of prime importance to check the quality of the weld being produced online owing to its obvious advantages.

Authors in their earlier studies investigated the role of the tool pin profile on tensile strength of welded joints by considering the parameters of (i) Tool rotational speed, (ii) Weld traverse speed and (iii) Plunge depth [5].

Acoustic Emission is the class of phenomena whereby transient elastic waves are generated by the rapid release of energy from a localized source or source within a material, or the transient elastic wave generated when a material undergoes deformation [6]. Acoustic Emission is a non – destructive technique different from other non- destructive testing techniques in two regards: (i) It pertains to the origin of the signal. Instead of supplying energy to the object under examination, AE simply listens for the energy released from the object followed by amplification and quantification of AE signals. AE tests are often performed on structures while in operation, as this provides adequate loading for propagating defects and thereby triggering AE and (ii) AE testing deals with dynamic processes or changes in a material. This is particularly meaningful because only active features (e.g., crack growth) are highlighted [7]. An advantage of using AE as a process monitoring tool is that the frequency range of AE is much higher than that of machine vibrations and ambient acoustic noise. Sources generating AE in different materials are unique due to the mechanical properties of materials [8].

Sharma et.al used Acoustic Emission for online monitoring of Electron Beam Welding of a Ti6Al4V alloy. They used the principle of AE analysis based on the fact of formation of defects in a material accompanied by dissipation of energy in the form of acoustic waves [9]. Beattie et.al conducted AE tests on aircraft Halon Fire Bottles to determine the presence of cracks based on AE counts [10]. Shiroishi et.al investigated defect detection methodologies for rolling element bearings through sensor signature analysis [11]. Dunegan used AE for measuring surface roughness. RMS or the Average Signal Level (ASL) of the AE signal generated was used for understanding of the process and measurement of roughness [12]. Craig et.al, in the review of the smart metallic structures program, focusing on the health monitoring system architecture (specifically the sensors, processors and analysis algorithms contained therein), used AE and fiber optic sensors to provide damage detection and strain monitoring in metallic structures [13]. Lee et.al studied the tensile behavior of low carbon steel welds by means of AE. During the study, AE characteristics were observed in the base metal, heat affected zone, and weld metal of a welded joint. AE energy rate was measured during the tensile tests. [14]. Miettinen et.al, used AE in monitoring of extremely slow rotating bearings [15]. Beggan et.al used AE to predict surface quality. During their study, AE signals were used to monitor tool wear and tool breakage thereby correlating them with surface quality [16]. Jaap H.HEIDA et.al, made investigations into the applicability of AE and ultrasonic inspection to monitor cure related

defects in thick resin transfer molded products. [17]. J. Kopač et.al studied the use of AE during drilling of carbon C15E steel and nodular gray iron (G40). During the study, they performed tool wear monitoring of the drilling process [18]. Akop'yan et.al made studies on correlation between parameters of AE signals and the degree of corrosion damage in Aluminum Alloys [19]. Xiaozhi Chen et.al used AE for tool condition monitoring based on wavelet analysis. These studies were made using a tungsten carbide tool to machine mild steel [20]. Sundaram et.al used AE in the study of change in geometry of flank wear in single point cutting tools [21].
Based on the literature, AE has been successfully adopted in many fields, including laser welding, health monitoring of aerospace structures, tool wear during drilling, turning, milling, etc. However, it has also been observed that there is little published information available about the use of AE to monitor FSW. Authors in their earlier studies have established the applicability of Acoustic Emission to monitor FSW. During establishing the relationship between FSW input process parameters and output AE parameters, one of the FSW process parameters were varied continuously and AE signals recorded. As the FSW input parameters varied the pattern of AE signal also varied [22].
The authors in their earlier work also concluded that there is a sudden increase in the values of AE parameters namely RMS, Amplitude, Energy whereas there is a sudden decrease in the value of Counts whenever there is the occurrence of a defect. This has been verified by comparing the X- Ray radiography image and pattern of AE signals. Authors have also developed a model for online monitoring of FSW that helps in taking necessary corrective action to produce defect free welds [23].
The motivation to understand the reason for the behavior of AE parameters has resulted in repeated experiments and carryout the time domain analysis of a few significant AE parameters obtained during welding.

Experimentation

Experimental work requires conducting experiments and sensing AE signals, studying the behavior of AE parameters derived from the AE signals during welding, and based on threshold values of the AE parameters distinguish between bad and good welds categorized by the presence or absence of defects by comparing with X- Ray radiography test results. The base material used in this study is Aluminum alloy AA2024-T3, a copper based alloy widely used for lower wing skins and the fuselage structure of commercial aircraft. Aluminum Plates of 5 mm thick were cut to a dimension of 300 mm × 75 mm with square edges to prepare a butt joint. Plates were fixed firmly on the FSW machine using suitable clamps. Welding was carried out using a conical threaded tool made of HSS with 60-65 HRC. Based on the similar lines of experiments conducted earlier by the authors, the optimal values of process parameters selected used in this study are tool rotational speed of 1000 rpm, traverse speed of 100 mm/ min and plunge depth in the range of 4.83 mm to 4.87 mm [5]. Three welds (Weld1, Weld 2, and Weld 3) were prepared using the optimal process parameters and AE signals recorded. The input parameters for producing the three welds are presented in Table I.

The AE sensor (Wide band, Piezoelectric transducer, operating frequency: 100 kHz – 2000 kHz, gain: 40 ±1dB) was mounted on the work piece by using a suitable couplant followed by connecting the sensor cable to a computer through a preamplifier with a band pass filter (Gain-40 dB, filter: 125 kHz - High Pass). The location where the sensor was fixed was confirmed by the Hsu-Nielsen method as per ASTM E 976-05. AE data were acquired for the welds produced using selected input process parameters and stored in the computer using an AE data acquisition system setup (make: MISTRAS, Physical Acoustic Corporation, USA), in Figure.2. The acquired AE data were processed to determine various AE parameters using AEwin software. The important AE parameters considered during the study are RMS, Energy,

Amplitude and Counts. Welded joints were subjected to X- Ray radiography testing to locate the position of defects produced during welding. Tensile specimens were prepared per ASTM - E8 / E8M-08 by choosing the portion of the welded joint where there were no defects revealed from the X –Ray radiography images. Tensile tests were carried out using an Electronic Tensometer (model: PC 2000, make: Kudale Instruments, Pune, India). The results of tests are summarized in Table II.

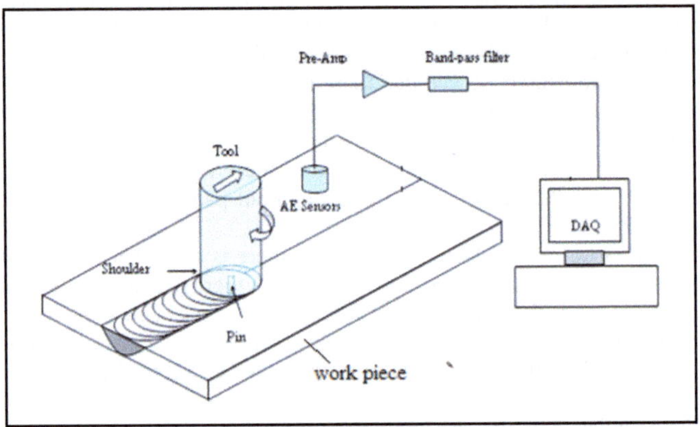

Figure 2. Schematic diagram showing FSW with the AE data acquisition system.
(Courtesy: MISTRAS, Physical Acoustic Corporation, USA)

Results and discussion

Three welds (weld 1, weld 2, and weld 3) were subjected to X- ray radiography. Out of three welds, weld 1 was found to contain a mix of both defect and non-defect regions. Conversely, welds 2 and 3 were found to be defect free having joint efficiencies of 80 and 85 % respectively. Hence, weld 1, having both defect and no defect locations, has been considered for time domain analysis. A photograph of the FSW joint (weld 1), prepared using the selected optimal process parameters, is shown in Figure 3.

Figure 3. Photograph of FSW joint (weld 1).

In this study, the time taken for complete plunge and dwell is about 40 sec. After plunging, the tool traverses with the set speed resulting in welding until the preset length of 200 mm weld is obtained. Radiography tests showed defects in weld 1 at a few locations. The screen shots of the signal pattern of AE parameters acquired in preparation of the weld 1 are shown in Figure 4. The individual 3D histogram of AE signal of RMS, Energy, Amplitude and Counts are shown in Figures 5 to 8 respectively. The RMS, Energy, Amplitude and Counts were found to be stable during 50 to 160 sec of welding for weld 1. Also, between 120 sec and 130 sec, there was a sudden increase in RMS, Energy, Amplitude and a sudden decrease in Counts revealing an occurrence of a defect subsequently confirmed by radiography. This has been shown as encircled marks on the screen shot of Figure 4. The details of variation in AE parameters, along with joint efficiency at non defect region of welds of AA2024-T3, are presented in Table II. Based on the lower and higher values of AE parameters (Table II), lower and higher threshold limits were set to distinguish the formation of good and bad welds presented in Table III [22].

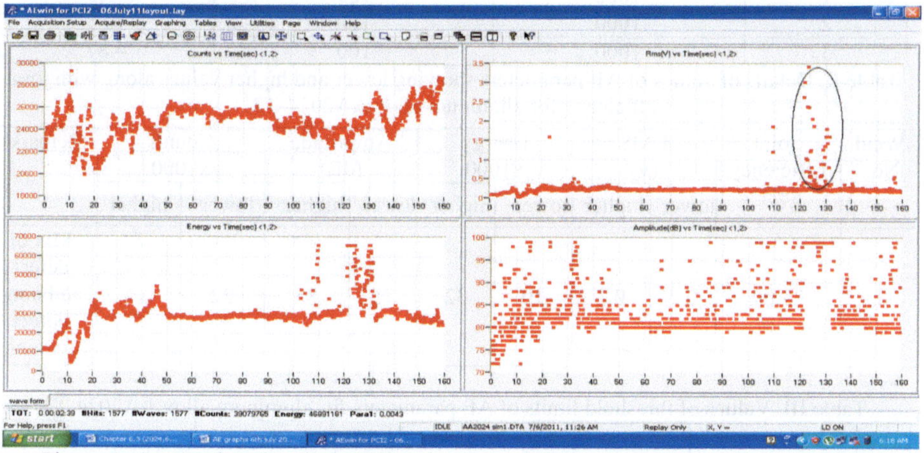

Figure 4. Screen shot showing AE signal pattern of different parameters during welding.

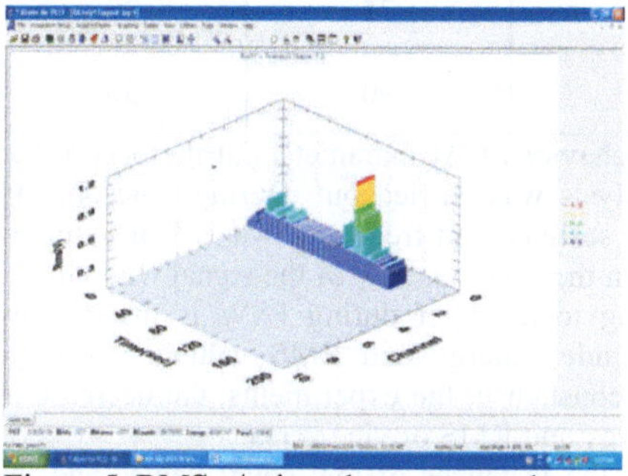
Figure.5 RMS v/s time data generated

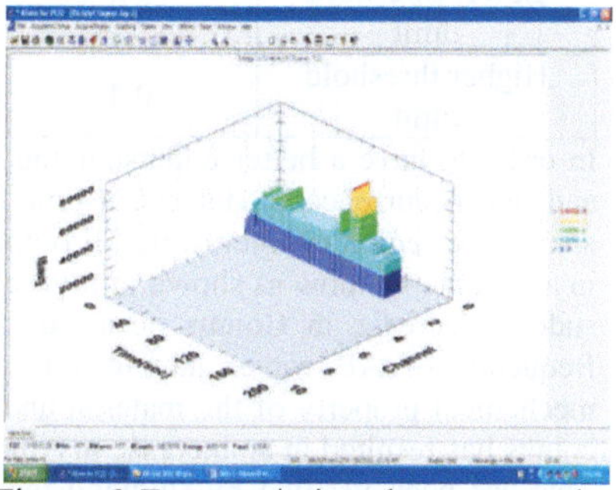
Figure. 6 Energy v/s time data generated

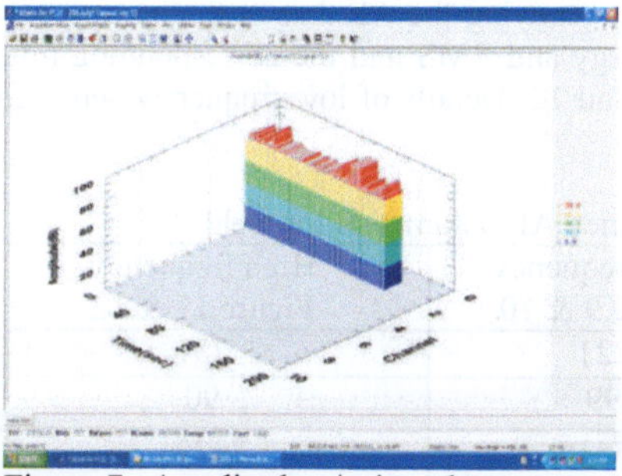
Figure.7 Amplitude v/s time data generated

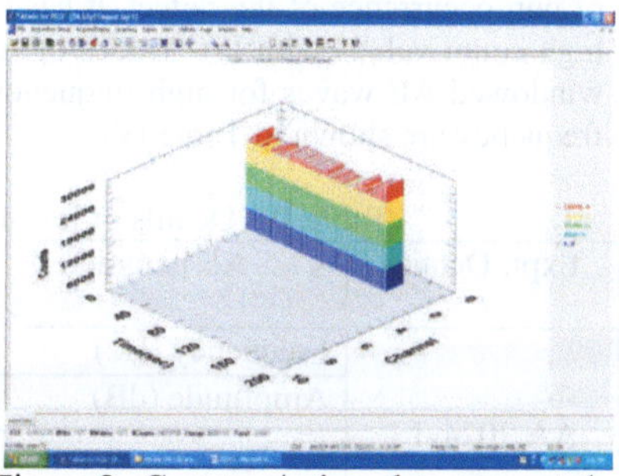
Figure.8 Counts v/s time data generated

Table I. Process parameters used during welding of AA 2024-T3.

Weld No.	Rotation Speed (rpm)	Traverse Speed (mm/min)	Plunge depth (mm)
1	1000	100	4.83
2	1000	100	4.85
3	1000	100	4.87

Table II. Details of values of AE parameters showing lower and higher values along with joint efficiency for aluminum alloy AA2024-T3.

Weld No.	Joint Efficiency (%)	RMS (v)		Energy x1000		Amplitude (dB)		Counts x1000		Remarks
		lower	higher	lower	higher	lower	higher	lower	higher	
1	75.00	0.2	0.3	10	32	75	85	22	24	Mix of defect and non-defect regions

Table III. Values of threshold limits of AE parameters for aluminum alloy AA2024-T3.

	AE parameters			
	RMS (v)	Energy x1000	Amplitude (dB)	Counts x 1000
Lower threshold limit	0.2	25	75	20
Higher threshold limit	0.4	70	90	30

In order to have a better understanding about the behavior of AE parameters at the occurrence and non-occurrence of defects, a time domain analysis was carried out. During time domain analysis, a region of interest in AE output plots was selected and frequency wave forms studied in a suitable window as shown in Figures 9 to 12. In the time window of the signal wherein the sudden decrease in Counts occurred, corresponding to a defect during FSW is due to low frequency of AE waves having very high Amplitude, Energy and RMS values. Since the mechanical property of the material and sensor are constant in the experiments, the decrease in counts has resulted in an increase in Amplitude, Energy and RMS or vice versa.

This has been verified by studying the AE data at the instances of low count values and the corresponding time windowed AE waves for low frequency, Figures 9 and 10. Also, at the point of non-occurrence of the defect, it has been verified by studying the AE data at the instances of high count values with low values of Amplitude, Energy and RMS and the corresponding time windowed AE waves for high frequency, Figures 11 and 12. Details of low frequency and high frequency are shown in Table IV.

Table IV. Details of low and high frequency AE parameters for weld

Expt. Details	AE parameter	Low frequency Figure.9 & 10	High frequency Figure.11 & 12
AA 2024-T3	Frequency (Hz)	221	231
	Amplitude (dB)	99	90
	Energy	65535	30796
	RMS(v)	0.2554	0.241
	Counts	22073	23115

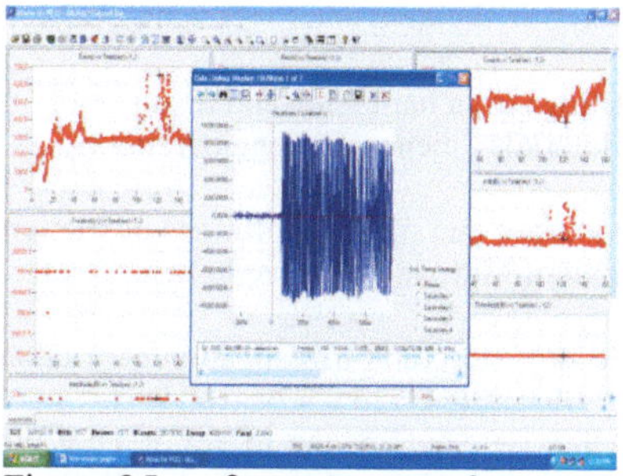

Figure. 9 Low frequency wave form

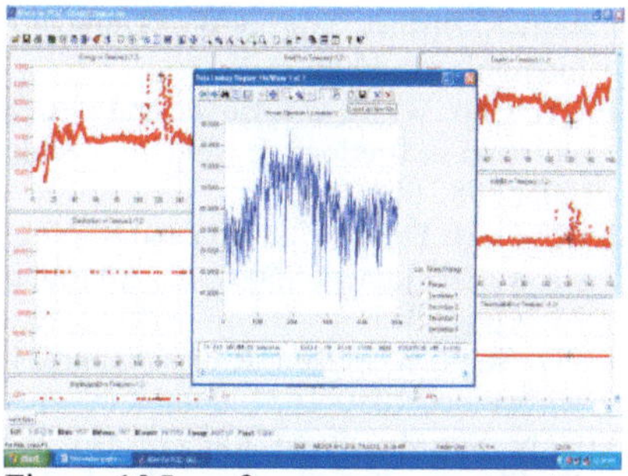

Figure. 10 Low frequency power spectrum

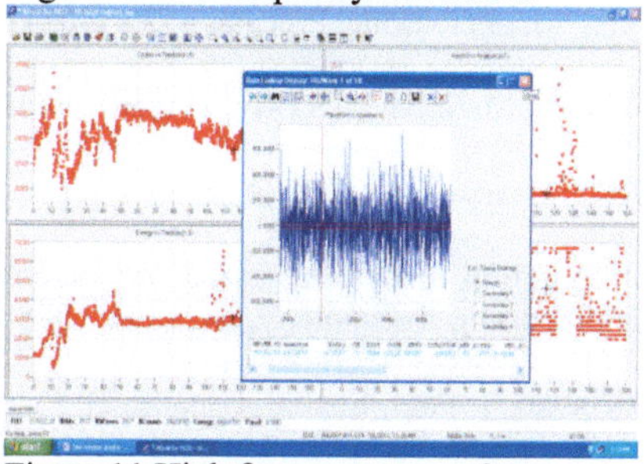

Figure.11 High frequency wave form

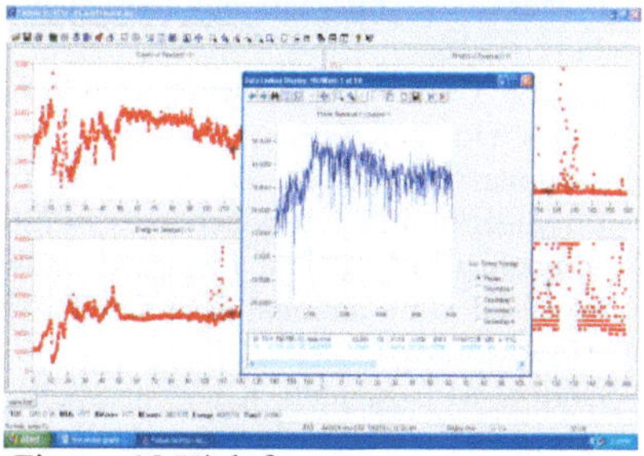

Figure. 12 High frequency power spectrum

Conclusions

- AE can be effectively used for online monitoring of FSW by suitably setting the threshold values of AE parameters to distinguish between good and bad welds.
- Time window analysis was found useful in better understanding of behavior of AE parameters in defect and defect free regions.
- The threshold values of AE parameters considered for distinguishing between good and bad weld are specific to the material being welded, thickness of material, and input process parameters considered.
- The outcome of results can be extended to other materials of different thickness during online monitoring of FSW using AE after making repeatability and reliability studies.
- The proposed work results in the inspection of a welded joint during the process of welding itself. Thus, the productivity of the process can be expected to increase substantially.
- The inspection of a welded joint using X-Ray radiography or any other method can be totally avoided once the relationship is established between selected AE parameters and quality of weld.

Acknowledgement

The authors wish to express sincere thanks to Aeronautical Research & Development Board (ARDB), New Delhi, for financial support rendered through R & D project No: DARO/08/2031473/M/I. Authors are thankful to Prof. Satish V Kailas, Department of Mechanical Engineering, Indian Institute of Science (IISc), Bangalore for extending facilities of computerized FSW setup to carry out experiments required for this investigation.

References

[1] Rajiv S. Mishra, Murray W. Mahoney-editors. Friction Stir Welding and Processing, ASM International, (2007)1-5.

[2] J. H. Cho, W. M, D E Boyce and P R Dawson. Modeling strain hardening and texture evolution in friction stir welding of stainless steel. Materials Science and Engineering A, 2005; 398: 146–163.

[3] H. Liu. H. Fujii, M. Maeda and K. Nogi - Tensile properties and fracture locations of friction stir welded joints of 6061–T6 Aluminum alloy. Journal of Materials Science Letters, 2003; 22:1061- 1063.

[4] J.Q. Su, T W Nelson, R Mishra and M Mahoney. Micro structural investigation of friction stir welded 7050-T651 aluminum. Acta Materialia, 2003; 51(3): 713 – 729.

[5] C. N. Suresha, B. M. Rajaprakash, and Sarala Upadhya. A Study of the Effect of Tool Pin Profiles on Tensile Strength of Welded Joints Produced Using Friction Stir Welding Process, International Journal of Materials and Manufacturing Processes, 26 (2011), 1111–1116.

[6] Mark F. Carlos. A E: Heeding the Warning Sounds from Materials, ASTM Standardization News, October 2003.

[7] Carpinteri. Structural damage diagnosis and life-time assessment by A E monitoring-Engineering Fracture Mechanics 74 (2007), 273–289.

[8] Xiaozhi Chen. Beizhi Li. A E method for tool condition monitoring based on wavelet analysis: International Journal of Advanced Manufacturing Technology,33(2007), 968–976.

[9] Sharma, M.I. Junaidh, K.K. Purushothaman, C.P. Kotwal, J. Paul, Shalini Tripathi, B. Pant and A.S. Sankaranarayanan. Online Monitoring of Electron Beam Welding of Ti6Al4V Alloy through Acoustic Emission. Proc. National Seminar on Non-Destructive Evaluation Dec. 7 - 9, 2006. [10] Beattie.A. G, Shurtleff. W. W. Results of Acoustic Emission Tests on Halon Fire Bottles-
report, Sandia National Laboratories.1996.

[11] Shiroishi. J, Li .Y, Liang .S, Kurfess .T,Danyluk.S. Bearing Condition Diagnostics via Vibration and Acoustic Emission Measurements-report. Georgia Institute of Technology Atlanta, Georgia, 1997.

[12] Dunegan. H.L. An Acoustic Emission Technique for Measuring Surface Roughness-a report. December 1998.

[13] Craig B. Van Way, Jayanth N. Kudva. Structural Health Monitoring of Full scale Components Using Acoustic Emission And Fiber Optic Sensors- Paper presented at the RTO AVT Specialists' Meeting on "Exploitation of Structural Loads/Health Data for Reduced Life Cycle Costs", held in Brussels, Belgium, and published in RTO May 1998, MP-7:11-12.

[14] Lee .C. S. Huh. J .H. Ll .D. M and Shin D. H. Acoustic Emission Behavior during Tensile Tests of Low Carbon Steel Welds. ISIJ International, Vol. 39 No. 4;1999: 365-370.

[15] Miettinen, J., Pataniitty, P- Acoustic Emission in Monitoring Extremely Slowly Rotating Rolling Bearing. Proceedings of COMADEM '99. Oxford, Cox moor Publishing Company. 1999. ISBN 1-901892-13-1. 289-29.

[16] C. Beggan, M. Woulfe, P. Young and G. Byrne. Using Acoustic Emission to Predict Surface Quality, Int J Adv Manuf Technol ,1999; 15:737–742.

[17] Jaap H. Heida and Walter G.J. 'HART, National Aerospace Laboratory NLR, Amsterdam, The Netherlands- Acoustic Emission and Ultrasonic Inspection for the Monitoring of Cure Related Defects in thick RTM Products, ECNDT - We.1.7.2,2006.

[18] J. Kopač, S. Sali. Acoustic emission in drilling carbon steel and nodular gray iron. Journal of Achievements in Materials and Manufacturing Engineering, Volume 19 Issue 1 November 2006:91-95.
[19] V. A. Akop'yan. Correlations between Parameters of Acoustic-Emission Signals and Corrosion Damage in Aluminum Alloys, ISSN 1061-8309, Russian Journal of Nondestructive Testing, 2007, Vol. 43, No. 6: 390–396.
[20] Xiaozhi Chen . Beizhi Li . Acoustic emission method for tool condition monitoring based on wavelet analysis- Int J Adv Manuf Technol .2007: 968 –976.
[21] Sundaram.S, Senthilkumar.P, Kumaravel .A and Manoharan. N. Study of Flank Wear in Single Point Cutting Tool Using Acoustic Emission Sensor Techniques. ARPN Journal of Engineering and Applied Sciences. VOL. 3, NO. 4, AUGUST 2008:32-36.
[22] B.M.Rajaprakash, C.N.Suresha, Sarala Upadhya, application of acoustic emission technique to monitor friction stir welding process to produce defect free welds, Friction Stir Welding and Processing VII, Proceedings of TMS 2013, Chapter 3.
[23] C.N. Suresha, "Studies on The Performance of the Welded Joints Produced by Friction Stir Welding Process using Acoustic Emission Technique" Ph.D thesis, Bangalore University, INDIA, 2013.

Friction Stir Welding and Processing VIII
Edited by: Rajiv S. Mishra, Murray W. Mahoney, Yutaka Sato, and Yuri Hovanski
TMS (The Minerals, Metals & Materials Society), 2015

ON THE MATERIAL BEHAVIOR AT TOOL/WORKPIECE INTERFACE DURING FRICTION STIR WELDING: A CFD BASED NUMERICAL STUDY

Gaoqiang Chen[1,2], Qingyu Shi[1], Zhili Feng[2]

[1] Department of Mechanical Engineering, Tsinghua University;
1 Tsinghuayuan Park, Haidian, Beijing 100084, PRC
[2] Materials Science & Technology Division, Oak Ridge National Laboratory.
1 Bethel Valley Road, Oak Ridge, TN 37831, USA (fengz@ornl.gov)

Keywords: Friction stir welding; Tool/workpiece interface; CFD; Material flow.

Abstract

The material behavior at the interface between the welding tool and the workpiece in friction stir welding (FSW) is a critical phenomenon governing heat generation and material flow. Previous computational modeling studies tend to simplify the material behavior at the interface, generally assuming simple stick or frictional slip conditions. In this paper, a new approach is developed that allows for dynamic friction between tool and workpiece, based on considerations of relationship between frictional force and material velocity at the interface. This new approach is implemented in a computational fluid dynamics (CFD)-based model to simulate the heat transfer and material flow. The simulated temperature field and velocity distribution are validated by experiments. The simulation results show that material velocity on tool/workpiece interface is significantly different from the tool velocity and increase with the interfacial frictional stress. It is found that when frictional stress is not large enough to enable large extent plastic flow, asymmetrical material flow appears. Material on the advancing side tends to flow at lower velocity than that at retreating side.

Introduction

Fundamental issues during friction stir welding (FSW) such as heat generation, temperature distribution and material flow are essential for optimization of welding parameters and tool geometry. A huge number of experimental and computational studies have been conducted to understand these fundamental issues. [1][2] In these studies, the transient material behavior at the tool/workpiece (T/W) interface during FSW has been taken a critical phenomenon governing both heat generation and material flow. Though previous experimental evidences showed that the interfacial contact condition was mixing of sticking and slip condition during the process [3][4], experimental studies gave little information on the distribution of material behavior on the interface, such as the material velocity. Most of previous computational and analytical modeling studies tend to simplify the material behavior at the interface, generally assuming simple stick [5-9] or frictional slip [5][10-11] conditions. Fraction of slip assumed in the studies range from 0% to 99% [5-11] in different welding conditions. The material behavior at the interface is but still unclear in the current studies on FSW.
The objective of this numerical study is to quantitatively study the material behavior at the T/W interface. In this paper, a new approach is developed that allows for dynamic friction between tool and workpiece, based on considerations of relationship between frictional force and material velocity at the interface. This new approach is implemented in a computational fluid dynamics

(CFD)-based model to simulate the heat transfer and material flow. The simulated distributions of temperature and material velocity are analyzed and validated by experimental results.

Experiments

FSW butt joining of two AA2024-T4 sheets is carried out. The dimensions of each workpiece are 145mm×55mm×3mm (length × width × thickness). The FSW tool is made by H13 steel. The tool shoulder is 13mm in diameter, and length of the conical pin is 2.4mm. The diameter of pin is 4mm at root and 3.5mm at tip. The FSW tool rotates at 1600rpm and travels at 20mm/min. A plunging depth of 0.5mm is used. No tilt angle is employed in the experiment. The length of the weld is 100mm. In order to measure the temperature on workpiece during welding, two K-type thermal couples (TC1 and TC2) are embedded in the work plates before welding. TC1 is embedded on the advancing side, while TC2 is on the retreating side. The thermal couples are located 3mm in the mid-thickness plane from the welding centerline and 37.5mm from the welding starting position. A specimen is cut after welding for macrostructure examination. The specimen is grounded, polished and etched with Keller's solution (95ml water, 1.5ml hydrochloric acid, 2.5ml nitric acid, 1ml hydrofluoric acid) for 1min. After that, the macrostructure of specimen is observed by optical microscope.

Model descriptions

A. Governing equations

Basic variables describing the heat flow and material flow during FSW are temperature and material velocity, which are obtained by solving by the governing equations. The conservation equations of mass and momentum for incompressible single-phase flow are given by,

$$\nabla \cdot (\rho \vec{v}) = 0 \tag{1}$$

$$\nabla \cdot (\rho \vec{v} \vec{v}) = -\nabla p + \nabla \cdot \left(\mu(\nabla \vec{v} + \nabla \vec{v}^T)\right) \tag{2}$$

where ρ is density, μ is viscosity, p is pressure and $\vec{v}$ is fluid velocity.
The energy conservation equation is given by,

$$\nabla \cdot (\vec{v} H) = \nabla \cdot (k \nabla T + S_V) \tag{3}$$

where H is enthalpy, T is temperature, k is thermal conductivity and S_V is a spatial source term regarding the heat generation due to plastic deformation.

B. Geometric model

Figure 1 illustrates the geometry of the CFD-based numerical simulation model. In the model, the FSW tool shape is explicitly reflected by the shape of tool-workpiece interface. The 'Fluid zone' is meshed using hexagonal grids. A relatively fine mesh is used in the vicinity of the tool in order to capture the rapid change of thermo-mechanical variables in this area.

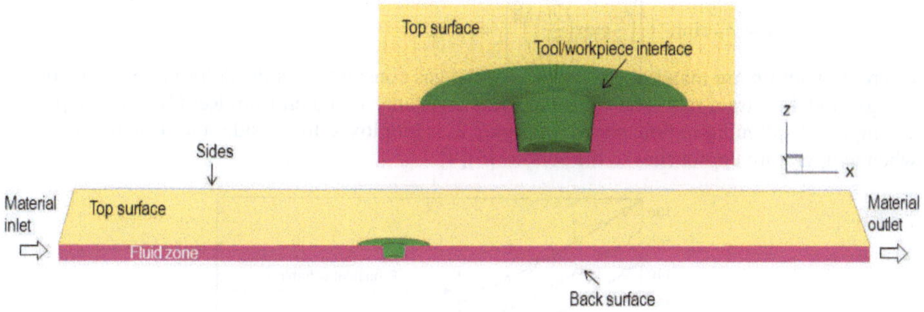

Figure 1 Geometric model.

C. Friction between tool and workpiece

In order to analyze the dynamic friction behaviors between FSW tool and workpiece, the frictional shear stress is applied at the T/W interface. The expression of the shear stress dependent on relative velocity between tool and workpiece, suggested by Ref. [13] is given by,

$$\vec{\tau}(\vec{v}_{relative}) = T_f \cdot \frac{\vec{v}_{relative}}{|\vec{v}_{relative}|} \cdot \delta(\vec{v}_{relative}) \quad (4)$$

where T_f is magnitude of maximum allowed shear stress at the T/W interface, $\vec{v}_{relative}$ is the relative velocity between tool and workpiece at the interface. The function $\delta(\vec{v}_{relative})$ is introduced to simplify the dynamic analysis when the relative velocity is close to 0 (See Ref [13] for details) . T_f is taken as constant over the T/W interface in this study.

D. Heat generation

Heat generation due to friction is taken as a heat flux at the T/W interface. The frictional heat flux at the interface is determined by the frictional shear stress and magnitude relative velocity between tool and workpiece, given as,

$$q_f = \eta \cdot |\vec{\tau}| \cdot |\vec{v}_{relative}| \quad (5)$$

where $\vec{\tau}$ is the shear stress at the T/W interface, given by Equ. (4) and $\vec{v}_{relative}$ is the relative velocity between tool and workpiece. η is mechanical efficiency, which is taken as 0.7 in the model.

The heat generated from plastic deformation is considered as a volumetric heat flux, taken as a source term in Equ. (3). The heat flux is calculated by the product of the flow stress and the plastic deformation rate, given as,

$$q_p = f \cdot \sigma \cdot \dot{\varepsilon} \quad (6)$$

where $f = 0.6$ is the fraction of plastic work dissipated as heat, σ is flow stress and $\dot{\varepsilon}$ is strain rate.

E. Material properties

The density of AA2024 is taken as 2700 kg/m^3 [14]. The temperature dependent thermal conductivity and heat capacity for AA2024 are taken from Ref [14]**Error! Reference source not found.**. The deformation resistance of aluminum alloys is both temperature and strain rate dependent. In the simulation model, material is considered as non-Newtonian fluid, the viscosity of which is temperature and strain rate dependent. The viscosity was calculated based on the following formulation [6],

$$\mu = \frac{\sigma}{3\dot{\varepsilon}} \quad (7)$$

where σ is flow stress and $\dot{\varepsilon}$ is strain rate. The expression of flow stress is proposed by Sheppard and Wright [15]**Error! Reference source not found.**,

$$\sigma = \sigma_P \sinh^{-1}\left[\left(\frac{\dot{\varepsilon}}{A}\exp\left(\frac{Q}{RT}\right)\right)^{\frac{1}{n}}\right] \tag{8}$$

where A, $\sigma_P$and n are material constants, R is the gas constant, Q is the deformation activation energy, and $\dot{\varepsilon}$ is strain rate. The values of these constants are taken from Ref [16]. In the paper, an empirical softening region, shown in Figure 2, is employed to consider the softening of metal when temperature approaches to the solidus [6][8].

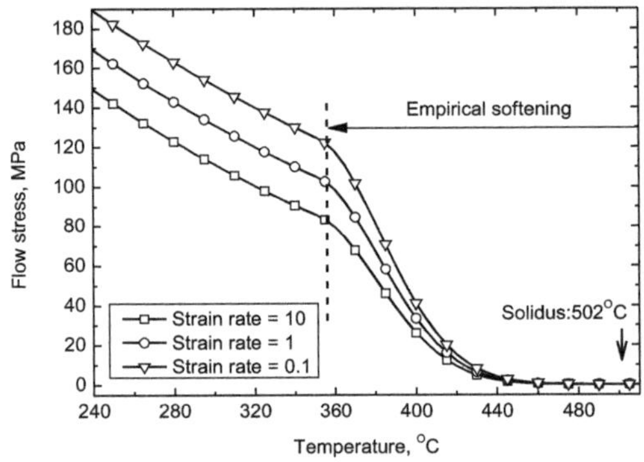

Figure 2 Flow stress of AA2024 in the model.

F. Boundary conditions

The FSW tool travels along the welding centerline during the real-world FSW, which would cause the shape change of the calculation domain due to translation of the T/W interface and thus increase the complexity of the models. In order to simplify the simulation models, an 'Inlet-outlet method' is adopted to include the welding speed by specifying material flow through the workpiece zone instead of changing the relative positions of tool and workpiece. Specifically, as shown in Figure 1, the material is assumed to flow into the 'Fluid zone' from 'Material inlet' at the welding speed and flow out of the domain from the 'Material outlet'. At 'Tool/workpiece interface', frictional shear stresses, defined by Equ. (4) and heat flux, defined by Equ. (5) due to friction, is applied. The 'Back surface' and 'Sides' of the work piece are assumed to contact with the fixtures. The contact heat transfer coefficient on 'Back surface' and 'Sides' is defined as 500 W/(m^2 K) [17] to consider the heat loss to the fixtures. 'Top surface' is taken as free surface and no shear stress is applied on it, where the contact heat transfer coefficient is assumed as 30 W/(m^2 K)[18].

Results & Discussion

Temperature and material velocity on the workpiece (welding parameters - 1600rpm and 20mm/min) are simulation with different interfacial shear stresses. Numerical results with stick condition on T/W interface are also calculated. A commercial CFD package, ANSYS Fluent 15.0, is used in the calculation. The initial temperature is set to 300K. After 2000 times steps (40s), a quasi-steady state is considered to be reached. The following analysis is based on the numerical solutions in a quasi-steady state.

A. Temperature field

The simulated temperature field on the workpiece zone is shown in Figure 3. As frictional stress exists at the T/W interface and large amount of heat is generated, temperature is relatively high around the tool. At the top surface of the workpiece, the temperature contour lines are close to circular as the welding speed is as slow as 20mm/min. It is shown in Figure 4 that the temperature distribution under the tool. The maximum temperature is less than 420°C with the shear stress of 12MPa, which is lower than the other 3 cases. The area with high temperature expands with the interfacial shear stress. Figure 5 compares the simulated and measured temperature evolution curves on advancing side and retreating side. Good agreement between the measured and simulated temperature could be found when the shear stress is 18MPa or the contact condition is 'stick'.

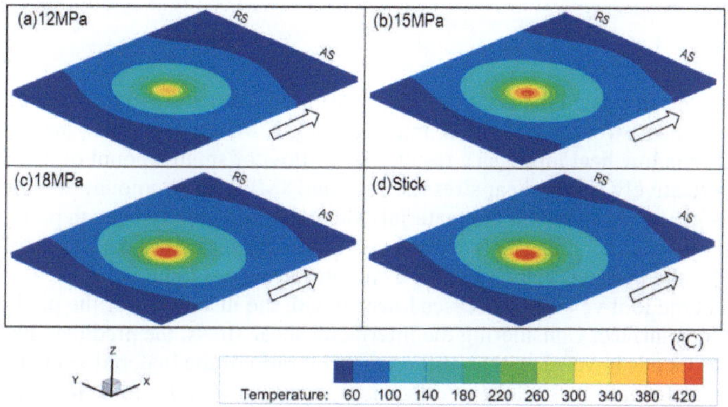

Figure 3 Simulated temperature fields with different interfacial stresses.

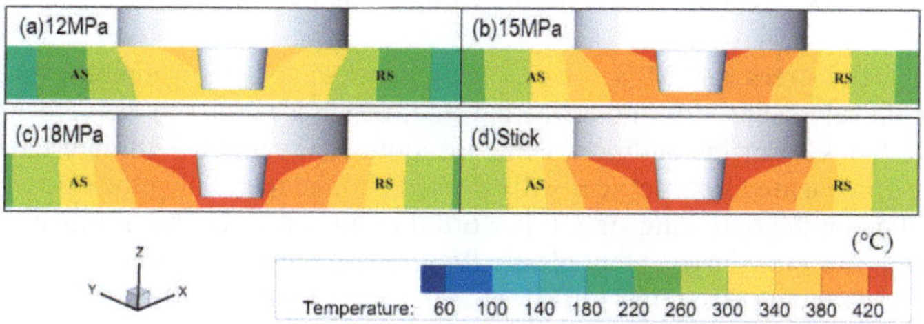

Figure 4 Simulated temperature fields under the tool.

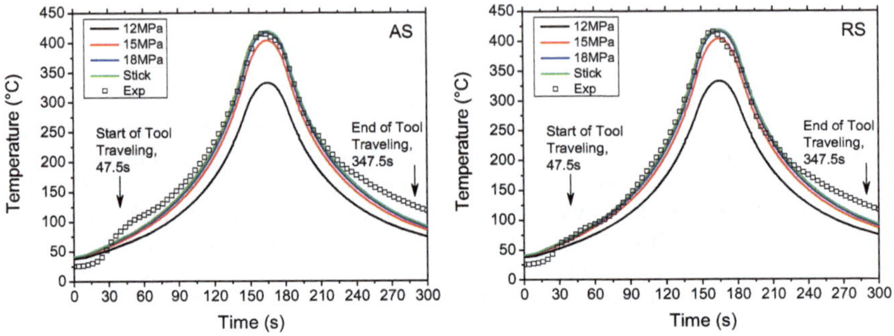

Figure 5 Comparison of measured and simulated temperatures.

B. Material velocity

Figure 6 shows the simulated material velocity distribution under the tool. In the cases with relatively lower shear stresses at the T/W interface, e.g. 12MPa and 15MPa, material is not well softened due to the low heat input; as a result, plastic flow of small amount of material is enabled. With relatively larger shear stresses, such as 18MPa, large amount of material flows as enough heat is generated to soften the material. The material velocity of each point on T/W interface under different conditions is plotted versus the distance from the tool axis in Figure 7. With the 'stick', the material velocity at the T/W interface is set to be the same as the tool velocity, so that the tool velocity increases linearly with the distance over the pin bottom, pin side and shoulder surface. Considering the interfacial shear stress, the predicted material velocity at the T/W interface is lower than that of the tool. The ratio of the material velocity to the tool velocity is not uniform over the T/W interface and strongly depends on the interfacial shear stress. The interfacial material velocity at the interface tends to increase with the interfacial frictional stress. The material velocity is close to welding speed at the periphery of shoulder as the temperature is lower and the material is constrained by the surrounding materials.

When the interfacial stress is 18MPa, the predicted material velocity is almost symmetrically distributed on the advancing side and retreating side, which is similar to the velocity distribution simulated with 'stick' contact condition; while the simulated velocity with interfacial shear stress of 12MPa and 15MPa are asymmetrically distributed and the material velocity on the retreating side is larger than on the retreating side. It is worth noting that when the frictional shear stress is not large enough to enable large extent plastic flow, asymmetrical material flow appears and material on the advancing side tends to flow at lower velocity than that on the retreating side.

Compared to the macrostructure shown in Figure 8, the predicted stir zones are not big enough with the shear stress being 12MPa or 15MPa. The predicted velocity distribution when the interfacial shear stress is 18MPa reflects the shape of the stir zone in the macrostructure better than that predicted by 'stick' condition, because the stir zone does not exceed the shoulder periphery in the macrostructure.

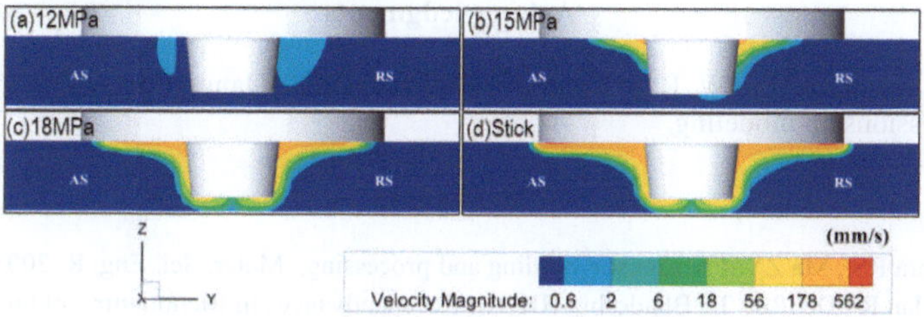

Figure 6 Simulated material velocity.

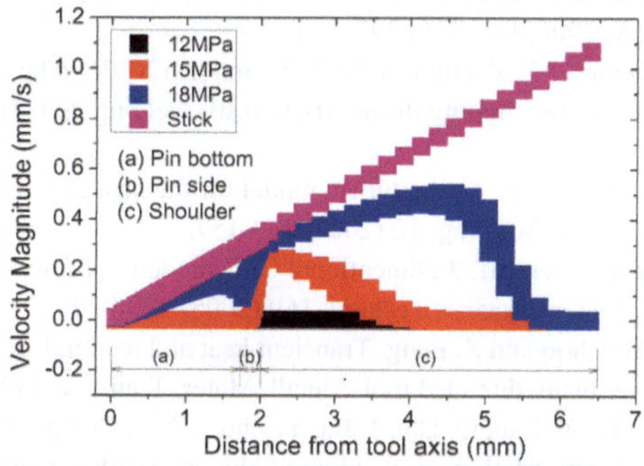

Figure 7 Velocity Magnitude at the tool/workpiece interface.

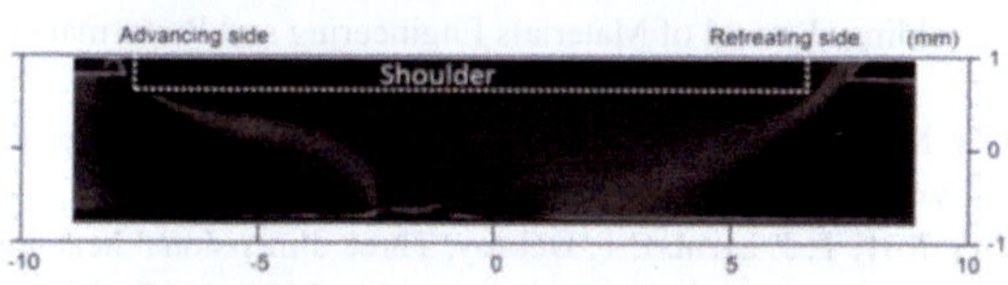

Figure 8 Macrostructure of friction stir weld.

Conclusions

Considering the relationship between frictional force and material velocity at the interface, the material behavior during FSW is analyzed via CFD based models. The following conclusions could be drawn:

(1) Good agreement could be found between the measured and simulated temperature when the shear stress is 18MPa or the contact condition is 'stick'. The predicted velocity distribution using interfacial shear stress of 18MPa reflects the stir zone shape better than that using 'stick' contact condition.

(2) The simulation results show that material velocity on tool/workpiece interface is significantly different from the tool velocity and strongly depends on the interfacial frictional stress.

(3) When frictional stress is not large enough to enable large extent plastic flow, asymmetrical material flow appears. Material on the advancing side tends to flow at lower velocity than that at retreating side.

Acknowledgments

The authors are grateful to W. R. Schwarz, C. Paglicco and K. Dang, from ANSYS INC. for useful discussions on modeling.

References

[1] Mishra RS , Ma ZY. Friction stir welding and processing, Mater. Sci. Eng. R, 2005 (50) 1–78.
[2] Nandan R , DebRoy T , Bhadeshia HKDH. Recent advances in friction-stir welding – process, weldment structure and properties, Prog. Mater. Sci., 2008(53) 980–1023.
[3] Schneider J , Beshears R , Nunes AC. Interfacial sticking and slipping in the friction stir welding process', Mater. Sci. Eng. A, 2006(435–436) 297–304.
[4] Qian J W, Li J L, Xiong J T, Zhang F S, Li W Y, and Lin X. Periodic variation of torque and its relations to interfacial sticking and slipping during friction stir welding. Sci. Technol. Weld. Joining. , 2012(4) 338-341.
[5] H. Schmidt, J. Hattel, J. Wert, An analytical model for the heat generation in friction stir welding, Modeling Simul. Mater. Sci. Eng. 12 (2004) 143-157.
[6] P. A. Colegrove, H. R. Shercliff, 3 Dimentional CFD modeling of flow round a threaded friction stir welding tool profile, J. Mater. Process. Technol. 169 (2005) 320-327.
[7] Z. Yu, W. Zhang, H. Choo and Z. Feng. Transient heat and mateiral flow modeling of friction stir welding of Magnesium alloy using threaded tool. Metall. Mater. Trans. A. 43A (2012) 724-737
[8] G. Chen, Q. Shi, Y. Li, Y. Sun, Q. Dai, J. Jia, Y. Zhu, J. Wu. Computational fluid dynamics studies on heat generation during friction stir welding of aluminum alloy. Comput. Mater. Sci. 79, (2013) 540–546.
[9] L. Shi, C. S. Wu, H. J. Liu. Modeling the Material Flow and Heat Transfer in Reverse Dual-Rotation Friction Stir Welding. Journal of Materials Engineering and Performance. 2014(23): 2918-2929
[10] A. Bastier; M. H. Maitournam; K. Dang Van; F. Roger. Steady state thermomechanical modelling of friction stir welding. Volume 11, Issue 3 (01 May 2006), pp. 278-288.
[11] R. Nandan, G. G. Roy, T. J. Lienert, T. Debroy, Three dimensional heat and material flow during friction stir welding of mild steel, Metall. Mater. Trans.. 37A (2006) 1247-1259.
[12] Hosein Atharifar, Dechao Lin, Radovan Kovacevic. Numerical and Experimental Investigations on the Loads Carried by the Tool During Friction Stir Welding. Journal of Materials Engineering and Performance. 2009(18) 339-350
[13] Mostaghel N., Davis T. Representations of coulomb friction for dynamic analysis. Earthquake engineering and structural dynamics, 1997, 26(5): 85-88.
[14] Mills, Kenneth C. (2002). Recommended Values of Thermophysical Properties for Selected Commercial Alloys. Woodhead Publishing. Online version available at:http://app.knovel.com/hotlink/toc/id:kpRVTPSCA1/recommended-values-thermophysical
[15] T. Sheppard and D.S. Wright, Determination of flow stress. Pt. 1. Constitutive equation for aluminum alloys at elevated temperatures, Metal. Technol., 1979, 6, p 215–223
[16] K. E. Tello, A. P. Gerlich, P. F. Mendez. Constants for hot deformation constitutive models for recent experimental data. Sci Technol Weld Joining 2010; 15: 260.
[17] M. Yu, W. Y. Li, J. L. Li, Y. J. Chao. Modelling of entire friction stir welding process by explicit finite element method. Materials Science and Technology. 2012, 28(7) 812-817
[18] H. Palaniswamy, G. Ngaile, T. Altan. Finite element simulation of magnesium alloy sheet forming at elevated temperatures. J. Mater. Process. Tech., 2004, 146: 52–60.

Friction Stir Welding and Processing VIII
Edited by: Rajiv S. Mishra, Murray W. Mahoney, Yutaka Sato, and Yuri Hovanski
TMS (The Minerals, Metals & Materials Society), 2015

FRICTION STIR WELDING OF AZ31B MAGNESIUM ALLOY WITH 6061-T6 ALUMINUM ALLOY: INFLUENCE OF PROCESSING PARAMETERS ON MICROSTRUCTURE AND MECHANICAL PROPERTIES

B. Mansoor[1*], A. Dorbane[1], G. Ayoub[2], A. Imad[3]

[1]Mechanical Engineering Program, Texas A&M University at Qatar, Doha, Qatar.
[2]Department of Mechanical Engineering, American University of Beirut, Beirut, Lebanon.
[3]Laboratoir de Mécanique de Lille, Ecole Polytech'Lille, University of Lille, Lille, France

Keywords: FSW, Dissimilar Material Joining, AZ31, 6061 Al, Intermetallics.

Abstract

The success of Friction Stir Welding (FSW) in joining light metal alloys has inspired attempts to further exploit its potential for joining materials which differ in chemical composition, structure, and/or properties. The FSW of relatively soft (e.g., Al/Mg) and hard (e.g., Fe/Ni) combinations of alloys is of particular interest in automotive and aerospace applications. However, joining of dissimilar alloys presents several unique challenges that include the different deformation behaviors, formation of detrimental intermetallic compounds, and differences in physical properties such as thermal conductivity. These factors lead to amplified asymmetry in both heat generation and material flow and consequently lead to the formation of a heterogeneous weld. In this work, a dissimilar metal joint was created between twin roll cast AZ31B magnesium alloy and Al 6061-T6 aluminum alloy plates by FSW. The main aim here is to investigate the effect of key process parameters such as tool rotation speed and welding speed on microstructural evolution and mechanical properties of the resulting heterogeneous joint. A detailed microstructural analysis was carried out to understand the composition of the intermetallic phases generated in the stirred zone and their impact on microhardness and over-all mechanical properties of the weld. Our key finding was that, weld configuration with placing the aluminum alloy plate on the advancing side resulted in a sound, defect free joint compared to the alternate configuration.

Introduction

The use of light metal alloys such as aluminum (Al), magnesium (Mg) and titanium (Ti) alloys in automotive vehicles for light weighting can significantly increase fuel efficiency and cut harmful CO_2 emissions [1, 2]. However, to enable promote the manufacture and integration of parts made from light-weight alternative materials, one key objective is to develop reliable dissimilar material joining technologies. The ability to build parts from a variety of lightweight materials and to join them is a major technological challenge for the transportation industry in its push to increase the use of advanced lightweight materials.

Joining Al alloys to Mg alloys poses one such specific challenge. In this regard, conventional fusion based welding processes such Gas Tungsten Arc Welding (GTAW) [3, 4], Electron Beam Welding (EBW) [4, 5], Laser Beam Welding (LBW) [6], have been applied to join Al alloys to Mg alloys. However, they produce coarse grains, large and continuous intermetallic regions in the weld zone accompanied by a large Heat Affected Zone (HAZ) in the base metals [7]. Also, certain other joining processes involving the use of Zn based filler materials such as brazing [8, 9], diffusion bonding [10, 11], cold metal transfer MIG welding [12, 13] and conventional MIG and TIG (with filler metal) welding [14-16]. However, the formation of different types and continuous layers of intermetallic compounds deteriorates the mechanical properties of the joint. Therefore, it can be argued that the fundamental problem with all dissimilar metal joining techniques of Al alloy to Mg alloy involving solidification see the formation of intermetallic compounds. The lack of control over the size, type and distribution of hard and brittle intermetallics has a detrimental effect on the joint strength.

Solid state welding processes which provide relatively controllable reaction times and heat inputs are found to be promising for dissimilar welding of Mg alloys to and Al alloys and result in a high strength joint [7]. Researchers have considered techniques such as linear friction welding [17], ultrasonic welding [18, 19], resistance spot welding [20] and FSW [21-30] have been studied. In this regard, FSW a solid state joining process that in addition to frictional heat input involves stirring and plastic deformation of the joint region has shown promising initial results. Also, FSW has been studied to join magnesium or aluminum alloys to steels [31, 32] and titanium [33, 34], again with proming results. In this work, we have attempted to study the effect of process parameters on the weld zone microstructure and the formation, morphology, size and distribution of the intermetallic region and its subsequent impact on mechanical properties.

Experimental Procedure

FS welds were performed using a Gantry Friction Stir Welding Machine of type FSW-LM-08, which was designed to weld aluminum alloys with maximum thickness of 10 mm. The tool used for the present study had a 10 mm tool shoulder diameter and a threaded pin with maximum diameter of 3.6 mm and minimum diameter of 2.9 mm. All the welds runs were performed using a tool tilt angle of 3°. The FSW coupons were all 250 x 50 x 3 mm strips sheared from the base material sheets and the edges were cleaned to avoid any prior in-homogeneities in weld. The base material were Al6061-T6 aluminum and AZ31B magnesium, both were obtained from commercial vendors. The nominal chemical composition of the two materials are presented in table I.

Table I: Nominal composition and mechanical properties of Al and Mg alloys.

Alloys	Wt. %	Al	Mg	Si	Cu	Cr	Fe	Zn	Ti	Ca	YS (MPa)	UTS (MPa)
Al6061-T6	Min	95.8	0.8	0.4	0.15	0.04	-	-	-	-	275	350
	Max	98.6	1.2	0.8	0.4	0.35	0.7	0.25	0.15	-		
AZ31B	Min	2.5	0.2	-	-	-	-	0.6	-	-	150	300
	Max	3.5	1	0.1	0.05	-	0.005	1.4	-	0.04		

The samples for microstructure examination of as-received and welded joints were prepared according to the standard metallographic preparation. The FSW plates were cut in the cross-section using TECHCUT 5" precision sectioning machine, then mounted using TechPress 2™ hydraulic-pneumatic automatic mounting press and black phenolic hot mounting resin. The mounted samples were ground and polished utilizing MetPrep 4™ Grinder/Polisher and silicon carbide abrasive paper with grits of 320, 600 and 1200 with rotating the sample 90° between each step. The polishing procedure was performed using Spec-cloth and 1 μm diamond polycrystalline solution with alcohol based BlueLube lubricant, followed by 0.04 μm colloidal silica and same lubricant mentioned before. The etching was performed using a standard acetic-picral solution to reveal the microstructure of the magnesium side of the joint. The microstructure was observed using Zeiss AxioVert 40 MAT optical light microscope equipped with an ERc5s camera. Scanning election microscopy was carried to better examine the fine grain structure and intermetallics formed in the stir zone. In addition, EDS analysis was implemented to investigate the chemical composition of the intermetallic. In this preliminary work reported, we only carried out Vicker's micro-hardness measurements to estimate the relative strength the weld. The microhardness measurements were done in the thickness plane perpendicular to the weld line at a distance of 2 mm from the top of the weld.

Results and Discussion

In this section we present the initial findings of our work. Figure 1 (a-b) present the pictures of two welded coupons. Figure 1 (a) shows a dissimilar weld of Al and Mg performed using tool rotation speed of 1600 rpm and 250 mm/min with placing Mg in the advancing side. While Figure 1 (b) shows the appearance of FS welded plates using 1400 rpm and 500 mm/min welding tool parameters with placing Al in the advancing side. Although, the welding parameters were different for the two cases but it can be seen that al plate configuration has a profound effect on the over-all welding. It was found that when the Al plate is placed in the Advancing Side (AS), a clear uniform defect-free weld zone can be obtained, as shown in Figure 1 (b). This result was further verified by examination of the weld zone under a microscopy.

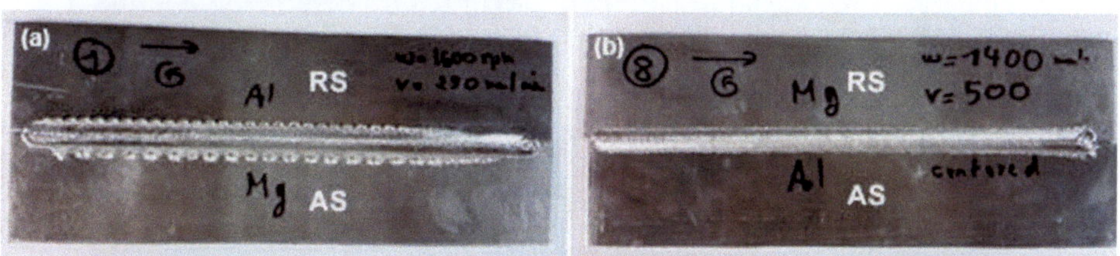

Figure 1. Top view of two welds : (a) Dissimilar FSW of Al and Mg with parameters of 1600 rpm and 250 mm/min, Mg in advancing side; (b): With parameters of 1400 rpm and 500 mm/min, Al in advancing side.

The structure and morphology of the weld was observed under an optical microscope. The weld zone cross-sections perpendicular to the tool traverse direction were observed to distinguish between the two welded materials, Al 6061-T6 and AZ31B. Figure 2 presents a heterogeneous FSW joint of aluminum to magnesium. As already observed in Figure 1, when aluminum was placed in the Retreating Side (RS), a poor quality weld was obtained and many defects such as the voids and excessive tool penetration were readily observed. These finding is well supported by the available literature on dissimilar joining of Al alloys to Mg alloys [21-30].

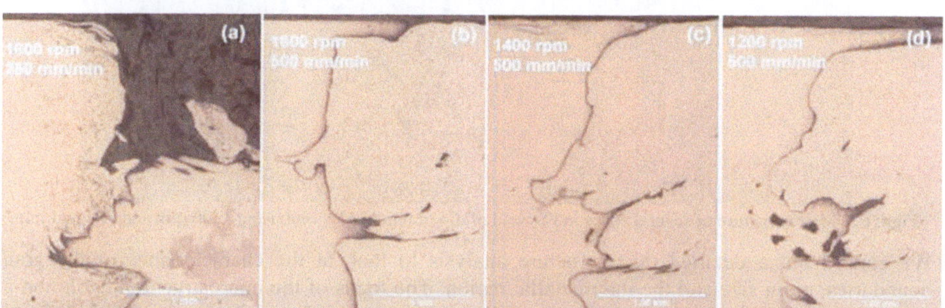

Figure 2. Cross section of weld zone performed using at: (a) 1600 rpm rotational speed and 250 mm/min welding speed (aluminum on the RS). (b-d) Constant welding speed of 500 mm/min and varying rotational speeds to 1600, 1400, 1200 rpm (aluminum on the AS).

Figure 2 (b) through (d) presents an Al/Mg dissimilar FSW weld with aluminum being placed in the advancing side. The influence of the tool rotational speed on the weld zone integrity was studied by keep the welding speed constant at 500 mm/min and reducing the tool rotation speed to 1600 rpm to 1200 rpm. It is observed that there were fewer defects (b, c) as compared to the previous weld (aluminum on RS). However, some internal defects such as worm hole, can be still be observed in the aluminum part, and as the tool rotational speed was decreased to 1200 rpm, again a larger number of voids were observed when comparing to the two previous cases. At this point, it was concluded that tool the rotation speed of 1600 rpm and tool advance speed of 500 mm/min produced the optimum weld.

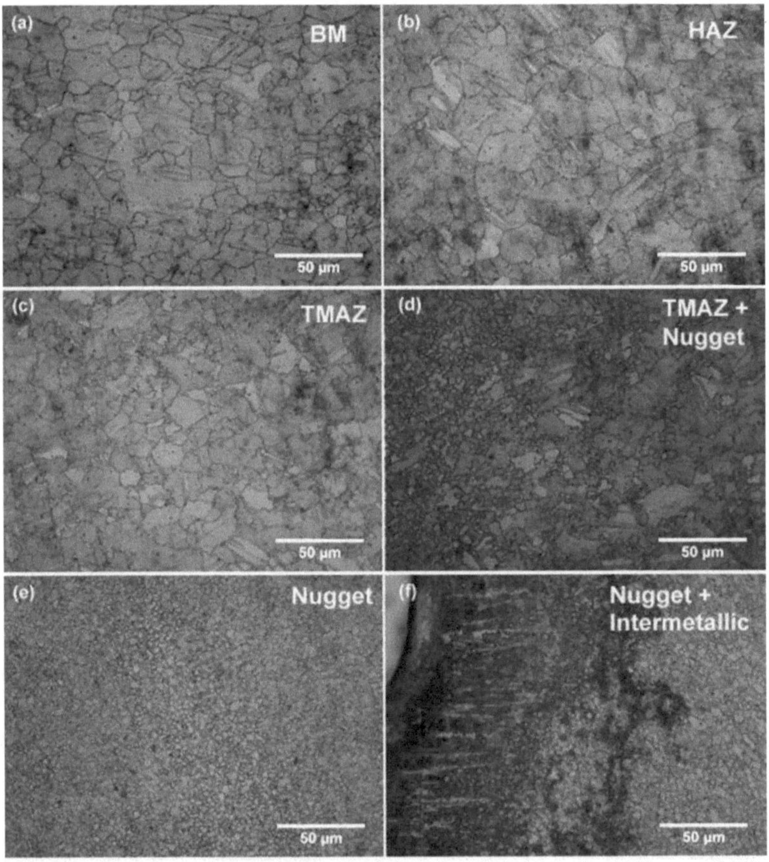

Figure 3. Microstructure of several zones in FS weld of dissimilar Al Mg joint using 1600 rpm and 500 mm/min.

We carried out a detailed microstructure analysis to look at the characteristics of the grain boundaries, grain size and the intermetallic region. The focus of the microstructural evolution in the intial stage of this study was magnesium alloy side of the weld zone. Figure 3 presents different microstructures obtained from several zones in the retreating side of the weld. These zones were identified to be the Base Metal (BM) (Figure 3 (a)), HAZ (Figure 6 (b)) and Thermo-Mechanically Affected Zone (TMAZ) (Figure 3 (c)). The heat affected zone shows larger grains compared to the base metal, while the thermo-mechanically affected zone shown in Figure 3 (d), shows some degree of grain refinement. A very small grain size was observed in the weld nugget as shown in Figure 3 (e). The strong degree of grain refinement (from 20 microns as-received to ~ 1 micron) in Mg rich side of the TMAZ and nugget region was also observed by several other researchers in dissimilar FSW of Al alloy to Mg alloy [21-30]. Figure 3 (f) presents the microstructure of the nugget and the intermetallic region. The microstructure of the intermetallic was not readily resolved with the optical micrographs. Figure 4 presents the SEM micrograph of the dissimilar FSW weld performed using 1600 rpm and 500 mm/min with aluminum placed in the advancing side. Figure 4 (a) shows the interaction between AZ31B magnesium alloy and 6061-T6 aluminum alloy with a void defect in both magnesium side and aluminum side. Figure 4 (b) is a magnified area where the intermetallic of magnesium is interacting with aluminum with the absence of defects in the joint.

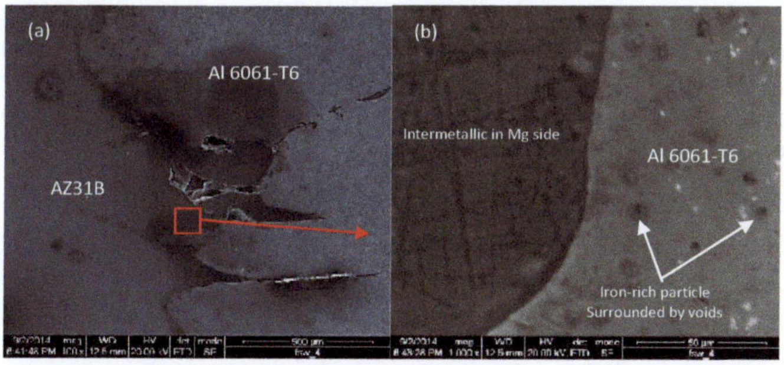

Figure 4. SEM micrograph of intermetallic region and nugget zone in the FSW joint obtained by using 1600 rpm rotational speed and a welding speed of 500 mm/mi, (a) low resolution image shows the over-all morphology of the joint, (b) the transition region between the Al and the Mg intermetallic side of the joint.

Figure 5 presents the SEM observations of dissimilar Al/Mg welded plates performed with 1400 rpm and 500 mm/min with aluminum placed in the advancing side. Figure 5 (a) shows the AZ31B and Al 6061-T6, the dark area is the intermetallic formed in the magnesium side. Figure 5 (b) a magnified area taken from Figure 5 (a) and highlighted with a red rectangle. It also shows an iron-

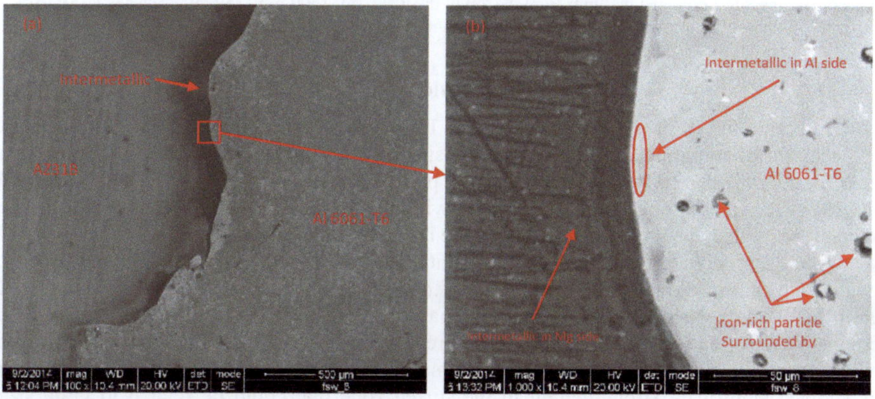

Figure 5. SEM micrographs of FSW sample at 1400 rpm and 500 mm/min. (a) low resolution image shows the over-all morphology of the joint, (b) the transition region between the Al and the Mg intermetallic side of the joint.

rich particle surrounded by voids and micro iron rich phases with the same phenomena in the aluminum side. From these SEM analyses we observe that the voids are initially around the intermetallic phases or iron-rich phases of Al6061-T6. This could be a reason for the presence of the micro-defects in the joint. Also the interaction between Al6061-T6 and AZ31B alloys was defect-free in the most regions. The intermetallic formed in the interface between Al and Mg were reported by McLean et al. [35] and the same observation was reported by Kostka et al. [36] when they studied the microstructure of friction stirred welded AA6040 and AZ31B.

Micro-hardness tests were performed for 4 samples with different welding parameters. It is observed in Figure 6 that the microhardness values in the magnesium part are between 55HV and 65HV. Then under the tool it starts to increase to reach values between 70 and 80 HV. The intermetallic phases showed values between 90 and 110 HV in the center of the weld, these values were results of the brittle nature of the intermetallic formed for to the dynamic recrystallization. The same trend of this increase of the hardness in the middle of the weld was obtained by Somasekharan et al. [37] when they studied the FSW of Az31B-H24 and Al6061-T6, they attributed this trend to the grain size reduction and presence of intermetallics. In the aluminum side and under the tool shoulder, the hardness values are found to be between 70 and 120 HV. Then it reaches 135 HV in the base metal.

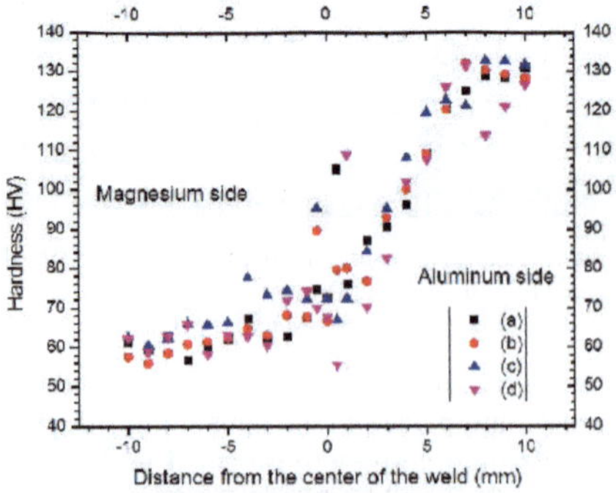

Figure 6. Hardness test obtained in 2 mm from the top of the weld. (a): 1600 rpm, 250 mm/min, Mg (AS); (b): 1600 rpm, 500 mm/min, Al (AS); (c): 1400 rpm, 500 mm/min, Al (AS); (d): 1200 rpm, 500 mm/min, Al (AS).

Conclusions

The twin roll cast AZ31B magnesium and Al6061-T6 aluminum were successfully friction stirred welded using different parameters and material positioning. It was found that placing magnesium (lower YS material) on the advancing side gave poor results with various defects in the weld region. However, performing dissimilar FSW with aluminum (higher YS material) on the advancing side produced better joints. Based on the intial results reported here, the rotational speed of the tool was varied and optimum welding conditions were obtained. It was also conclude that the magnesium rich side of the nugget zone showed evidence of extensive grain refinement and the grain size was reduced to ~ 1 micron. It has been observed that different intermetallic compounds were formed in the stir zone. The nature and morphology of these intermetallic layers were different on the aluminum rich and magnesium rich side of the weld zone. The microhardness results showed significant strengthening in the stirred zone that could be related to grain size reduction. A detailed TEM and EDS analysis and tensile characterization would be carried out in future to fully understand the influence of the intermetallics on mechanical properties of the joint.

Acknowledgment

The authors acknowledge the support provided by Qatar National Research Fund under NPRP grant # 4-1063-2-397 for this work. QNRF is a constituent member of the Qatar Foundation.

References

1. Pollock, T.M., *Weight Loss with Magnesium Alloys.* Science, 2010. **328**(5981): p. 986-987.
2. Mathaudhu, S.N. and E.A. Nyberg, *MAGNESIUM ALLOYS IN U.S. MILITARY APPLICATIONS: PAST, CURRENT AND FUTURE SOLUTIONS*, in *Magnesium Technology 2010*, S.R. Agnew, et al., Editors. 2010. p. 27-32.
3. Liu, P., et al., *Microstructure characteristics in TIG welded joint of Mg/Al dissimilar materials.* Materials Letters, 2007. **61**(6): p. 1288-1291.
4. Ben-Artzy, A., et al. *Joining of light hybrid constructions made of magnesium and aluminum alloys.* in *TMS Annual Meeting*. 2002.
5. Chi, C.-T., et al., *Aluminum element effect for electron beam welding of similar and dissimilar magnesium–aluminum–zinc alloys.* Scripta Materialia, 2007. **56**(9): p. 733-736.
6. Bannour, S., et al., *The influence of position in overlap joints of Mg and Al alloys on microstructure and hardness of laser welds.* J. Laser Appl., 2013. **25**(3).
7. Liu, L., D. Ren, and F. Liu, *A Review of Dissimilar Welding Techniques for Magnesium Alloys to Aluminum Alloys.* Materials, 2014. **7**(5): p. 3735-3757.
8. Liu, L., J. Tan, and X. Liu, *Reactive brazing of Al alloy to Mg alloy using zinc-based brazing alloy.* Materials Letters, 2007. **61**(11–12): p. 2373-2377.
9. Wang, Z., H. Wang, and L. Liu, *Study on low temperature brazing of magnesium alloy to aluminum alloy using Sn–xZn solders.* Materials & Design, 2012. **39**(0): p. 14-19.
10. Zhao, L.M. and Z.D. Zhang, *Effect of Zn alloy interlayer on interface microstructure and strength of diffusion-bonded Mg–Al joints.* Scripta Materialia, 2008. **58**(4): p. 283-286.
11. Liu, L.M., L.M. Zhao, and R.Z. Xu, *Effect of interlayer composition on the microstructure and strength of diffusion bonded Mg/Al joint.* Materials & Design, 2009. **30**(10): p. 4548-4551.
12. Shang, J., et al., *Microstructure characteristics and mechanical properties of cold metal transfer welding Mg/Al dissimilar metals.* Materials & Design, 2012. **34**(0): p. 559-565.
13. Wang, J., J. Feng, and Y. Wang, *Microstructure of Al–Mg dissimilar weld made by cold metal transfer MIG welding.* Materials Science and Technology, 2008. **24**(7): p. 827-831.
14. Zhang, H.T. and J.Q. Song, *Microstructural evolution of aluminum/magnesium lap joints welded using MIG process with zinc foil as an interlayer.* Materials Letters, 2011. **65**(21–22): p. 3292-3294.
15. Liu, F., D. Ren, and L. Liu, *Effect of Al foils interlayer on microstructures and mechanical properties of Mg–Al butt joints welded by gas tungsten arc welding filling with Zn filler metal.* Materials & Design, 2013. **46**(0): p. 419-425.
16. Liu, F., Z. Zhang, and L. Liu, *Microstructure evolution of Al/Mg butt joints welded by gas tungsten arc with Zn filler metal.* Materials Characterization, 2012. **69**(0): p. 84-89.
17. Bhamji, I., et al., *Linear friction welding of aluminium to magnesium.* Science and Technology of Welding and Joining, 2012. **17**(5): p. 368-374.
18. Panteli, A., et al., *The effect of high strain rate deformation on intermetallic reaction during ultrasonic welding aluminium to magnesium.* Materials Science and Engineering: A, 2012. **556**(0): p. 31-42.
19. Robson, J., A. Panteli, and P.B. Prangnell, *Modelling intermetallic phase formation in dissimilar metal ultrasonic welding of aluminium and magnesium alloys.* Sci. Technol. Weld. Join., 2012. **17**(6): p. 447-453.

20. Hayat, F., *The effects of the welding current on heat input, nugget geometry, and the mechanical and fractural properties of resistance spot welding on Mg/Al dissimilar materials.* Mater. Des., 2011. **32**(4): p. 2476-2484.
21. Mofid, M.A., et al., *Submerged friction-stir welding (SFSW) underwater and under liquid nitrogen: An improved method to join Al alloys to Mg alloys.* Metallurgical and Materials Transactions A: Physical Metallurgy and Materials Science, 2012. **43**(13): p. 5106-5114.
22. Morishige, T., et al., *Dissimilar welding of Al and Mg alloys by FSW.* Materials Transactions, 2008. **49**(5): p. 1129-31.
23. Cao, X., C. Garnier, and P. Wanjara. *Tensile strength of friction stir spot welded dissimilar AA5754-to-AZ31B alloys.* in *9th International Conference on Trends in Welding Research, June 4, 2012 - June 8, 2012*. 2013. Chicago, IL, United states: ASM International.
24. Zettler, R., et al. *A study on dissimilar friction stir welds between Al and Mg alloys.* in *7th International Conference on Trends in Welding Research, May 16, 2005 - May 20, 2005*. 2005. Pine Mountain, GA, United states: ASM International.
25. Cabibbo, M., et al., *Microstructural aspects of a fiction stir welded joint of magnesium AZ31 and aluminium 6060 alloys.* Metallurgia Italiana, 2012(10): p. 13-18.
26. Hunt, F., et al. *Friction stir welding of dissimilar magnesium alloys for automotive applications*. in *2007 World Congress, April 16, 2007 - April 19, 2007*. 2007. Detroit, MI, United states: SAE International.
27. Zettler, R., et al., *Dissimilar Al to Mg alloy friction stir welds.* Advanced Engineering Materials, 2006. **8**(5): p. 415-21.
28. Firouzdor, V. and S. Kou, *Al-to-mg friction stir welding: effect of material position, travel speed, and rotation speed.* Metallurgical and Materials Transactions A (Physical Metallurgy and Materials Science), 2010. **41**(11): p. 2914-35.
29. Malarvizhi, S. and V. Balasubramanian, *Influences of tool shoulder diameter to plate thickness ratio (D/T) on stir zone formation and tensile properties of friction stir welded dissimilar joints of AA6061 aluminum–AZ31B magnesium alloys.* Materials & Design, 2012. **40**(0): p. 453-460.
30. Sato, Y.S., et al., *Constitutional liquation during dissimilar friction stir welding of Al and Mg alloys.* Scripta Materialia, 2004. **50**(9): p. 1233-1236.
31. Jana, S., Y. Hovanski, and G.J. Grant, *Friction Stir Lap Welding of Magnesium Alloy to Steel: A Preliminary Investigation.* Metallurgical and Materials Transactions A, 2010. **41**(12): p. 3173-3182.
32. Watanabe, T., H. Takayama, and A. Yanagisawa, *Joining of aluminum alloy to steel by friction stir welding.* Journal of Materials Processing Technology, 2006. **178**(1–3): p. 342-349.
33. Dressler, U., G. Biallas, and U. Alfaro Mercado, *Friction stir welding of titanium alloy TiAl6V4 to aluminium alloy AA2024-T3.* Materials Science and Engineering: A, 2009. **526**(1–2): p. 113-117.
34. Aonuma, M. and K. Nakata, *Dissimilar metal joining of ZK60 magnesium alloy and titanium by friction stir welding.* Materials Science and Engineering: B, 2012. **177**(7): p. 543-548.
35. McLean, A.A., et al., *Friction stir welding of magnesium alloy AZ31B to aluminium alloy 5083.* Science and Technology of Welding & Joining, 2003. **8**(6): p. 462-464.
36. Kostka, A., et al., *Microstructure of friction stir welding of aluminium alloy to magnesium alloy.* Scripta Materialia, 2009. **60**(11): p. 953-956.
37. Somasekharan, A. and L. Murr, *Microstructures in friction-stir welded dissimilar magnesium alloys and magnesium alloys to 6061-T6 aluminum alloy.* Materials Characterization, 2004. **52**(1): p. 49-64.

FRICTION STIR WELDING AND PROCESSING VIII

Poster Session

Friction Stir Welding and Processing VIII
Edited by: Rajiv S. Mishra, Murray W. Mahoney, Yutaka Sato, and Yuri Hovanski
TMS (The Minerals, Metals & Materials Society), 2015

ASSESSMENT OF FRICTION STIR WELD QUALITY BY ANALYZING THE WELD BEAD SURFACE USING BOTH DIGITAL IMAGE PROCESSING AND ACOUSTIC EMISSION TECHNIQUES

R. Rajashekar[1], B.M. Rajaprakash[2], Sarala Upadhya[3]

[1,2,3]Mechanical Engineering Department, University Visvesvaraya College of Engineering, K.R. Circle, Bangalore – 560 001, Karnataka State, India

Keywords: Friction Stir Welding, Banded texture, Digital Image Processing, Statistical Image Parameters, Digital Image Correlation, Acoustic Emission.

Abstract

The success of Friction Stir Welding (FSW) depends upon how well the tool shoulder interacts with base material, generating required frictional heat, controlling flow of material to ensure proper containment and consolidation resulting in a defect free weld. In the present work, the weld bead surface of AA6082-T6 specimens produced by shoulders of different geometry (Flat and Concentric rings) with selected welding parameters are assessed by using digital image processing. The images of the weld bead surface are captured, processed, and analyzed by extracting first and second order statistical image parameters along good and defect weld regions in comparison with X-Ray Radiography. The variations in statistical image parameters of weld bead images are observed, enabling digital image correlation to assess weld quality. The results obtained are also verified by Acoustic Emission (AE) data acquired during FSW, showing variations similar to that of image data along good and defect weld regions.

Introduction

The aluminum alloys generally used in the aircraft and automobile industries show low weldability by traditional fusion welding process. An alternate, improved way for satisfactorily joining aluminum alloys is provided by the development of FSW by The Welding Institute (TWI), UK [1]. FSW is an environment friendly, solid-state joining process in which materials are thermomechanically processed using no consumables and resulting in welds free from melting related defects. The two different modes of material flow, i.e., "Pin-driven material flow" and "Shoulder-driven material flow" observed during FSW, merges together with adequate temperature and hydrostatic pressure generated from the shoulder contact with the base material. The shoulder, under the action of optimal axial load, transfers the material from the leading edge into weld cavity, preventing the material from flowing out of the weld cavity to form defect free welds [2].

Tool shoulders are designed to produce heat (through friction and material deformation) to the surface and subsurface regions of the workpiece. Among the key functions of the tool shoulder is the control of material to ensure proper containment and consolidation. The shoulder profile is designed to increase or decrease the contact surface area and increase or decrease the amount of heat input. This also changes the amount of deformation experienced by the material at the top of the weld. Features like scrolls, ridges or knurling, grooves, and concentric circles are machined onto the tool shoulder profile (concave, flat, and convex). These features help direct traction forces inwardly toward the probe improving weld closure by preventing the plasticized material from being expelled [3].

Studies on the weld surface of FSW of aluminum alloys produced using tools of different shoulders showed that properties in the weld zone are influenced by the tool shoulder geometry. The efficiency of the shoulder can be evaluated by the amount of flash produced along with the surface appearance and uniformity of the weld bead [4]. The combined experimental and numerical investigation of texture patterns in friction stir welds along different cross sections found that a banded texture of a weld has close ties to the microstructure of the weld region and the bands have a strong effect on the mechanical behavior of the welded joints [5].

The images of weld beads formed at different welding conditions such as tool pin failure and lesser pin depth were processed and analyzed to propose a methodology for online condition monitoring of FSW using image processing techniques. The presence of defects and the variations in weld quality are identified by studying a) the texture pattern of the weld bead image from a histogram plot, b) variation in gray level from line profile plot and c) uniformity of weld bead contours from contour plot [6].

The study of the texture forms an important clue for the recognition of the surfaces processed by different manufacturing process [7]. Image texture can be quantitatively evaluated using properties such as fineness, smoothness, coarseness, granulation, etc., using statistical methods. This involves calculation of properties based on the gray tones of the specimens following different orders based on the number of gray values considered for analysis and the type of relationship used [8,9]. The first order statistics involves computation and extraction of features such as Mean, Standard deviation, Entropy, etc. from the histogram of the images. The higher order statistics use Gray Level Co-occurrence Matrix (GLCM), also known as the gray-level spatial dependence matrix which considers the spatial relationship of pixels to examine the image texture by features such as Contrast, Correlation, Energy, etc. [10].

AE data acquired during FSW using different weld parameters, materials (similar and dissimilar) were studied in comparison with X-ray radiography results to determine the threshold values of AE signals. Based on the threshold values obtained, a model to monitor online FSW of similar and dissimilar materials is developed, which helped in assessing the weld quality and to modify the welding process parameters accordingly to obtain quality welds [11].

The authors, in their earlier publication, reported the use of machine vision without interfacing the use of AE signals during welding [12]. In the present investigation, studies on the use of machine vision and AE in assessing weld quality being produced are dealt with.

The above studies infer that a banded texture from a friction stir weld surface, formed by the tool shoulder interaction, can be viewed as characteristic features relevant for the inspection and evaluation of FSW. The low cost, non-contact, non destructive machine vision technology can be applied to assess FSW quality by analyzing weld bead images. In this investigation, the goal is to improve the methodology used for weld quality assessment using an image processing technique [8]. Weld bead surfaces were produced by different methods including using either selected weld parameters a flat shoulder, or a flat shoulder with concentric rings. Images from these surfaces were processed to quantify the image parameters by evaluating first and second order statistical parameters from histogram and GLCM respectively. The variations in statistical parameters, i.e., Entropy, Contrast and Energy, which quantify weld bead texture by its roughness and uniformity, were studied along good and defect weld regions identified by X-Ray radiography results to develop a criteria for weld quality assessment. Further, the performance of good welds was assessed by conducting tensile tests along regions of good weld showing variations in image data. The results obtained are verified with AE data, one of the techniques employed in monitoring FSW [11]. This improves the reliability of the machine vision technique for weld quality assessment.

Experimental Details

In this study, commercially available aluminum 6082-T6 alloy plates of dimensions 300mm x75mm x5 mm are used for preparing butt welded joints on an ETA stir welding machine at IISc., Bangalore. The workpieces are clamped rigidly on the backing plate with an AE sensor mounted on the workpiece as shown in Fig. 1. The FSW trials are carried out with two tools of different shoulder geometry (flat and flat shoulder with concentric rings) having a frustum shaped threaded pin made of tool steel. The configurations of the tools are shown in Table I. By adopting the optimum welding parameters for AA6082-T6 used in the literature, two sets of specimens are welded per tool using the parameters shown in Table II [11]. The first set of specimens was comprised of welding plates without holes across the thickness. The welds are identified as FS1 and CN1 and were produced using tools with either a Flat shoulder or a Flat shoulder with concentric rings respectively. The second specimens (FS2 and CN2), are welded with plates having holes across the thickness at a distance of 10mm from the butting edge to induce defects and also to study variations in weld texture formed during material consolidation in the hole region by the two shoulder geometries considered in the present work.

Figure 1. Clamping of workpieces on ETA FSW machine with an AE sensor.

The standard PCI-2 based AE testing equipment of MISTRAS, Physical Acoustics Corporation, U.S.A has been used to acquire AE signals during FSW. The signals are processed and analyzed as the component is subjected to continuous or repetitive stress during welding. Discontinuities in components release energy as the component is subjected to an external stimulus (change in pressure, load, or temperature) which travels in the form of high-frequency stress waves or Acoustic emission. The most detectible Acoustic emissions take place when a loaded material undergoes plastic deformation or when a material is loaded at or near its yield stress. The stress waves or oscillations are received with the use of sensors (transducers) that in turn convert the energy into a voltage. This voltage is electronically amplified and with the use of timing circuits is further processed as AE signal data. The most widely used signal measurement parameters are Counts, Amplitude, and the Energy. Amplitude (A) is the highest peak voltage attained by an AE Waveform. This is a very important parameter since it directly determines the detectability of the AE event. Counts (N) are the threshold-crossing pulses. Counts depend on the magnitude of the source event, but they also depend strongly on the acoustic properties and reverberant nature of the specimen and the sensor. Energy (E) is the measured area under the rectified signal envelope. Energy (MARSE) is often preferred over counts because it is sensitive to amplitude as well as duration and it is less dependent on threshold setting and operating frequency.

Table I. FSW Tool dimensions and different shoulder configurations.

Tool Dimensions (mm)		Tool type	
		Flat shoulder with conical threaded pin	Concentric rings shoulder features with conical threaded pin
Pin length	4.7		
Pin Diameter	D=6 d=4		
Shoulder Diameter	20		
Shoulder Length	25		
LH Thread pitch	1		

The weld bead images are captured off-line using a high resolution digital camera (Nikon D90) with uniform lighting. The color images are converted to gray scale image, cropped along the weld bead and processed by applying various filters to reduce the noise and subsequently enhanced by stretching the contrast by histogram equalization using the Image Processing Tool Box of MATLAB. Several Regions of Interests (ROI) are defined along good and defect weld regions by comparing with X-Ray radiography results. The image parameters are quantified by extracting statistical image parameters for each ROI. The statistical methods involve calculation of image properties based on the gray tones and they are of a different order based on the number of gray values considered for analysis and the type of relationship used. The "First order" statistics involves computation and extraction of features such as Mean, Standard deviation and Entropy, from the histograms of the images. The "Second order" statistics involves computation and extraction of features such as Contrast, Correlation, Energy and Homogeneity based on the GLCM. The GLCM characterize the texture of an image by calculating how often pairs of pixels with specific gray values and in a specified spatial relationship occur in an image.

The tensile tests of defect free welds, as revealed by X-Ray Radiography, is conducted to study the variations in the strength of weld along the regions of image data variations. The base material AA6082-T6 is a medium strength alloy with excellent corrosion resistance and good mechanical properties (Yield strength: 291 N/mm^2, Ultimate tensile strength: 305 N/mm^2). The tensile test specimens are prepared as per ASTM-E8/EM-08 standards. The specimens are cut using wire EDM as per the dimensions along the direction perpendicular to the tool traverse and tested on an Electronic Tensometer (Make: Kudale Instruments-PUNE, INDIA, Model: PC 2000, Capacity: 20 KN).

Table II. Weld plates and selected weld parameters used to weld the specimens.

Workpiece	Weld specimens		Welding Speed (mm/min)	Tool Rotation Speed (rpm)	Plunge Depth (mm)	Tool Tilt angle
	Flat shoulder	Concentric rings shoulder				
Plate without holes across thickness of plate	FS1	CN1	100	1000	4.85	2^0
Plate with holes across thickness of plate	FS2	CN2				

Results and Discussion

X-Ray Radiography of FS1 reveals no defects. The processed weld bead image of FS1 and equal sizes of ROI defined along the weld bead image are shown in Fig. 2. The AE data acquired during welding of FS1, showing marginal variations along the weld, is shown in Fig. 3. The weld bead texture is quantified by extracting first order and second order statistical image parameters from histogram and GLCM of ROI respectively. Among the several first order statistical parameters: Mean, Standard deviation and Entropy for assessing weld quality, Entropy is found to be significant. Similarly among second order statistical parameters: Contrast, Correlation, Energy and Homogeneity for assessing weld quality, Contrast and Energy are found to be significant. The ROI 3, 5, 8 and 13 showing large variations in the image data are considered for the analysis. The tensile strength obtained along these ROI is correlated with image parameters to assess weld quality (Fig. 4).

The image of FS2 welded by butting the edges of the plates with holes across the cross section using a flat shoulder tool and selected welding parameters has been processed and analyzed. The processed image of weld bead FS2 is shown in Fig. 5. The X-ray radiography of FS2 reveals defects around the regions of holes. The AE signal data acquired during welding of FS2 shows abrupt variations along the weld when a defect forms due to the presence of drilled holes across the thickness of plate, Fig. 6. The equal sizes of ROI defined along good and defect weld regions are shown superimposed on the processed weld bead image in Fig. 5. The weld bead texture is quantified along ROI by extracting significant first order and second order statistical image parameters from histogram and GLCM of ROI respectively (Fig. 7). The variations in statistical image parameters along ROI are studied to understand the occurrence of defects through image and AE signal data to set criteria for weld quality assessment.

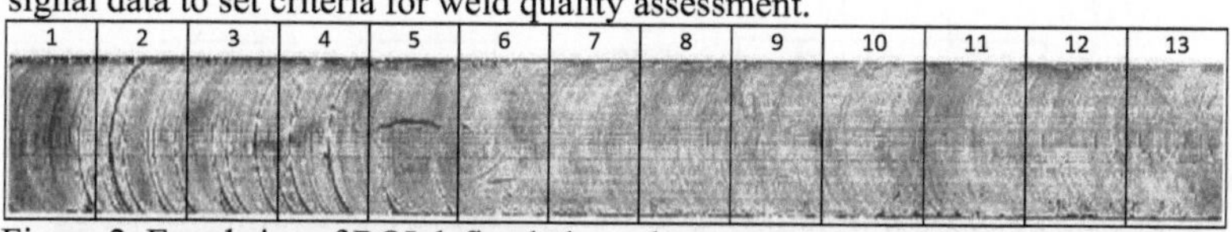

Figure 2. Equal size of ROI defined along the processed weld bead image of FS1.

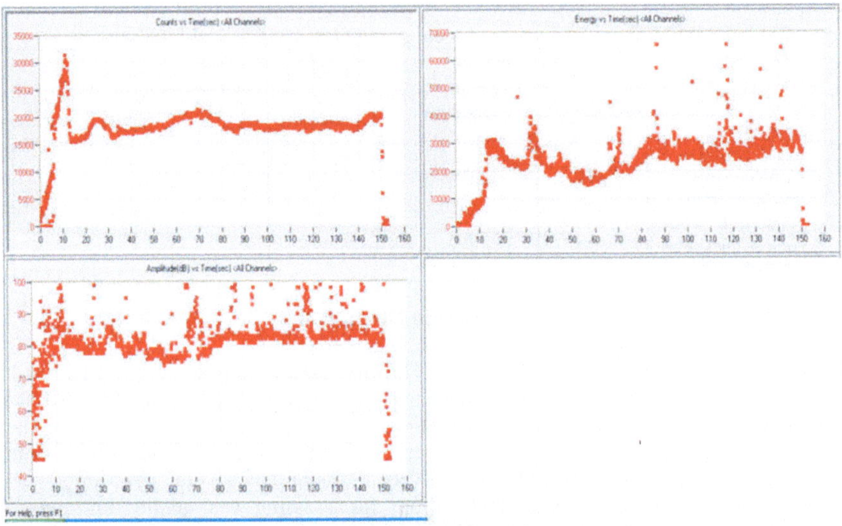

Figure 3. AE signal data (Energy, Counts and Amplitude) acquired during FSW of FS1.

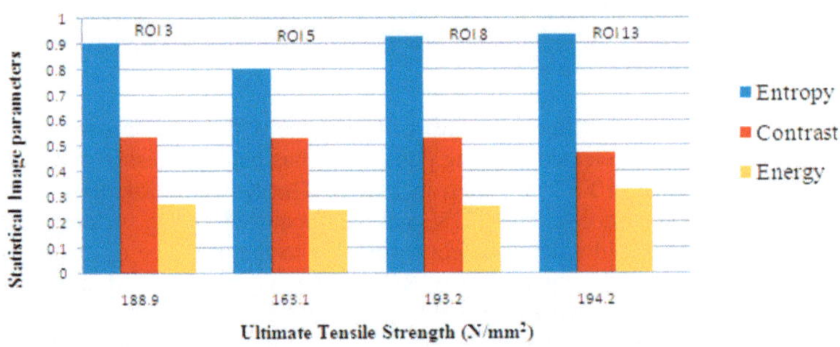

Figure 4. Variations in statistical image parameters along ROI of different tensile strength of FS1.

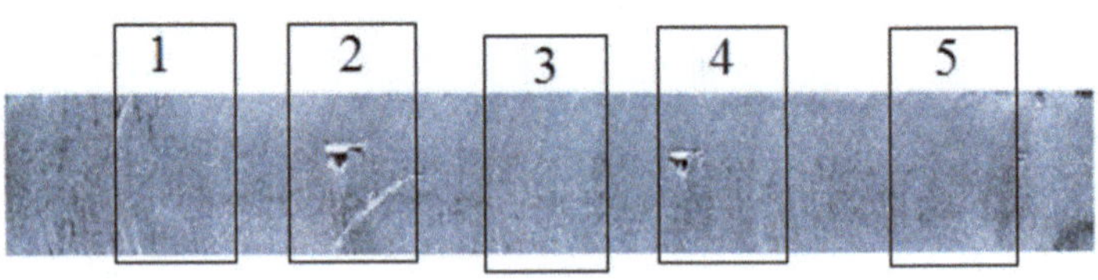

Figure 5. The ROI along good (1,3,5) and defect (2,4) weld regions of processed weld bead image FS2.

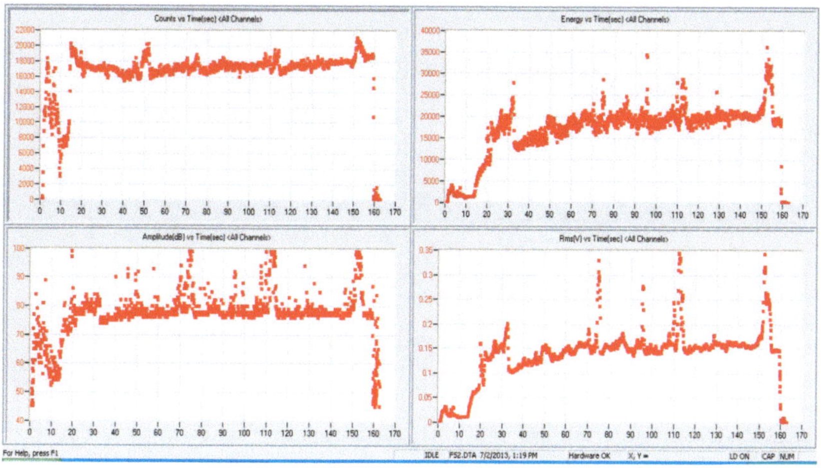

Figure 6. AE signal data: Counts, Energy, RMS and Amplitude acquired during FSW of FS2.

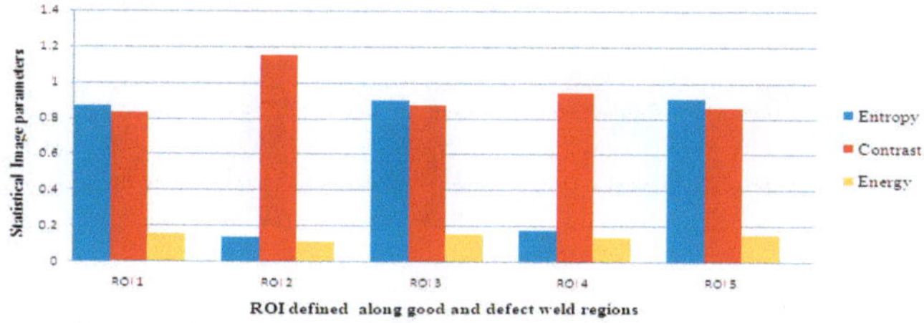

Figure 7. Variations in statistical image parameters of ROI defined along good (1,3,5) and defect weld regions (2,4) of FS2.

The processed image of CN1 welded plates without holes across the cross section using a flat shoulder tool with concentric rings and selected welding parameters is shown in Fig. 8, X-ray radiography of the specimen reveals no defect in the weld.AE signal data acquired during FSW of CN1 are shown in Fig. 9. The equal sizes of ROI, defined along the processed weld bead image, are shown superimposed on the processed weld bead image in Fig. 8. The weld bead texture is quantified by extracting first order and second order statistical image parameters of ROI. The ROI 3, 7, 9 and 11, which shown large variations in image data, are considered for analysis. The tensile test is conducted by extracting the specimens along each ROI. The tensile strength obtained is correlated with image parameters to assess the weld quality (Fig. 10).

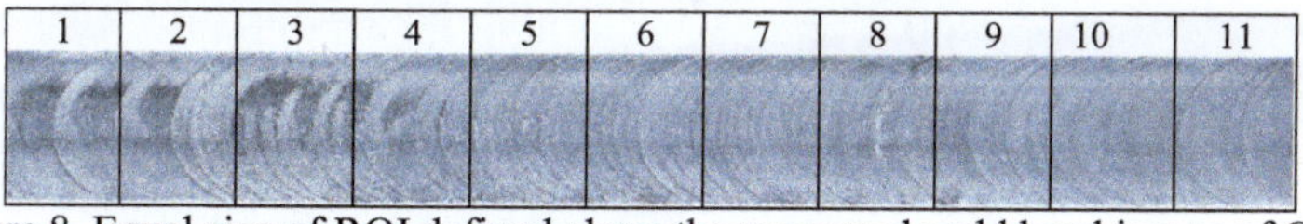

Figure 8. Equal size of ROI defined along the processed weld bead image of CN1.

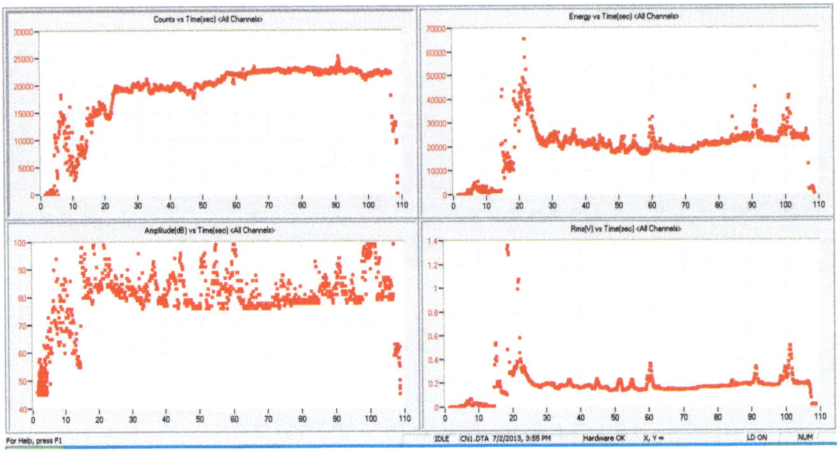

Figure 9. AE signal data: Counts, Energy, RMS and Amplitude acquired during FSW of CN1.

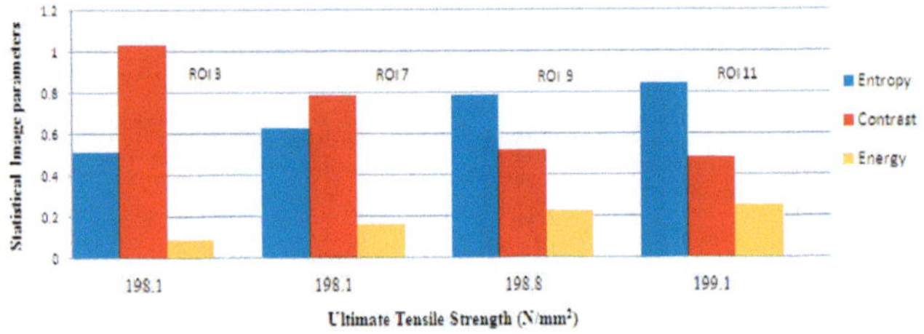

Figure 10. Variations in statistical image parameters along ROI of different tensile strength of CN1.

The processed image of CN2 welded by butting the edges of the plates with holes across the cross section using a flat shoulder tool with concentric rings and selected welding parameters is shown in Fig. 11. X-ray radiography of CN2 reveals defect around the regions of holes. AE signal data, acquired during FSW of CN2, shows significant variations whenever there is a defect along the weld, Fig. 12. The equal sizes of ROI defined along good and defect weld regions superimposed on the processed weld bead image is shown in Fig. 11.The weld bead texture has been quantified by processing ROI to extract first and second order statistical image parameters, Fig. 13. Variations in textural image parameters are studied to understand the occurrence of defects during FSW using a flat shoulder tool with concentric rings through image and AE signal data to set criteria for weld quality assessment.

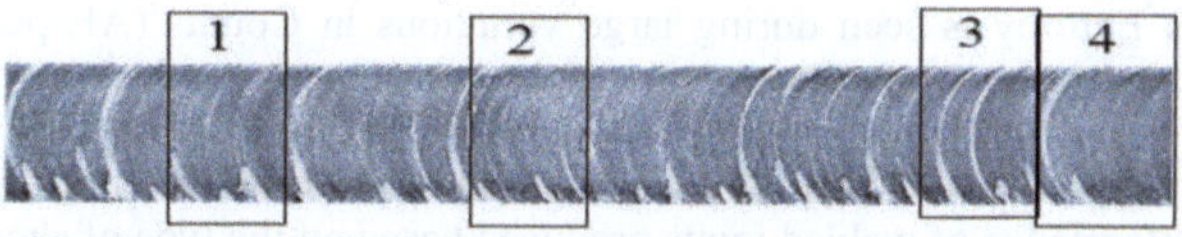

Figure 11. The ROI along good (1,4) and defect (2,3) weld regions of processed weld bead image of CN2.

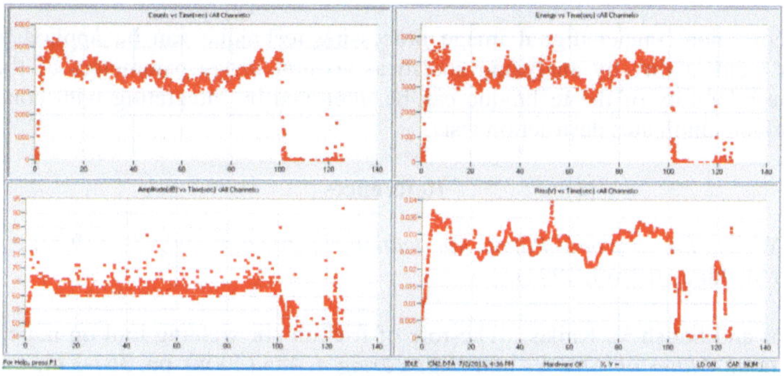

Figure 12. AE signal data: Counts, Energy, RMS and Amplitude acquired during FSW of CN2.

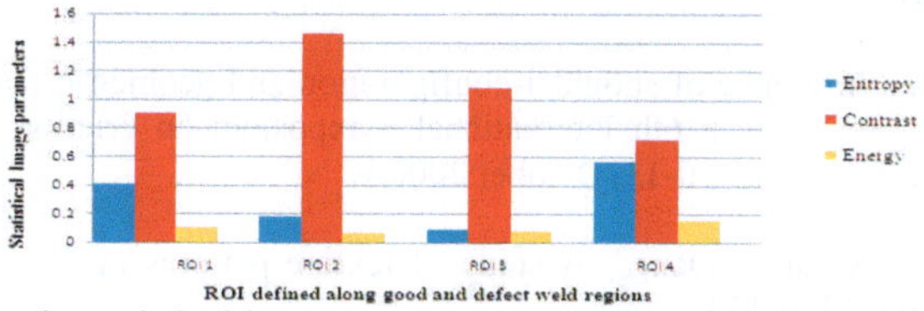

Figure 13. Variations in statistical image parameters of ROI defined along good (1,4) and defect (2,3) weld regions of CN2

Conclusions

In this work, a typical non-contact digital image processing technique is applied to assess the friction stir weld quality by analyzing weld bead surface texture produced using two different shoulder geometries at constant weld parameters. The results obtained are verified with AE data acquired during FSW using a contact AE sensor. From the results derived the following conclusions are drawn:

1. The good weld regions showed more uniformity in the bead surface with reduced roughness compared to that of defect weld regions. This is indicated by the increased values of Entropy and Energy and reduced values of Contrast of image.

2. The increase in the weld strength of defect free welds is observed along the regions showing an increase in uniformity and reduced roughness due to proper interaction of the tool shoulder.

3. An increase in Entropy is seen during large variations in Counts (AE parameter), between 15,000 to 25,000, indicating proper interaction of the tool shoulder in the good weld regions.

4. Shoulder geometry affects the type of surface texture produced and in turn can be used for comparison of performance of welded joints produced based on the type of shoulder geometry.

5. The presence of shoulder features resulted in improvement of strength from 165 to 200N/mm^2.

6. A low cost, non-contact digital image processing technique can be applied to assess the friction stir weld quality by analyzing statistical textural image parameters of the weld bead surface. The reliability of the technique can be improved by integrating with contact type AE technique which eliminates destruction tests.

References

1. Thomas W. M., *International Patent Application No PCT/GB92/02203* (10 June, Gothenburg, Friction Stir Welding at TWI, 1993).

2. K. Kumar and Satish V. Kailas, "The role of friction stir welding tool on material flow and weld formation," *Materials Science and Engineering A,* 485 (2008), pp. 367-374.

3. Rajiv S. Mishra and Murray W. Mahoney, *Friction Stir Welding and Processing,* (ASM International®, 2007)

4. Dwight. A. et al., "Influence of shoulder configuration and geometric features on FSW track properties" (Paper presented at 6th International Symposium on Friction Stir Welding Saint-Sauveur, Nr Montréal, Canada, 10-13, October 2006).

5. Shaowen XU and Xiaomin Deng, "A study of texture patterns in friction stir welds," *Acta Materialia,* 56 (2008), 1326-1341.

6. S. Muthukumaran et al., "Condition monitoring of first mode of metal transfer in friction stir welding by image processing techniques," *International Journal of Advanced Manufacturing Technology,* 36 (2008), 484-489.

7. B. Dhanasekar and B. Ramamoorthy, "Evaluation of Surface Roughness Using a Image Processing and Machine Vision System," *Journal of Metrology Society of India,* 21, (2006), 9-15.

8. Robert M.Haralick, K.Shanmugam and Itshak Dinstein, "Textural Features for Image Classification," *IEEE transactions on Systems, Man and Cybernetics,*1973, no. 6: 610-621.

9. K. Venkatramana and B.Ramamoorthy, "Statistical methods to compare the texture features of machined surfaces," *Pattern Recognition,* 29, (9) (1996), 1447-1459.

10. E.S. Gadelmawla, "A vision system for surface roughness characterization using the gray level co-occurrence matrix," *NDT&E International*, 37, (2004), 577–588.

11. C.N. Suresha, "Studies on Performance of Welded Joints Produced by Friction Stir Welding process using Acoustic Emission Technique" (PhD thesis, Bangalore University, 2013).

12. R. Rajashekar and B.M. Rajaprakash, " Application of Digital Image Processing for Assessing welds produced by Friction Stir Welding process using Flat and Concave shoulder geometry" (Paper presented at International Conference on Friction based Processes, Indian Institute of Science, Bangalore, Karnataka, India, 3-5, September 2014).

Friction Stir Welding and Processing VIII
Edited by: Rajiv S. Mishra, Murray W. Mahoney, Yutaka Sato, and Yuri Hovanski
TMS (The Minerals, Metals & Materials Society), 2015

Development of FSW Simulation Model-Effect of Tool Shape on Plastic Flow

Yurika Miyake[1], Fumikazu Miyasaka[1], Shuhei Matsuzawa[1], Shunta Murao[1], Kenta Mitsufuji[1], Shinnosuke Ogawa[1]

[1]Adaptive Machine Systems, Osaka University, Osaka 565-0871, Japan

Keywords: FSW, MPS method, Particle methods, Numerical analysis, Onion ring, Mesh less.

Abstract

The friction stir welding (FSW) is known as non-melting joining. Numerical analysis methods for FSW also have been developed. In these models, general grid methods have traditionally been used. However, there are some problems when these methods are employed. Calculation of the advection term for both momentum and temperature needs technical attention. To analyze a few substances of different physical properties such as the phase transformation on the bonded interface or the dissimilar joining, some complex process is required. In this study, these problems are avoided by adopting particle methods for FSW simulation. Lagrange approach is mainly used, so this particle methods calculate mass transfer and surface deformation more easily than general grid methods. The effectiveness of this method is verified by plastic flow around the tool examined by particle trace. As a result, phase transformation and tool's shape change are taken into account for the analysis of the material flow around the tool.

Introduction

The friction stir welding (FSW) is developed at the Welding Institute in 1991 [1] and has been used widely in the field of industry. Numerical analysis models for FSW also have been developed and there are many papers about it [2]-[6]. Traditionally, FDM and FEM are used for the analysis method. However, the metal around tool is deformed and stirred very intensively during FSW so that large deformation of surface of the work piece takes place. These methods using mesh or grid are difficult to handle such surface deformation. Therefore, in this paper, particle methods without mesh or grid proposed for analysis of FSW. In this method, the behavior of metal being stirred can be described directly as moving of particles. The particles can move keeping variables such as momentum, mass and enthalpy, therefore, the advection term or substance transfer can be dealt with very simply and automatically. Moreover, the mass conservation is also satisfied automatically.

This study has shown the usability of particle methods so far by performing the analysis of various FSW simulation models such as FSSW, dissimilar FSW, and general FSW. In this paper, two models which have different shapes of probe are employed. The influence of the difference in the shape of probe can be appeared by the analysis results.

Numerical model

In this paper, the plastic flow of metal during FSW is approximately described as highly viscous fluid. The governing equations for plastic flow are equations of Navier-Stokes and mass conversation. These equations are respectively given by:

$$\frac{Du}{Dt} = -\frac{1}{\rho}\nabla P + \nu\nabla^2\boldsymbol{u} + \boldsymbol{g} \qquad (1)$$

$$\frac{D\rho}{Dt} = 0 \tag{2}$$

where ρ is density, P is pressure, ν is dynamic viscosity, $\boldsymbol{u}$ is velocity and $\boldsymbol{g}$ is gravity acceleration. The way to decide the value of ν is described later in Eq. (5). *D/Dt* denotes Lagrangian derivative which means a derivative with respect to a moving coordinate system. In this paper, Lagrangian formulation is employed to these equations so that the advection term, which is likely to be cause of numerical error, does not need to be calculated in the particle methods.

In FSW, the heat transfer is quite important factor. This thermal reaction is governed by the thermal diffusion equation as follows:

$$\frac{DT}{Dt} = \frac{k}{\rho C_p}\nabla^2 T + \frac{Q_h}{\rho C_p} \tag{3}$$

Here C_p is specific heat, k is thermal conductivity, T is temperature, and Q_h is heat generation per unit volume. The definition of Q_h is discussed later in Eq. (10).

The distribution of viscosity depends on temperature and equivalent strain rate. The viscosity of metal as plastic flow, which is employed in Navier-Stokes equations, is generally described as follows [10]:

$$\eta = \frac{\sigma(\dot{\varepsilon},T)}{3\dot{\varepsilon}} \tag{4}$$

$$\nu = \frac{\eta}{\rho} \tag{5}$$

where η is viscosity, $\dot{\varepsilon}$ is equivalent strain rate and σ is equivalent flow stress. Dynamic viscosity which is need for Eq. (1) is obtained by Eq. (5).

Using velocity distribution, the equivalent strain rate at each calculation point is obtained as follows:

$$\dot{\varepsilon} = \left(\frac{2}{3}\dot{\varepsilon_{ij}}\dot{\varepsilon_{ij}}\right)^{1/2} \tag{6}$$

$$\dot{\varepsilon_{ij}} = \frac{1}{2}\left(\frac{\partial u_i}{\partial x_j} + \frac{\partial u_j}{\partial x_i}\right) \tag{7}$$

Therefore, technically, when the value of equivalent flow stress is obtained, then the viscosity is obtained by Eq. (4), (6) and (7). Regardless of temperature and equivalent strain rate, the value of equivalent flow stress depends on material. In this paper, equivalent flow stress is approximated by equations below [9]:

$$\sigma = \frac{1}{\alpha}\ln\left\{\left(\frac{Z}{A}\right)^{\frac{1}{n}} + \left[\left(\frac{Z}{A}\right)^{\frac{2}{n}} + 1\right]^{\frac{1}{2}}\right\} \tag{8}$$

$$Z = \dot{\varepsilon}\exp\left(\frac{Q}{RT}\right) \tag{9}$$

where R is gas constant, α, A, n, Q are material constants. In this paper, A1100 is chosen for material of work piece. Material constants including α, A, n, Q for these aluminum are summarized in Table 1.

During FSW, heat is generated due to plastic deformation. In this paper, this heat generation is calculated as follows:

$$Q_h = \sigma\dot{\varepsilon} \cdot 0.9 \tag{10}$$

Eq. (10) means that 90 % of the work for plastic deformation is transformed into heat.

Table 1 Material constants for A1100

Material	A1100
Density[kg/m^3]	2710
thermal conductivity[W/m·K]	234
specific heat[J/Kg·K]	900
α[MPa^{-1}]	0.045
A	exp(24.67)
N	5.66
Q	158300

Particle methods

Particle methods are developed for hydrodynamics problem. In particle methods, the fluid moves as particles in terms of variables from Lagrangian formulation. Unlike FEM with both nodes and element, only nodes (particles) are used in particle methods so that there is no need to care about distortion of elements. Therefore, the behavior of fluid with large deformation could be analyzed without any complex operations.

There are some traditional particle methods such as MPS method [7] and SPH method [8]. In this paper, MPS method is adopted, but it is expected that SPH method or other particle methods also can get similar numerical results. In MPS method, to solve Eq. (1) and (2), the differential operators at particle i such as ∇P_i and $\nabla^2 \boldsymbol{u}_i$ are discretized by interaction among other particles j within a certain radius of r. ∇P_i and $\nabla^2 \boldsymbol{u}_i$ of are respectively expressed as follows [7]:

$$\nabla P_i = \frac{d}{W}\sum_{j\neq i}\left[\frac{P_j - P_i}{|\boldsymbol{r}_{ij}{}^2|}\boldsymbol{r}_{ij} w(|\boldsymbol{r}_{ij}|)\right] \tag{11}$$

$$\nabla^2 \boldsymbol{u}_i = \frac{2d}{W}\sum_{j\neq i}\left[\frac{\boldsymbol{u}_j - \boldsymbol{u}_i}{|\boldsymbol{r}_{ij}{}^2|} w(|\boldsymbol{r}_{ij}|)\right] \tag{12}$$

Here P_i and $\boldsymbol{u}_i$ are respectively pressure and velocity at particle i, $\boldsymbol{r}_{ij}$ is distance between particle i and j, and d is a number of space dimension (in this paper, d=3). Other variables are presented as follows [8]:

$$w(|\boldsymbol{r}_{ij}|) = \begin{cases} \frac{r}{|\boldsymbol{r}_{ij}|} - 1 & (0 \leq |\boldsymbol{r}_{ij}| \leq r) \\ 0 & (r < |\boldsymbol{r}_{ij}|) \end{cases} \tag{13}$$

$$W = \sum_{j\neq i} w(|\boldsymbol{r}_{ij}|) \tag{14}$$

In MPS method, Eq. (13) is called weight function. It should be noticed that the interactions among particles are weighted averaged in Eq. (11) and (12) using Eq. (13) and (14). The shorter the distance between particles is, the stronger the interaction of the particles is. In general, the value of r is about two times the minimum distance between particles at initial placement.

Using Eq. (12), the diffusion term in Eq. (1) is usually computed explicitly. However, as we see later, the distribution of viscosity differs entirely from linear and has quite large value. In such case, it is well known that explicit calculation tends to get some errors. Therefore in this paper, the diffusion term in Eq. (1) is resolved implicitly.

Analyzed model

The illustration of analyzed model is shown in Fig. 1. The initial particle arrangement which correspond to Fig.1 is shown in Fig.2. The work piece is handled as highly viscous fluid though only edge of it is handled as rigid solid. In Fig.2, yellow particles represent the "rigid particles" and red and blue particles represent "fluid particles". The red and blue particles are the same material which colored only to ease to clear the movement of each particles. There are two models which difference is the shape of the probe. Model 1 has a cylindrical probe shown in Fig.1 (b) and Fig.2 (a). Model 2 has a truncated conical probe shown in Fig.1 (c) and Fig. 2 (b). In both models, the tool has an angle of three degrees. The actual shape of the probe has little taper and probe has thread in usual experiment. In this paper, however, the vertical velocity which corresponds to the effect of thread is added on the lateral face particles of probe. This additional velocity is presented as follows:

$$u_z = \Delta L \times n_s \tag{15}$$

where u_z is additional velocity, ΔL is the pitch of thread, n_s is rotating speed.

The process of FSW includes three phase. First phase is plunging process in which the tool is plunged into work piece until the bottom of shoulder is in contact with surface of work piece. The second phase is dwelling process in which the tool keeps only rotating in a few seconds. In this process the temperature develop enough to sir the work piece. The third phase is welding process in which the tool starts to go on the joint line and after a while the plastic flow and temperature field becomes quasi-stationary state.

In this paper, however, only the welding process is reported. In this calculation, besides, the initial temperature of whole model is equal to room temperature and the initial velocity field is zero.

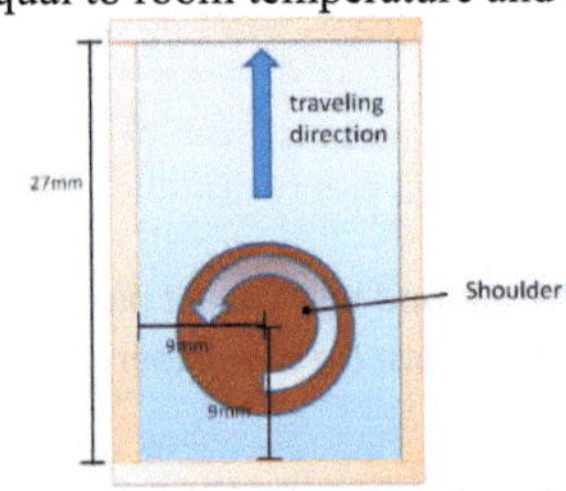

Fig.1 (a) Plane view of analyzed model

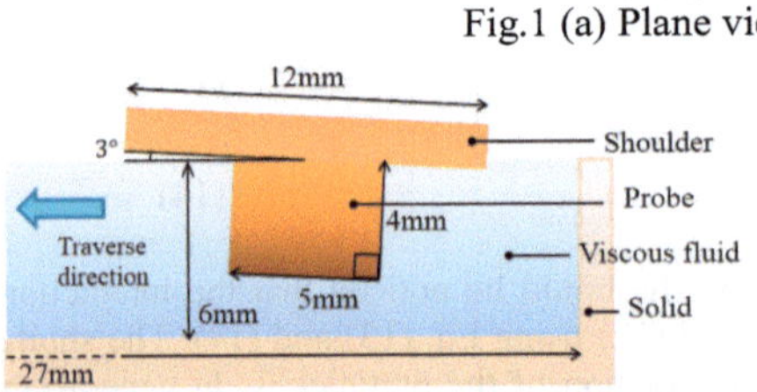

Fig.1 (b) Front view of analyzed Model 1

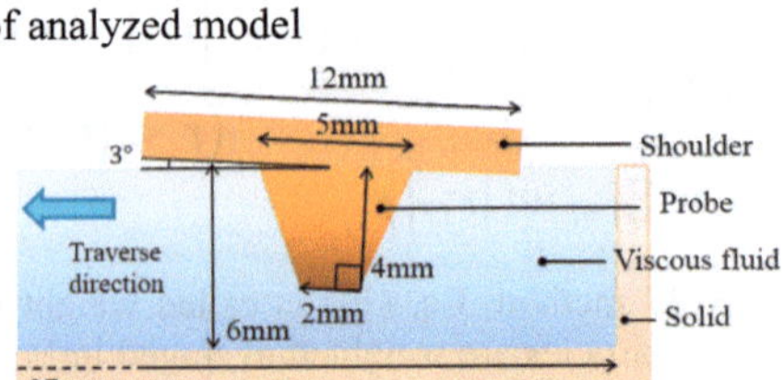

Fig.1 (c) Front view of analyzed Model 2

Fig. 1 Illustration of analyzed Model

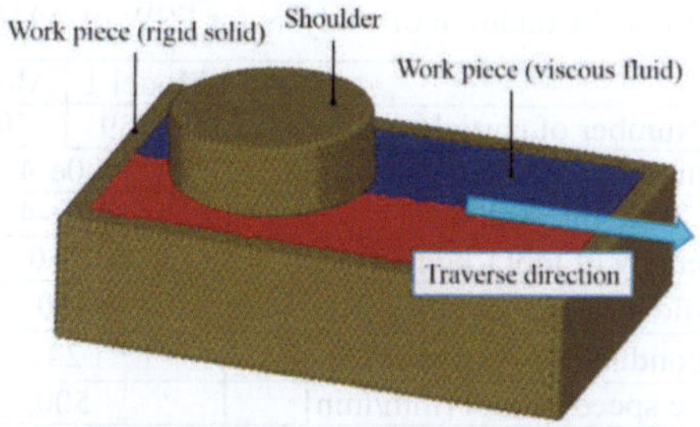

Fig.2 (a) Whole particles in model

Fig.2 (b) Tool particles in Model 1 Fig.2 (c) Tool particles in Model 2

Fig. 2 Initial particle arrangement in the calculation

Analyzed results

Using two different models in Fig. 1 (b) and Fig. 1 (c), the influence of the shape of the probe on the material flow around the tool can be investigated. In this calculation, A1100 is employed as the material of the work piece.

In Fig. 4, the velocity field of two models is shown at 0.82 seconds on the cross section same as the view of Fig. 3. The velocity field is intensely turbulent in Fig. 4 (a) about Model 1. The turbulent flow does not appear in Fig. 4 (b) about Model 2. Furthermore, circulating flow appears in Fig. 4 (a). Fig. 5 (a) indicates the magnified view around circulating flow of Fig. 4 (a). Fig. 5 (b) indicates the magnified view of Fig. 4 (b) which is the same location of Fig. 5 (a). The figures about Model 1 show that the flow rate is so increased below the shoulder that some materials cannot move into the horizontal direction and downward material flow is generated. The radius of the truncated conical probe is smaller than that of the cylindrical probe unless below the shoulder, therefore there is not enough flowable space to make downward flow.

The temperature distributions are respectively shown in Fig. 6 at the same time in Fig. 4. Below the shoulder, the distributions look almost same. However, the distributions around the tip of probe are different. The temperature in Fig. 6 (b) is lower than that in Fig. 6 (a).

With respect to the velocity field and temperature distribution, the analyzed result is qualitatively correct to the following theory. The velocity increases in proportion to the radius by the definition of angular velocity as follows:

$$u = r\omega \tag{16}$$

The velocity influences the equivalent strain rate, equivalent flow stress and temperature increases as seen from Eq. (3) to Eq. (10).

Table 2 Condition of analysis for FSW of A1100

	Model 1	Model 2
Number of particles	70759	70746
Initial minimum particle distance[mm]	5.0e-4	
Time interval[sec]	1.0e-4	
Density of tool [kg/m^3]	2710	
Specific heat of tool [J/Kg·K]	460	
Thermal conductivity of tool[W/m·K]	24	
Traverse speed of tool [mm/min]	500	
Rotation speed of tool [rpm]	500	
Tilt angle of tool [degree]	3	
Pitch of tread on probe [mm]	0.7	
heat-transfer coefficient of air [Wm^{-2}K^{-1}]	50	
Room temperature [K]	293	

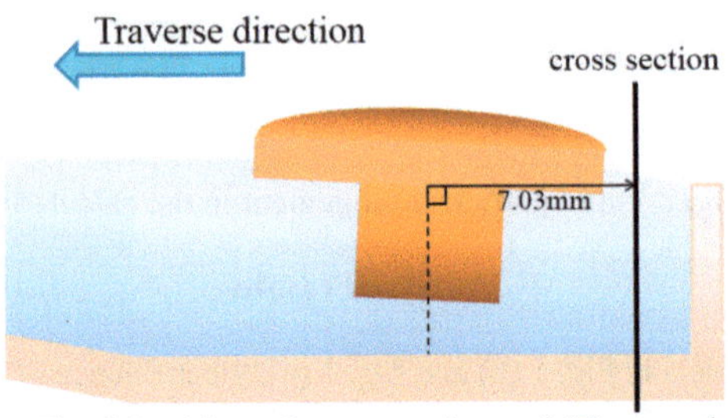

Fig. 3 Position of cross section at 0.82 seconds

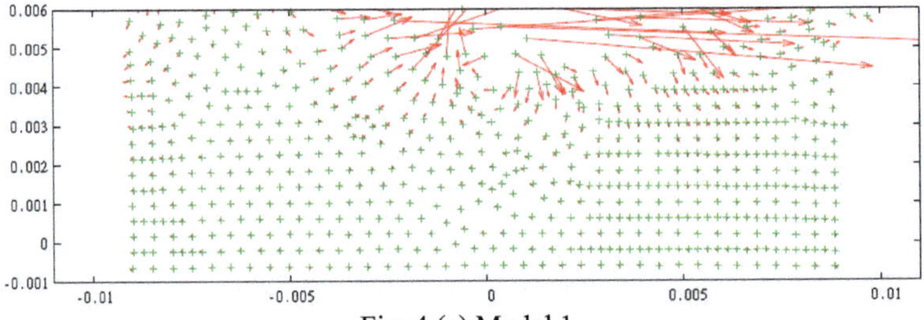

Fig. 4 (a) Model 1

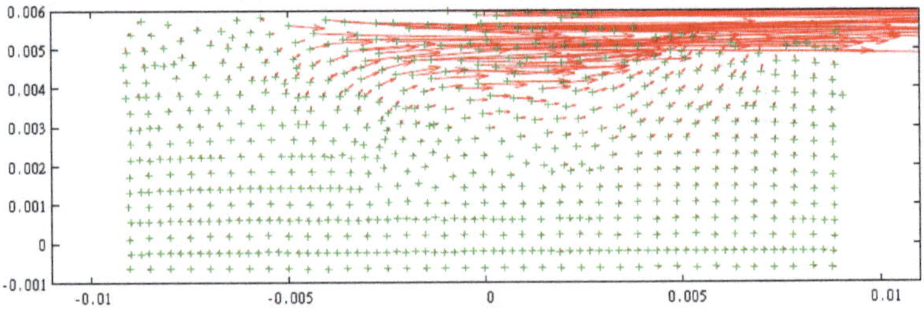

Fig. 4 (b) Model 2

Fig. 4 The velocity field at 8.02 seconds on the cross section in Fig. 3

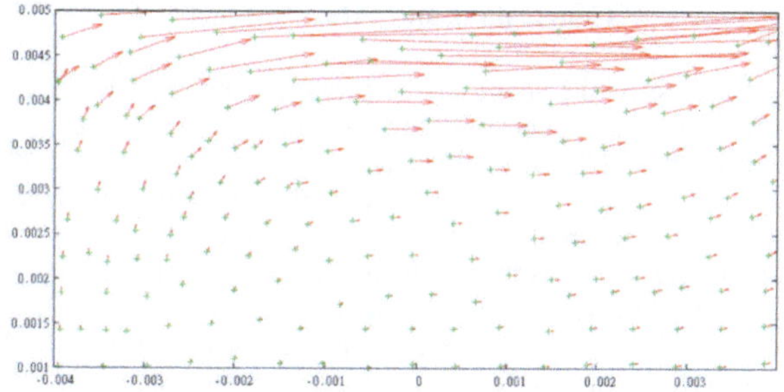

Fig. 5 (a) Model 1

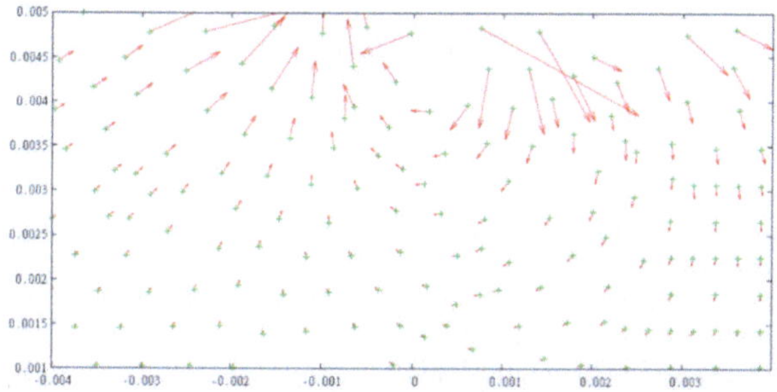

Fig. 5 (b) Model 2

Fig. 5 The magnified views of the velocity field in Fig. 4

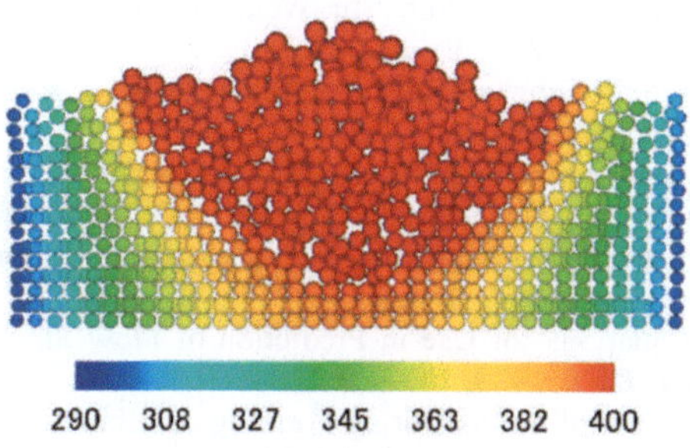

Fig. 6 (a) Model 1

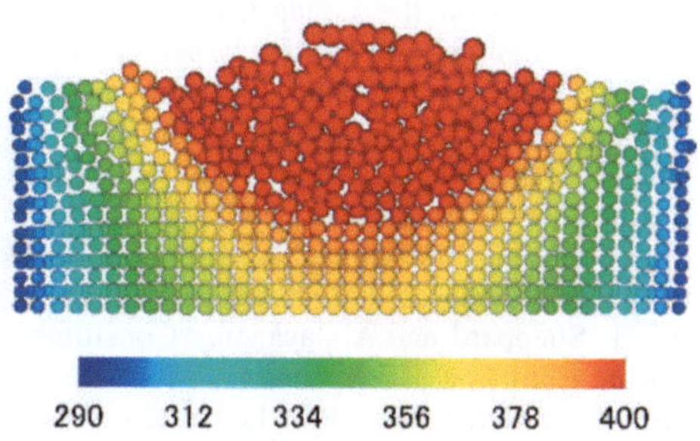

Fig. 6 (b) Model 2

Fig. 6 Temperature distributions at 0.82 seconds on the cross section in Fig. 3 [K]

Conclusions and future subjects

In this paper, the influence of the shape of the probe on the material flow around the tool is investigated. From the analyzed results about the model of the cylindrical probe, it is shown that the velocity field is turbulent and the heat widely spreads. However, the velocity field by truncated conical probe is not turbulent and the temperature of the tip of the probe is lower than that by cylindrical probe. Moreover, the circulating flow appears in the velocity field by cylindrical probe. This indicates the onion ring, which is formed behind the tool, is affected by the tool shape.

In this paper, the work piece is represented as highly viscous fluid. The phase transformation is not taken into account in this model. Therefore, this model cannot handle the defects such as tunneling defects. In future, the material of the work piece should be represented by highly viscous liquid and viscous elastomer to consider the phase transformation.

References

[1]: Thomas, M. W., Nicholas, J., Needham, J. C., Murch, M. G., Templesmith, P. and Dawes, C. J. "Friction stir butt welding.", GB Pat. Application 9125978.8, December 1991; US Pat. 5460317, October 1995.

[2]: P. Ulysse, "Three-dimensional modeling of the friction stir-welding process", *International Journal of Machine Tools & Manufacture* 42 (2002) 1549–1557.

[3]: S Guerdoux and L Fourment, "A 3D numerical simulation of different phases of friction stir welding", Modelling Simul. Mater. Sci. Eng. 17 (2009) 075001 (32pp).

[4]: M Song and R Kovacevic, "Numerical and experimental study of the heat transfer process in friction stir welding", *Proceedings of the Institution of Mechanical Engineers*, Part B: Journal of Engineering Manufacture 217 (1), pp. 73-85.

[5]: Hosein Atharifar, Dechao Lin, and Radovan Kovacevic, "Numerical and Experimental Investigations on the Loads Carried by the Tool During Friction Stir Welding", *Journal of Materials Engineering and Performance.*, 18 (4), 2009, p 339–350.

[6]: H Schmidt and J Hattel, "A local model for the thermomechanical conditions in friction stir welding", *Modelling Simul. Mater. Sci. Eng.* 13 (2005) 77–93.

[7]:S. Koshizuka, Y. Oka, "Moving-Particle Semi-implicit Method for Fragmentation of Incompressible Fluid" *Nucl. Sci. Eng. Soc.*, vol. 123, pp. 421-434, 1995.

[8]: Gingold,R.A.& Monaghan, J. J. ," Smoothed particle hydrodynamics – theory and application to non-spherical stars" *Mon. Not. R. astr. Soc.* 181, (1977), 375-389.

[9]: T. Sheppard and A. Jackson, "Constitutive Equations for Use in Prediction of Flow Stress During Extrusion of Aluminum Alloys", *Mater. Sci. Tech.*, 1979, 13 (3), p 203–209.

[10]: O.C. Zienkiewicz, P.C. Jain, and E. Onate, "Flow of Solids During Forming and Extrusion: Some Aspects of Numerical Solutions", *Int. J. Solid Struct.*, 1978, 14, p 15–38.

Friction Stir Welding and Processing VIII
Edited by: Rajiv S. Mishra, Murray W. Mahoney, Yutaka Sato, and Yuri Hovanski
TMS (The Minerals, Metals & Materials Society), 2015

TEMPERATURE DISTRIBUTION AND WELDING DISTORTION MEASUREMENTS AFTER FSW OF Al 6082-T6 SHEETS

Iurii A. Golubev[1], Evgenii V. Chernikov[1], Anton A. Naumov[1], Vesselin G. Michailov[1]

[1]St.Petersburg State Polytechnic University. Address: 29 Polytechnicheskaya street, St.Petersburg, Russia, 195252; Phone +78122972095, Fax +78125526080

Keywords: Friction Stir Welding, temperature distribution, welding distortion, high-strength Al alloys.

Abstract

The method of Friction Stir Welding (FSW) has a number of advantages whereas it takes place in solid state without melting. Following FSW, a fine recrystallized equiaxed structure is formed in the welded area under combined temperature and severe plastic deformation. The first experiments for FSW of an Al 6082 T6 alloy was to weld with different parameters in order to measure the temperature distribution in directions parallel and perpendicular to the welding direction. The developed earlier finite element model for temperature calculations during FSW was checked using experimental data. The second experiments for an Al 6082 T6 alloy was to friction stir weld with different tool rotation speeds in order to determine the influence of the mentioned parameter on welding distortion. The 3-D geometry of sheets before and after friction stir welding was measured by means of a precise robotic system. The results of measurements were compared between each other and shown in the present research.

Introduction

FSW was invented at The Welding Institute (TWI) of the UK in 1991 as a solid-state joining technique [1]. FSW has number of advantages compared to the conventional welding methods such as a reduced heat affected zone and elimination of porosity, hot cracks, harmful evaporation and UV emissions. The equiaxed and relatively homogeneous microstructure with fine grain size is usually observed during FSW in the nugget zone of the weld [2]. Such a microstructure improves mechanical properties of the weld compared to fusion welding methods. Also, for some materials, such as cast aluminum alloys, conventional steels, and NiAl bronze, the weld or friction stir processed zone microstructure has mechanical properties superior to the parent metal. This metallurgical aspect is the conception of Friction Stir Processing (FSP), a solid state thermo-mechanical metal processing technique. The purpose of FSP is to achieve microstructural modification by homogenization and grain refinement by means of severe plastic deformation in metallic alloys.

FSW/FSP processes have 4 stages (Figure 1):

1. *Plunging*. A non-consumable tool with a shoulder and a pin plunges into a metal. The heat generated by frictional rubbing of tool shoulder softens the metal.
2. *Dwelling*. The required elevated temperature (lower than the melting point) is achieved for subsequent metal stirring.
3. *Stirring*. After dwelling, the tool translates along the surface of the work piece.
4. *Pulling*. In the end of the process, the tool pulls in vertical direction and in the processed material, an exit hole occurs (except when a run-off plate or retractable pin options are used).

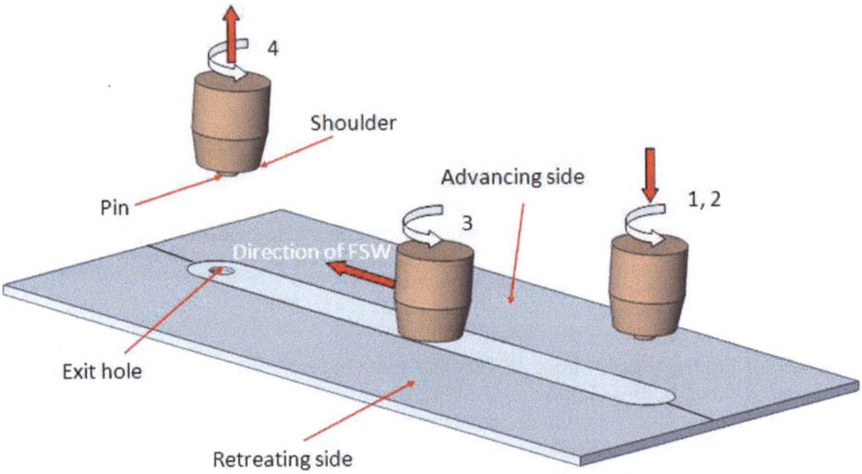

Figure 1. Schematic illustration of the FSW.

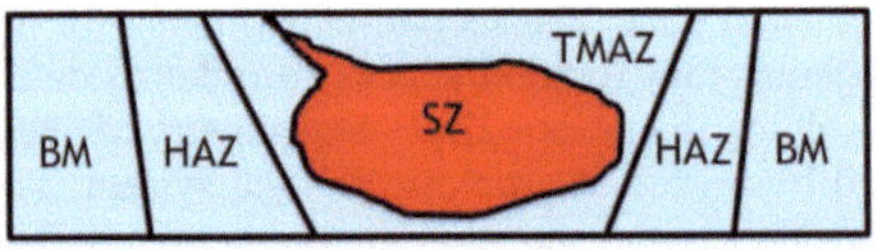

Figure 2. Schematic illustration of a welded/processed area microstructure formed after FSW/FSP.

Different zones are formed in the processed/welded area during FSW/FSP; zones are schematically presented in Figure 2 illustrating:

1. Stir zone (SZ) or weld nugget/weld metal – a fully dynamically recrystallized structure zone due to severe plastic deformation and frictional heating located directly where the tool translates;
2. Thermo-mechanically affected zone (TMAZ) – a unique transition zone with a highly deformed structure between the parent material and the weld nugget/weld metal but with insufficient heat and deformation to facilitate recrystallization;
3. Heat-affected zone (HAZ) - zone beyond the TMAZ which experiences a thermal cycle but does not undergo any plastic deformation.

Welds made by FSW have special features. For example, the side where the tool rotation opposes the material flow is named the advancing side (AS), while the other side is named the retreating side (RS). This effect leads to the non-uniform distribution of temperature in the welded area. The temperature is usually lower in the RS then in the AS. The advancing side is located on the left hand side in Figure 2 where the SZ comes to the surface. According to the non-uniform distribution of temperature in the welded area, the properties in the weld area also have a non-uniform character. Information on temperature distribution is necessary to predict the joint microstructure and mechanical properties.

The purpose of the present paper is to adjust the earlier developed finite element model for the temperature calculation during FSW [3] and to measure welding distortion after welding. The temperature measurement was done by means of thermocouples but only in the zones beside the weld nugget and the TMAZ. Numerical simulation enables an additional ability of temperature determination in the weld joints including the zone of plastic deformation and the weld nugget.

Joined plates by conventional fusion welding methods have distortion caused by residual stresses. Friction stir welding has less distortion compared to the conventional welding methods. This could be explained by the reduced HAZ and TMAZ compared to the conventional welding methods.

1. Experimental Procedure

Friction Stir Welding experiments were done with the aluminum alloy 6082-T6 on a modified CNC machine. The tool for welding has a pin with 4 mm diameter, 1.8 mm length and 12.5 mm diameter shoulder. The thickness of welded sheets was 2 mm. The rotation speed was varied from 710 to 1120 rpm with constant welding speed of 400 mm/min and the tilt angle set to 2^0.

1.1. Temperature Measurement Experiments During Plunging and Dwelling

The initial stages of Friction Sir Welding are plunging and dwelling. Temperature measurement experiments were done according to the scheme shown in Figure 3. Nine type K thermocouples were welded on the top surfaces of the sheets. Plunging speed was set to 8 mm/min with a dwell time of 2 sec in all experiments.

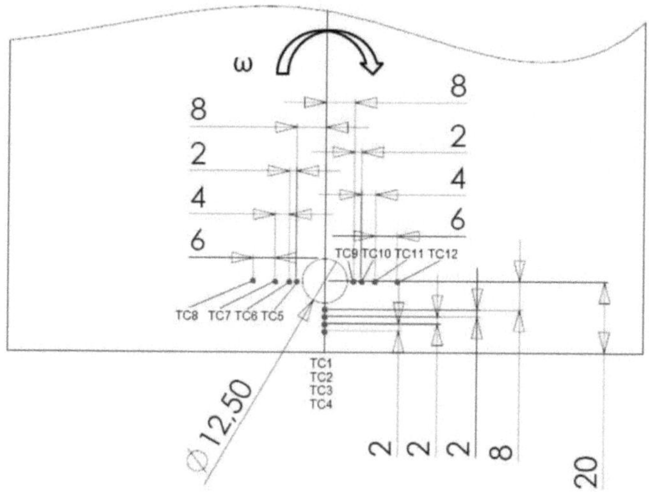

Figure 3. Schematic illustrating thermocouple positions.

Experiments were completed with rotation speeds of 710, 900, 1120 rpm. Temperature cycles for 710 rpm are presented in Figure 4.

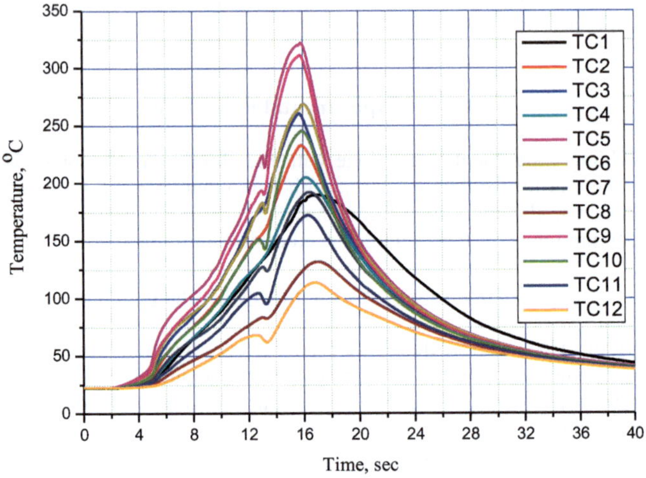

Figure 4. Temperature cycles during plunging and dwelling at 710 rpm.

Similar experiments were completed for 900 and 1120 rpm. The maximum temperature during plunging and dwelling was 345^{0}C. This peak temperature was reached during plunging and dwelling at 1120 rpm.

1.2. Temperature Measurement Experiments During FSW

Temperature measurement experiments during FSW were completed with similar weld parameters. Type K thermocouples were placed on both the advancing and retreating sides. The scheme of the thermocouple positions is shown in Figure 5.

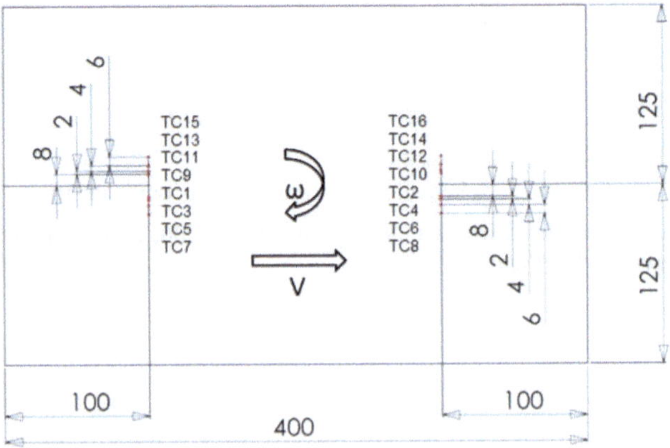

Figure 5. Schematic illustrating thermocouple positions during FSW.

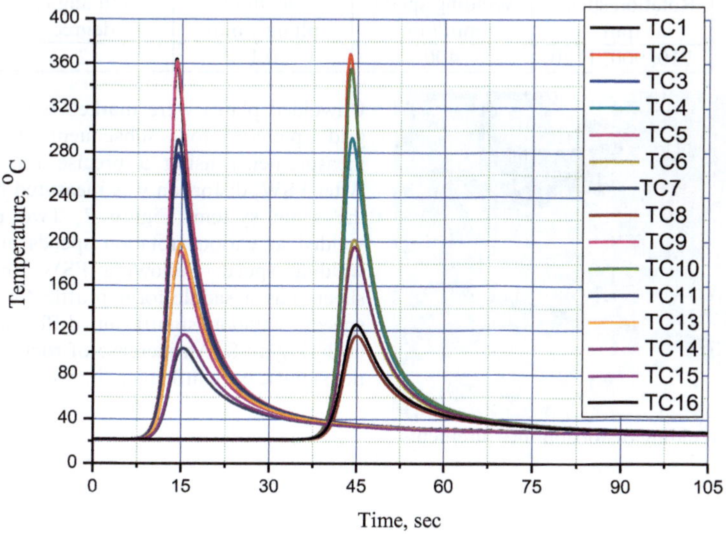

Figure 6. Temperature cycles during FSW at 1120 rpm.

It was noticed that temperature did not accumulate during welding stage. A maximum temperature of 370°C was reached on the advancing side during FSW at 1120 rpm. A minimum temperature of 312°C was measured during FSW on the retreating side at 710 rpm. The temperature difference between the advancing and the retreating sides decreases with increasing rotation speed.

Peak temperature measured from thermocouples № 9, 1, 15, 7 (Figure 5) was compared with the peak temperature calculated by means of an earlier finite element model developed for temperature calculations during FSW [3], Table I. Maximum absolute deviation is 12,7 % and mean absolute deviation is 8,4 %.

Table I. Temperature comparison between measurements and calculations.

TC №	Experiment	FEM
TC9	340	360
TC1	312	336
TC15	111	110
TC7	94	106

2. Welding Distortion Measurements

Two plates of Al 6082 T6 with dimensions 2x125x400 mm were friction stir welded. The FSW tool was made of structural steel with the geometry described above. The plates were placed on a backing plate and secured with clamps. The FSW parameters are listed in Table II.

Table II. FSW parameters

Rotation speed, rpm	Welding speed, mm/min	Depth of plunge, mm	Tilt angle, degree
710, 900, 1120	400	0.1	2

Figure 7. Welding distortion measurements.

Unwelded plates were marked with an 88 point grid pattern for subsequent initial shape measurements using a precise robotic system. After FSW, distortion was measured by means of the same system, Figure 7. Two plates were welded at various rotation speeds at a constant welding speed. Following FSW, the shape of sheets has a saddle-form profile for all rotation speeds without significant difference between each, Figure 8. An increase of rotation speed did not significantly influence the welding distortion.

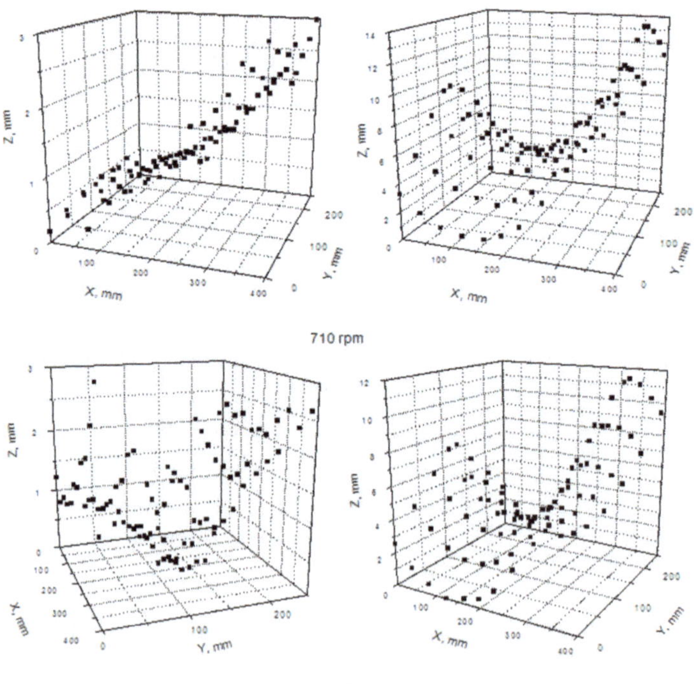

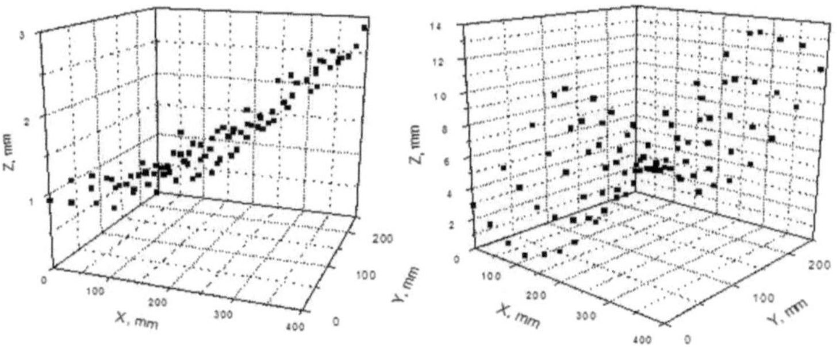

1120 rpm

Figure 8. Geometry of welded plates before and after FSW with different rotation speeds.

Conclusions

Temperature measurements were completed during different stages of FSW. The temperature difference between the advancing and retreating sides decreases with increasing rotation speed. The developed earlier finite element model for temperature calculation during FSW was checked using the experimental data obtained during the experiments. Peak temperature measured by the thermocouples was compared with peak temperature calculated by means of the earlier finite element model for temperature calculation during FSW. Maximum absolute deviation is 12,7 % and mean absolute deviation is 8,4 %. The increase of rotation speed from 710 rpm up to 1120 rpm does not significantly influence welding distortion.

ACKNOWLEDGEMENT

The research has been performed at St. Petersburg State Polytechnic University under the contract № 14.Z50.31.0018 with the Ministry of Education and Science of the Russian Federation.

REFERENCES

[1]. Thomas WM, Nicholas ED, Needham JC, Murch MG, Templesmith P, Dawes CJ (1991) G.B. Patent Application No. 9125978.8.

[2]. Mishra R.S., Mab Z.Y. Friction stir welding and processing. *Materials Science and Engineering*: R: Reports. – 2005, pp. 1-78.

[3]. Golubev Iurii, Naumov Anton, Michailov Vesselin. Developing Finite Element Model of the Friction Stir Welding for Temperature Calculation. *23rd International Conference on Metallurgy and Materials METAL 2014, Brno, Czech Republic.* In press.

AUTHOR INDEX
Friction Stir Welding and Processing VIII

SUBJECT INDEX
Friction Stir Welding and Processing VIII

N

O

P

R

S

T

W

MIX
Papier aus verantwortungsvollen Quellen
Paper from responsible sources
FSC® C105338

FSC
www.fsc.org

If you have any concerns about our products,
you can contact us on
ProductSafety@springernature.com

In case Publisher is established outside the EU,
the EU authorized representative is:
Springer Nature Customer Service Center GmbH
Europaplatz 3, 69115 Heidelberg, Germany

Printed by Libri Plureos GmbH
in Hamburg, Germany